Undergraduate Lecture Notes in Physics

Series Editors

Neil Ashby, University of Colorado, Boulder, CO, USA

William Brantley, Department of Physics, Furman University, Greenville, SC, USA

Matthew Deady, Physics Program, Bard College, Annandale-on-Hudson, NY, USA

Michael Fowler, Department of Physics, University of Virginia, Charlottesville, VA, USA

Morten Hjorth-Jensen, Department of Physics, University of Oslo, Oslo, Norway

Michael Inglis, Department of Physical Sciences, SUNY Suffolk County Community College, Selden, NY, USA

Barry Luokkala ⓘD, Department of Physics, Carnegie Mellon University, Pittsburgh, PA, USA

Undergraduate Lecture Notes in Physics (ULNP) publishes authoritative texts covering topics throughout pure and applied physics. Each title in the series is suitable as a basis for undergraduate instruction, typically containing practice problems, worked examples, chapter summaries, and suggestions for further reading.

ULNP titles must provide at least one of the following:

- An exceptionally clear and concise treatment of a standard undergraduate subject.
- A solid undergraduate-level introduction to a graduate, advanced, or non-standard subject.
- A novel perspective or an unusual approach to teaching a subject.

ULNP especially encourages new, original, and idiosyncratic approaches to physics teaching at the undergraduate level.

The purpose of ULNP is to provide intriguing, absorbing books that will continue to be the reader's preferred reference throughout their academic career.

Pieter Kok

A First Introduction to Quantum Physics

Second Edition

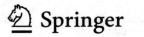

Pieter Kok
University of Sheffield
Sheffield, UK

ISSN 2192-4791   ISSN 2192-4805   (electronic)
Undergraduate Lecture Notes in Physics
ISBN 978-3-031-16164-3   ISBN 978-3-031-16165-0   (eBook)
https://doi.org/10.1007/978-3-031-16165-0

This Springer imprint is published by the registered company Springer Nature Switzerland AG
The registered company address is: Gewerbestrasse 11, 6330 Cham, Switzerland

Preface to the Second Edition

One of the benefits of teaching a course based on one's own lecture notes is seeing how students react to explanations, and discovering what are the main conceptual difficulties. In particular, I found that the students were struggling with the notion of a state space for a physical system; they had trouble understanding the origin of the minus sign in the matrix form of the beam splitter transformation; and they found it hard to relate the spin states in the x and y directions to the spin states in the z direction. These are three deeply connected concepts that are foundational to the understanding of the theory. In the second edition, I spend ample time on these three topics in the hope that the text has become clearer. In Sect. 2.1 I give several examples of classical state spaces to help get a feel for the concept, and in Sect. 2.4, I expand on the meaning of each number in the beam splitter matrix. I also give a classical description of a beam splitter as a half-reflecting piece of glass to help argue where the minus sign in the transformation matrix comes from.

In the few years since the first edition, I also came across a few topics that I thought might be relevant and interesting to new readers. I added a section to Chap. 2 on the Elitzur-Vaidman bomb detection thought an experiment, because it brings home the strange nature of quantum physics very dramatically. In Chap. 3, I corrected the omission of the magnitude of the spin-$\frac{1}{2}$ system, which is an important concept later when the reader encounters the addition of angular momentum. I also thought it would be good to add a second example of a physical observable, namely the polarisation of a photon. This will help the students spot the differences and similarities of two different observables, how the physics of a polarised photon is different from the electron spin but the maths is the same. I also moved the eigenvalue problem of the interaction Hamiltonian from Chap. 5 to Chap. 4, which leads to a more complete description earlier on, and allowed me to include the Zeeman effect of an atom in a magnetic field. This is an important effect that can be used to control the state of an atom, and the Zeeman effect was historically important in the discovery that sun spots are magnetic in origin.

In Chap. 6, I added the no-cloning theorem, because it is a foundational principle of quantum theory and is easy to understand at this level. It also leads directly to the idea of quantum key distribution, which is at the forefront of quantum technologies right now. I go in detail through the BB84 protocol and show how it allows two parties, Alice and Bob, to set up a secret key without an eavesdropper, Eve, gaining any knowledge of it. This topic naturally leads to the question of how we can extend quantum entanglement over large distances using quantum repeaters, which is one of the key technologies that will underpin the Quantum Internet of the future.

In Chap. 7, we saw how entanglement, interactions, and the density matrix are fundamentally related, and I added a section on how we can use the properties of the reduced density matrix to create measures for how much entanglement is present in a quantum state. In Chap. 9, I incorporated not only the EPR thought experiment but also the Einstein-Bohr debate on the photon and the box. These are still some of the most powerful demonstrations of how careful we must be when we reason about quantum uncertainty. I also moved the discussion of Schrödinger's cat from Chap. 10 to Chap. 9, which then allowed me to include a discussion on how to create cat-like states in the lab using opto-mechanical systems.

I included a brief overview of the history of quantum mechanics, which is full of interesting characters and stories. We often refer to the pioneers of quantum theory, so it is important that students have some familiarity with the history in order to participate in these shared stories. I collected some of the main results from the end of the nineteenth century to the end of the twentieth century to give a broad overview of the development of the field without getting too bogged down in the historical details.

As I delivered this course, I prepared extra problems and homework exercises, which are also included in the second edition.

And finally, I am greatly indebted to Nuno Peres, Ed Daw, and Alessandro Fedrizzi for numerous comments on the first edition, Marta Marchese for inspiring the section on opto-mechanical systems, and Mark Wilde for advice on entanglement measures. But my greatest thanks goes out to all the students who were subjected to my teaching, and who asked many questions. Their struggle helped me improve this book.

Sheffield, UK   Pieter Kok

Preface to the First Edition

Quantum mechanics is one of the crowning achievements of human thought. There is no theory that is more successful in predicting phenomena over such a wide range of situations—and with such accuracy—than quantum mechanics. From the basic principles of chemistry to the working of the semi-conductors in your mobile phone, and from the Big Bang to atomic clocks, quantum mechanics comes up with the goods. At the same time, we still have trouble pin-pointing exactly what the theory tells us about nature. Quantum mechanics is hard, but perhaps not as hard as you think. Let's compare it to another great theory of physics: electromagnetism.

When we teach electricity and magnetism in school and university, we start with simple problems involving point charges and line currents. We introduce Coulomb's law, the law of Biot and Savart, the Lorentz force, and so on. After working through some of the most important consequences of these laws, we finally arrive at Maxwell's equations. Advanced courses in electrodynamics then take over and explore the consequences of this unification, treating such topics as wave guides, gauge invariance, relativity, et cetera. The pedagogical route is going from the simple, tangible problems to the general and abstract theory. You need to know quite a bit of electromagnetism and vector calculus before you can appreciate the beauty of Maxwell's equations.

The situation in teaching quantum mechanics is generally quite different. Instead of simple experimentally motivated problems, a first course in quantum mechanics often takes a historical approach, describing Planck's solution of black body radiation, Einstein's explanation of the photoelectric effect and Bohr's model for the atom from 1913. This is then followed by the introduction of the Schrödinger equation. The problem is that appreciating Schrödinger's equation requires a degree of familiarity with the corresponding classical solutions that most students do not yet have at this stage. As a result, many drown in the mathematics of solving the Schrödinger equation and never come to appreciate the subtle and counterintuitive aspects of quantum mechanics as a fundamental theory of nature.

It does not have to be like this. We can develop the core principles of quantum mechanics based on very simple experiments and without requiring much prior mathematical knowledge. By exploring idealized behaviour of photons in interferometers, electron spins in magnetic fields, and the interaction of simple two-level atoms with light, we can put our finger quite precisely on the strange, puzzling and wonderful aspects of nature as described by quantum mechanics. We can then illustrate the theory with modern applications such as gravitational wave detection, magnetic resonance imaging, atomic clocks, quantum computing and teleportation, scanning tunnelling microscopy, and precision measurements.

Another reason to write this book was to make use of the wonderful possibilities that are offered by new media. Physics is an experimental science, and seeing how systems behave in interactive figures when you nudge them in the right way hopefully gives the reader an immediate connection between the experiments and the physical principles behind them. That is why I have included many interactive elements to accompany the text, which are available online. I firmly believe that replacing static figures on a page with interactive and animated content can be a great pedagogical tool when used correctly.

This book introduces quantum mechanics from simple experimental considerations, requiring only little mathematics at the outset. The key mathematical techniques such as complex numbers and matrix multiplication are introduced when needed, and are kept to the minimum necessary to understand the physics. However, a full appreciation of the theory requires that you also take a course on linear algebra. Sections labeled with 🖐 indicate topics that are not part of the core material in the book, but denote important applications of the theory. They are the reason we care about quantum mechanics. The first half of this book is devoted to the basic description of quantum systems. We introduce the state of a system, evolution operators and observables, and we learn how to calculate probabilities of measurement outcomes. The second half of the book deals with more advanced topics, including entanglement, decoherence, quantum systems moving in space, and a more in-depth treatment of uncertainty in quantum mechanics. We end this book with a chapter on the interpretation of quantum mechanics and what the theory says about reality. This is the most challenging chapter, and it relies heavily on all the material that has been developed in the preceding nine chapters.

I am greatly indebted to my colleague Antje Kohnle from the university of St. Andrews, who helped me navigate the pitfalls of interactive content, and without whom this book would have been much less readable. I also wish to thank Dan Browne, Mark Everitt, and Derek Raine for deep and extended discussions on how to organise a first course in quantum mechanics. Finally, I want to thank Rose, Xander, and Iris for their patience and support during the writing of this book.

Sheffield, UK   Pieter Kok

Contents

List of Figures

Chapter 1
Three Simple Experiments

In this chapter, we consider a series of simple experiments. By contemplating the meaning of the outcomes of these experiments we are forced to adopt some very counterintuitive conclusions about the behaviour of quantum particles.

1.1 The Purpose of Physical Theories

Since antiquity, people have tried to understand the world around them in terms of simple principles and mechanisms. The Greek philosopher Aristotle (who lived in the 4th century BCE in Athens, Greece) believed that all heavenly bodies moved in perfect circles around the Earth. The discrepancy of this basic principle with the observed movement of the planets led to increasingly complicated models, until Copernicus introduced a great simplification by assuming that the planets orbit the Sun instead. Galileo, Kepler, and Newton refined this theory further in the sixteenth and seventeenth century, with only Mercury's orbit resisting accurate description. Solving this last puzzle ultimately culminated in Einstein's theory of general relativity in the early twentieth century.

What makes science different from other human endeavours is that our theories about the world we live in must conform to the outcomes of our observations in well-designed and well-executed experiments. Particularly in physics, our experiments form the ultimate arbiter whether we are on the right track with our theories or not. A theory that can predict the outcomes of our experiments is considered successful.

However, it is not enough to just predict the motion of the planets, the behaviour of magnets, or how electrical components should be wired to build a radio. We want

Supplementary Information The online version contains supplementary material available at https://doi.org/10.1007/978-3-031-16165-0_1.

P. Kok, *A First Introduction to Quantum Physics*, Undergraduate Lecture Notes in Physics, https://doi.org/10.1007/978-3-031-16165-0_1

to know why planets, magnets, resistors, and capacitors behave the way they do. We naturally assume that there is an underlying microscopic world that determines the way resistors and capacitors respond to currents, and how magnets interact. Indeed, this has been an extraordinarily successful programme. Electricity and magnetism are explained by only four basic equations, called Maxwell's equations, and gravity is understood by a single equation, called Einstein's equation.

These equations tell us not only how to describe the behaviour of planets and magnets, but they give an explanation of that behaviour in terms of underlying physical "stuff". In the case of electricity and magnetism the underlying stuff is charges, currents, and electric and magnetic fields. In the case of gravity, the underlying stuff is space-time, which has properties like curvature. These fields and curved space are assumed to really exist, independent of whether we look at it or not. Similarly, we all believe that atoms exist, and that their collective motion causes directly observable phenomena such as air pressure and temperature. In other words, atoms are real. And so are curved space and electromagnetic fields.

In physics, we want to construct theories that explain a wide variety of phenomena based on a few types of physical objects, plus the rules that govern these objects. The objects are taken as really "there". This is called scientific realism, and it is what most scientists believe at heart.

In the beginning of the twentieth century, physicists came up with a new theory to describe the behaviour of atoms that was extraordinarily successful. It predicted new phenomena that were subsequently discovered, and there has not been a single credible experiment that contradicts it. I am talking, of course, about quantum physics.

But quantum physics is not like electrodynamics or general relativity. Simple scientific realism is difficult to maintain, and this has led to all sorts of seemingly fantastical claims about the nature of underlying reality. We will explore these difficulties, and in the process develop the basic structure of the theory. You will be able to perform calculations in quantum physics, and develop a clear picture of what we can and cannot say about the underlying reality of nature according to quantum physics.

1.2 Experiment 1: A Laser and A Detector

We start with a simple thought experiment. Consider the situation shown in Fig. 1.1, in which a laser is pointed at a photodetector. The laser has two settings, "HI" and "LO", corresponding to high and low output power, respectively. There is an interactive

Fig. 1.1 A laser and a detector (see supplementary material 1)

Fig. 1.2 The idea of photons (see supplementary material 2)

Laser   Detector

version of the figure available online—like most figures in this book—and it lets you change the laser power to see the effect on the detector. Let us say that the high output is about 1.0 mW, which is the intensity of a typical laser pointer. The photodetector converts light into a current, which is read out by the current meter. When the laser is set to the high power output, we see a steady current on the meter. The strength of the current is directly proportional to the intensity of the light.

Next, we reduce the power of the laser by switching to the setting "LO". We expect that continuously lowering the power output of the laser will continuously decrease the current in the detector. At first, this is indeed how the system behaves. However, it is an experimental fact that for very low intensities the current is no longer a continuous steady current, but rather comes in pronounced pulses. This is the first counterintuitive quantum mechanical result. In what follows we will explore the consequences of this fact.

You may imagine a constant beam of light from the laser to the detector, but there is a reason we have drawn no line in Fig. 1.1. At this stage we do not know what is happening between the laser and the detector, and we need to be very careful not to make any assumptions that do not have a so-called *operational* meaning in terms of light sources and detectors, that is, a meaning that relates directly to how things operate on a directly observable scale. You may say: "but I can see the light between the laser and the detector if I blow chalk dust in the beam", and you would be right. However, the chalk dust and your eye would then become a second detection system that we do not yet wish to consider. Having said this, it is customary to interpret the current pulses as small "light packets" traveling from the laser to the detector, shown in Fig. 1.2. These chunks of electromagnetic energy are commonly called photons.

At this stage it is important to remember that we don't really know anything about these photons, other than that they are defined as the cause of the pulses on the current meter. In particular, you should not assume that they behave like normal objects such as marbles or snooker balls. We need to perform more experiments to establish how they behave, and we will explore this in the coming sections.

1.3   Experiment 2: A Laser and A Beam Splitter

For our next experiment, we set the laser output again to "HI". We place a beam splitter between the laser and the detector, which consists of a piece of glass with a semi-reflective coating that lets half of the light through to the original detector. The other half is reflected by the beam splitter. We set up a second detector and current

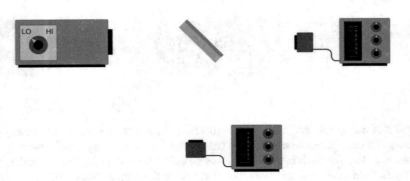

Fig. 1.3 A laser and a beam splitter (see supplementary material 3)

meter to monitor the light that is reflected by the beam splitter (shown in Fig. 1.3). The current created by each detector is half the current created by the detector in experiment 1 (Fig. 1.1). You may know already that the intensity of light is related to its energy, so this experiment demonstrates energy conservation of the beam splitter. It divides the intensity of the laser evenly over the two detectors.

Incidentally, you may have noticed that this situation is somewhat similar to the chalk dust in the laser beam: the chalk acts a little bit like a beam splitter, and your eye is the detector. However, using a beam splitter is much more accurate, since we can in principle precisely tune the reflectivity.

Next, we switch the laser setting from "HI" to "LO", and find that we again find current pulses. The pulses look exactly the same as in experiment 1, and the total number of pulses that we detect per second is also unchanged (reflecting the fact that the power output of the laser is the same as in 1). This is shown in Fig. 1.3. The pulses appear randomly in the two detectors. We cannot predict which detector will trigger a current pulse in advance. In other words, the probability that detector D_1 is triggered is $p_1 = \frac{1}{2}$, and the probability that detector D_2 is triggered is $p_2 = \frac{1}{2}$. The sum of the probabilities is $p_1 + p_2 = 1$, as it should be.

Moreover, at low enough intensity we never find a pulse in both detectors simultaneously. If we return to the mental picture of chunks of energy, we can now say that the photon triggers detector D_1 or detector D_2, but never both simultaneously. In other words, the photon is *indivisible*. We have experimentally established this as a physical property.

It looks like the photon really is behaving as a particle. What is a bit strange is that a static element such as a beam splitter (which is, after all, just a piece of glass) should introduce a probabilistic aspect to the experiment. On the other hand, how else could it be? Each photon is created independently, so there should not be any conspiracy between the photons to create a regular pattern of pulses in the detectors. Therefore, if the intensity of the light is to be divided evenly over the two detectors, each photon must make a random decision at the beam splitter. Or so it seems.

1.4   Experiment 3: The Mach–Zehnder Interferometer

For our final experiment we replace the detectors by mirrors, and recombine the two beams using a second beam splitter (see Fig. 1.4). The outgoing beams of this second beam splitter are then again monitored by detectors. The setup is shown above. When we set the laser to high intensity ("HI"), we can arrange the beam splitters and mirrors such that there is no signal in detector D_1, and all the light is detected by detector D_2. This is a well-known wave effect, called interference. According to the theory of optics, light is a wave, and the lengths of the two paths between the beam splitters are such that the wave transmitted from the top of BS2 has a phase that is exactly opposite to the phase of the reflected wave coming from the left of BS2 (see Fig. 1.5). The device is called a Mach–Zehnder interferometer.

When we reduce the power of the laser again all the way down to the single photon level (setting "LO"), the current in detector D_2 reduces until only single current pulses appear. Detector D_1 stays silent. This is consistent with experiment 2, where the signal also reduces to pulses in the current. However, this does not sit well with the mental image we developed earlier, in which photons are particles that choose randomly whether they will be reflected or transmitted at the beam splitter. How can a single particle cause interference such that detector D_1 remains dark?

We have established that photons are indivisible, so it can not be true that the photon splits up into two smaller parts at the first beam splitter and recombines at the second beam splitter to always trigger detector D_2. A natural question to ask is then: which of the two paths between the two beam splitters does the photon take, the top or the bottom? To answer this we modify our experiment.

The photodetectors we considered thus far are not the only way to measure light. In recent years people have spent a lot of effort to create so-called quantum non-demolition (QND) detectors. These have the benefit that they do not destroy the

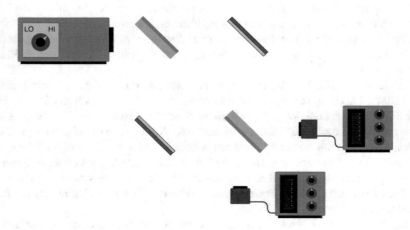

Fig. 1.4   A Mach–Zehnder interferometer (see supplementary material 4)

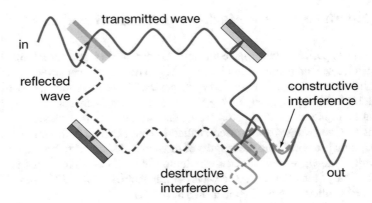

Fig. 1.5 Classical waves in a Mach–Zehnder interferometer. Waves appearing from the second beam splitter with opposite phase cause destructive interference, resulting in a complete absence of light in that output beam. All the light appears in the output with the constructive interference. The amplitude of the wave is the square root of the intensity

photon. A detailed description of these detectors is beyond the scope of this book, and we do not really need to know how they work. It is sufficient to note that they detect the presence of a photon without destroying it. In Fig. 1.6, we draw them as loops connected to current meters. A blip on the meter indicates that a photon passed through the loop. We can use QND detectors to find out whether the photon took the upper path or the lower path of the Mach–Zehnder interferometer by placing a loop in each path. This is shown in Fig. 1.5.

As you may have expected, we find that our photon passes either through loop A or through loop B, but never through both. In addition, after finding a photon in one of our QND detectors, we always find a single current pulse in one of our regular photodetectors. However, it is no longer exclusively detector D_2 that creates the current pulses. After many trials in this experiment, it becomes clear that each QND detector is triggered with equal probability, and so are the normal photodetectors. In other words, the interference pattern, where detector D_1 always remained dark, has disappeared!

So the observation which path the photon takes, A or B, will destroy the interference effect. This is one of the most fundamental aspects of quantum physics: *looking at a physical system changes its behaviour*. This conclusion is entirely determined by experimental findings, making it inescapable. Note that in order to look at which path the photon took, we had to interact with the photon via the loops of the QND detectors. So it is not our "consciousness" that made that photon behave differently, but the physical interaction that is necessary to "look" at the system, just like the chalk dust revealing a laser beam interacts with the light by reflecting parts of the beam towards your eye.

The big difference between classical systems and quantum systems is that in classical systems we can often "peel off" a small part from a bigger system (for

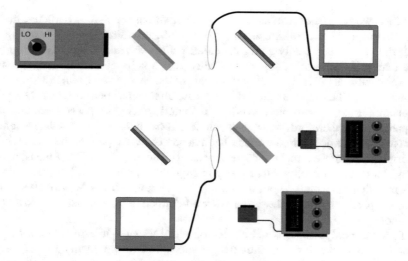

Fig. 1.6 A Mach–Zehnder interferometer with QND measurements of the photon path (see supplementary material 6)

example, reflect a little bit of light out of a beam using chalk) to detect how the large system is behaving. We typically assume that this does not disturb the large system very much. In contrast, a quantum system tends to be small—like a single photon—and we fundamentally cannot peel off part of it and leave the remainder undisturbed. As the experiments in this chapter have shown, we cannot assume that *unobserved* quantum systems behave just like classical systems.

1.5   The Breakdown of Classical Concepts

The three simple experiments described above show us that photons behave in a rather strange way. This forces us to abandon some classical notions that are deeply ingrained in our thinking about nature. Here, by classical we mean those natural phenomena that are well described by the great "classical" theories of physics, including Newton's mechanics, Maxwell's electromagnetism, Kelvin's thermodynamics, and even Einstein's special and general theory of relativity. We start with a summary of what we have learned about photons in our experiments:

- Photons are defined as the cause of current pulses in our photodetectors.
- Photons are indivisible.
- Single photons can exhibit interference when there is more than one path a photon can take.
- Interference disappears when we measure which path the photon takes in the interferometer.

The first two properties lead us to believe that photons are indeed particles, but the last two properties complicate our mental picture considerably.

If the photon is actually going via either path A or path B, then extracting this information reveals something about reality, which is by definition independent of our knowledge about it. Therefore, if the path of the photon was a real property of the photon, the QND measurement should have no effect on its behaviour on the second beam splitter. Yet it does have an effect: the interference disappears when we know its path. We are therefore forced to conclude that our photon does not take a definite path that we just do not know about, but that somehow the path of the photon is not determined before we make the measurement! The act of the QND measurement brings into reality the physical property of the photon path. Before the measurement it is meaningless to talk about the path of the photon. This is very different from classical particles, which always have a well-defined path, whether we know about it or not.

To avoid confusion, it is helpful to introduce a new name for objects like the photon. Instead of calling the photon a particle or a wave, we say that it is a quantum of light. The term "quantum" means "little amount". You see that quanta are neither particles nor waves, but share some characteristics of both. In the next chapter we will attempt to describe these objects more accurately using maths.

Our discussion demonstrates that classical concepts such as intrinsic properties of a system are no longer valid in quantum physics. This was summed up by John Wheeler[1] as follows:

No phenomenon is a physical phenomenon until it is an *observed* phenomenon.

It is very difficult to describe what happens to the photon, and ordinary language is almost inadequate. We will return to this in Chap. 10. Often people write that the photon takes both paths at the same time, but you can see from the experiments and their interpretation that this is not quite accurate. In the next chapter, we will make our discussion more precise and quantitative.

Exercises

1. What is the purpose of physical theories? Should they say something about the world beyond the correct prediction of measurement outcomes? Keep a record of your answer, so you can compare it to your answer to Exercise 1 of Chap. 10.
2. If the beam splitter in Fig. 1.3 is 70% reflective, instead of 50%, what would be the probabilities of finding the photon in detectors D_1 and D_2?
3. If both beam splitters in Fig. 1.4 are 70% reflective, instead of 50%, show that we can still have perfect destructive interference.
4. Imagine we send a high-intensity laser beam into the Mach–Zehnder interferometer with the QND detectors in Fig. 1.6. What do you expect will happen to the

---

[1] J. A. Wheeler, interview in *Cosmic Search*, **1** (4), Fall 1979.

classical interference pattern in the output detectors? You may assume that the QND detectors are perfectly responsive.

5. What will happen to the interference pattern in the output detectors of Fig. 1.4 if the photon has a weak interaction with the medium it travels through, such that the medium has a (partial) memory of the photon's trajectory?

6. What are the probabilities of detectors D_1 and D_2 firing in the Mach–Zehnder interferometer of Fig. 1.4 if there is a 50:50 chance that the first beam splitter is not there?

Chapter 2
Photons and Interference

In this chapter, we will give the mathematical description of the previous chapter's experiments. This will show us how we can calculate the probabilities of measurement outcomes in quantum mechanics. We need to introduce simple matrices and basic complex numbers.

2.1 The State of a Physical System

In physics, we often refer to the *state* of a physical system as the collection of all its physical properties. The state of a system is typically a mathematical description that we can use in calculations. As an example, consider the state of a classical non-relativistic point particle in a one dimensional box of length L. If we assume that there is no internal structure to the particle, the state of the particle can be summed up by the position x in the interval $[0, L]$, and it velocity v, which lies in the interval $(-\infty, +\infty)$. So the state of the particle is the combination of two numbers, (x, v). The collection (set) of all possible values pairs of x and v is called the *state space* of the particle.

The details of the state space are important, because it puts constraints on what physical processes are allowed. For example, we can have no physical process that results in values for x that are greater than L, since it would imply that the particle can end up outside the tube. We specifically set up the state space such that this is not possible. Consequently, the fact that the particle can have a positive velocity means that it must either bounce off the end of the tube, or get stuck there forever.

Supplementary Information The online version contains supplementary material available at https://doi.org/10.1007/978-3-031-16165-0_2.

P. Kok, *A First Introduction to Quantum Physics*, Undergraduate Lecture Notes in Physics, https://doi.org/10.1007/978-3-031-16165-0_2

Another example is a particle in a three-dimensional spherical cavity with radius a. Here, it is convenient to use spherical coordinates, and we know that the radial position r is confined to the interval $[0, a]$, namely within the sphere, but the polar angle θ and azimuthal angle ϕ are unconstrained and may take the values $\theta \in [0, \pi]$ and $\phi \in [0, 2\pi)$. The velocity of the particle now has three components, one for each cartesian coordinate, v_x, v_y, and v_z. Each coordinate can again take any real value between $-\infty$ and $+\infty$ again, so the total state space consists of all the six numbers

$$(r, \theta, \phi, v_x, v_y, v_z) \in [0, a] \times [0, \pi] \times [0, 2\pi) \times \mathbb{R}^3 , \tag{2.1}$$

where we used the notation $\mathbb{R}^3$ for the three-dimensional velocity space $(-\infty, +\infty) \times (-\infty, +\infty) \times (-\infty, +\infty)$. The variables that make up the state space are called *degrees of freedom*, since we can pick any allowed values for these variables and arrive at a unique and valid state of the particle.

If we have N classical particles in a volume, like a monatomic gas where each atom has no internal degree of freedom, we generally need $6N$ numbers to completely describe all the positions and velocities of the particles; three per particle for the position components in three-dimensional space, and three per particle for the velocity components in three-dimensional space. The position components may be constrained by the size of the volume, like in our previous examples, and the velocity components may be constrained by, for example, the speed of light in relativistic physical theories.

The idea of a state space does not only apply to particles. Any physical system has a state and a corresponding state space (see Fig. 2.1). The state consists of values for all the degrees of freedom that are necessary to uniquely determine the behaviour of the system. This defines an array of numbers. The state space is the set of all arrays with all possible combinations of valid numbers. As another example, consider a gas obeying the ideal gas law

$$pV = Nk_B T , \tag{2.2}$$

where p is the pressure, V is the volume of the gas, T is the temperature, N is the number of particles, and k_B is Boltzmann's constant. When the gas is in thermal equilibrium the ideal gas law completely specifies the physical behaviour. So our state space consists of the variables $p \in [0, \infty)$, $V \in [0, \infty)$, $N \in \mathbb{N}$, and $T \in (0, \infty)$. We exclude zero for T, since absolute zero can never be reached. Often, we consider

Fig. 2.1 The state of a system (see supplementary material 1)

Car   Photon

a constant number of particles N, which then means that with the constraint of the ideal gas law the ideal gas has two degrees of freedom. We often sketch this state space as pV diagrams.

Notice that we could have used the state space for N particles that we introduced above, but imposing the ideal gas law and thermal equilibrium will vastly cut down the number of different macro-states (where we consider only p, V, N, and T). It is just not possible to calculate anything thermodynamically meaningful in terms of all the individual particle positions and velocities when N is on the order of Avogadro's number.

One final example shows how we can extend the above notion of a state space to include probabilities. Consider two coins, each with heads (H) and tails (T). If we toss the coins there will be four possible outcomes:

$$\{HH, \ HT, \ TH, \ TT\}.$$

Often it is relatively straightforward to determine all possible outcomes, but the real question of interest is the distribution of the probabilities of these outcomes. This leads to four probabilities

$$\Pr(HH), \quad \Pr(HT), \quad \Pr(TH), \quad \text{and} \quad \Pr(TT). \tag{2.3}$$

Each probability is a number between 0 and 1, and the probabilities sum to 1:

$$\Pr(HH) + \Pr(HT) + \Pr(TH) + \Pr(TT) = 1. \tag{2.4}$$

The state space of the two coins (which may not be fair!) is then given by these four probabilities, along with the constraint that they sum to one.

In the remainder of this chapter, we will construct the state space of a photon in a beam splitter, and we will work out how we can calculate the probabilities of where the photon will end up. We wil need some extra mathematics involving vectors, matrices, and complex numbers, and we conclude the chapter with two examples. Our construction of the state space in quantum mechanics will not be as intuitive—at least initially!—as the examples we have given in this section, but it is necessary to obtain good agreement with experimentally observed results.

2.2 Photon Paths and Superpositions

In quantum mechanics, we have to take a little more care when talking about properties, because as we have seen, they are not independent of the measurements or "observations" of the system. Nevertheless, we do refer to the state of a quantum system as the description of the system that allows us to determine its physical behaviour.

Fig. 2.2 The photon can
enter a beam splitter from the
left and from the top, and it
exits the beam splitter to the
right and down to be
detected in photodetectors
D_1 and D_2, respectively

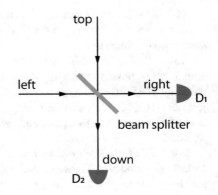

To explore the quantum state of a photon, consider the second experiment of the
previous chapter, where the photon enters the beam splitter from the left (see Fig. 2.2
for a more traditional "technical" diagram). The other way in which a photon can enter
the beam splitter and trigger the detectors is from the top (we will not be considering
photons coming in from other angles, since the outgoing beams for those inputs will
then no longer overlap). We say that the photon has two distinct quantum states,
indicated by the labels "left" or "top". We write this as

$$|\text{left}\rangle \quad \text{or} \quad |\text{top}\rangle \,. \tag{2.5}$$

The symbol "$|\cdot\rangle$" is called a "ket", and is often used to indicate a quantum state. This
is called "Dirac notation", after one of the founders of quantum mechanics, Paul
Dirac.[1] The label inside the ket may not have a mathematical meaning, and is mainly
an indicator as to which physical state the ket refers to.

We must distinguish the state of the photon before the beam splitter from the state
after the beam splitter, because the beam splitter will change the state of the photon.
After the beam splitter, there are again two paths that the photon can take, namely
"right" and "down". We also associate quantum states with the photon in these paths,
denoted by

$$|\text{right}\rangle \quad \text{or} \quad |\text{down}\rangle \,. \tag{2.6}$$

To distinguish clearly the state of the photon after the beam splitter (let's say at time
t_2) from the state of the photon before the beam splitter (at time $t_1 < t_2$) we can add
an extra label "2" as a subscript to the ket. We do not have to do this, but it is a
convenient reminder for now.

A photon entering the beam splitter from the left in state $|\text{left}\rangle_1$ will typically not
be in the exact state $|\text{right}\rangle_2$ or $|\text{down}\rangle_2$ after the beam splitter. To see why, consider
again the experiment with a single beam splitter. If the photon was always in the

---

[1] P. A. M. Dirac, *The Principles of Quantum Mechanics*, Oxford University Press, 1988.

Fig. 2.3 Interference of water waves

state $|\text{right}\rangle_2$, we would expect detector D_1 to fire every time. Similarly, if the photon was always in the state $|\text{down}\rangle_2$, we would expect detector D_2 to fire every time. However, we saw that the photons are detected randomly in detector D_1 and D_2, so the state of the photon must be some combination of $|\text{right}\rangle_2$ and $|\text{down}\rangle_2$.

Next, we note that the third experiment of the previous chapter (the Mach–Zehnder interferometer) gives rise to interference of the laser light, a phenomenon that is explained by the wave nature of light. We obtain interference by taking a *superposition* of two waves (see Fig. 2.3). A similar effect was observed using the photon, which always triggered detector D_2, and never D_1. It is therefore natural to suppose that the states $|\text{right}\rangle_2$ and $|\text{down}\rangle_2$ can be combined in a superposition. If we denote the state of the photon after the beam splitter by $|\text{photon}\rangle_2$, the superposition becomes

$$|\text{photon}\rangle_2 = a|\text{right}\rangle_2 + b|\text{down}\rangle_2 , \tag{2.7}$$

where a and b are numbers whose meaning we will return to later. The photon is neither in state $|\text{right}\rangle_2$, nor in the state $|\text{down}\rangle_2$, but in an entirely different state, namely $|\text{photon}\rangle_2$.

We could continue to develop the theory in terms of kets, but at this point we note a great mathematical convenience. The state of the photon behaves as a *vector*. We can add vectors to make another vector, and we can identify two special vectors that correspond to the distinct states $|\text{right}\rangle_2$ and $|\text{down}\rangle_2$. These vectors point in perpendicular directions, and they are called *orthogonal* vectors:

$$|\text{right}\rangle = \begin{pmatrix} 1 \\ 0 \end{pmatrix} \quad \text{and} \quad |\text{down}\rangle = \begin{pmatrix} 0 \\ 1 \end{pmatrix} . \tag{2.8}$$

Fig. 2.4 The state of the
photon as a vector

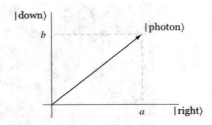

Consequently, the state of the photon after a beam splitter can be written as

$$|\text{photon}\rangle = a \begin{pmatrix} 1 \\ 0 \end{pmatrix} + b \begin{pmatrix} 0 \\ 1 \end{pmatrix} = \begin{pmatrix} a \\ b \end{pmatrix} \tag{2.9}$$

Equation (2.9) is equivalent to (2.7) and can be represented graphically as in Fig. 2.4. Note that the states $|\text{right}\rangle$ and $|\text{down}\rangle$ are orthogonal vectors of length 1. This means that any conceivable superposition of the photon paths can be decomposed into these two vectors. We call the vectors $|\text{right}\rangle$ and $|\text{down}\rangle$ the *basis vectors* in which any state vector $|\text{photon}\rangle$ can be written. The two basis vectors $|\text{right}\rangle$ and $|\text{down}\rangle$ together form a *basis* for the state space of a photon exiting a beam splitter. This state space consists of all the vectors determined by a and b in the state $|\text{photon}\rangle$. We will explore in due course exactly what are the allowed values for a and b.

It is the *direction* of the vector (with components a and b) that determines the superposition state, and we assume that the length of the vector remains unchanged. This places the restriction $a^2 + b^2 = 1$ on the state $|\text{photon}\rangle$. In the special case of the beam splitters in the previous chapter we have $a = b = 1/\sqrt{2}$, and the state of the photon is

$$|\text{photon}\rangle = \frac{1}{\sqrt{2}}|\text{right}\rangle + \frac{1}{\sqrt{2}}|\text{down}\rangle = \frac{1}{\sqrt{2}} \begin{pmatrix} 1 \\ 1 \end{pmatrix} \tag{2.10}$$

after the beam splitter.

If the state of the photon after a beam splitter can be in a superposition, then it makes sense that the state of the photon before the beam splitter can also be written as an arbitrary superposition of $|\text{left}\rangle_1$ and $|\text{top}\rangle_1$. Indeed, when we look at the second beam splitter of the Mach–Zehnder interferometer this is exactly what happens, since the input of the second beam splitter is in fact the output of a previous beam splitter. We are once more talking about a photon that can be in a superposition of two distinct states, and the mathematical description can again be given in terms of vectors:

$$|\text{left}\rangle_1 = \begin{pmatrix} 1 \\ 0 \end{pmatrix} \quad \text{and} \quad |\text{top}\rangle_1 = \begin{pmatrix} 0 \\ 1 \end{pmatrix}. \tag{2.11}$$

The label "1" on the ket determines again the stage in the experiment before the beam splitter. We could have added the same label to the vectors, but this will quickly become cumbersome and after a bit of practice it will be clear to which stage of the experiment a vector refers. If in doubt, you can always add the labels. The vectors |left⟩ and |top⟩ are the basis vectors for the state space of the photon prior to entering the beam splitter. The state spaces of the photon before and after the beam splitter are mathematically identical (i.e., two-dimensional vectors), but refer to different physical situations, with correspondingly different labels ("left" and "top", versus "right" and "down").

An important point about these vectors is that they are not vectors in the space that you and I occupy. Rather, they live in an abstract "space of possibilities" for the photon. The more possibilities there are for the photon (such as many paths), the more dimensions the space will have, because each distinct path has its own orthogonal vector. For now, we limit ourselves to two-dimensional spaces.

With the construction of the vectors |left⟩ and |top⟩—or |right⟩ and |down⟩—the state space of the photon in the Mach–Zehnder interferometer is determined by the values of the numbers a and b in (2.9). Before we determine these possible values, we have to take a brief mathematical digression.

2.3 *Mathematical Intermezzo*: Matrix Multiplication

Quantum mechanics is formulated in terms of matrices, with which you may not yet be familiar. Luckily, we will at first only need a few basic rules that are easy to explain. First, a matrix is a rectangular array of numbers, for example

$$\begin{pmatrix} 8 & 3 & 2 \\ 2 & 5 & 7 \end{pmatrix} \quad \text{or} \quad \begin{pmatrix} 12 & 0 \\ 1 & 13 \end{pmatrix}. \tag{2.12}$$

As you can see, a matrix has rows and columns. Sometimes a matrix has only one column or one row. Those matrices we refer to as vectors. For example:

$$\begin{pmatrix} 1 \\ 2 \end{pmatrix} \quad \text{or} \quad \begin{pmatrix} 2 & 4 & 3 \end{pmatrix}. \tag{2.13}$$

To understand the rest of this book, you need to know how to multiply two matrices. It is a very simple rule, based on the dot product (we will refer to it by its more general name, the scalar product). Suppose that we have two vectors, $\mathbf{a} = (2, 4)$ and $\mathbf{b} = \begin{pmatrix} 1 \\ 3 \end{pmatrix}$. The scalar product is

$$\mathbf{a} \cdot \mathbf{b} = \begin{pmatrix} 2 & 4 \end{pmatrix} \begin{pmatrix} 1 \\ 3 \end{pmatrix} = 2 \times 1 + 4 \times 3 = 14. \tag{2.14}$$

Fig. 2.5 Matrix
multiplication (see
supplementary material 5)
$$\begin{pmatrix} 2 & 3 \\ -1 & 4 \end{pmatrix} \begin{pmatrix} 1 & 1 \\ 5 & 2 \end{pmatrix} = \begin{pmatrix} \cdot & \cdot \\ \cdot & \cdot \end{pmatrix}$$

We have written the scalar product in matrix notation, with **a** a 1×2 row vector and
b a 2×1 column vector. Note that the length of the row of **a** must be equal to the
length of the column of **b** for the scalar product to work. The outcome of the salar
product is a "matrix" with one row and one column, more commonly known as a
number.

Matrix multiplication works in a very similar way to the scalar product (see
Fig. 2.5). Let us try to multiply the two matrices in (2.12). There are two possibilities:

$$\begin{pmatrix} 8 & 3 & 2 \\ 2 & 5 & 7 \end{pmatrix} \begin{pmatrix} 12 & 0 \\ 1 & 13 \end{pmatrix} \quad \text{or} \quad \begin{pmatrix} 12 & 0 \\ 1 & 13 \end{pmatrix} \begin{pmatrix} 8 & 3 & 2 \\ 2 & 5 & 7 \end{pmatrix}. \tag{2.15}$$

The length of the rows of the first matrix must equal the length of the columns of
the second, and therefore only the matrix multiplication on the right makes sense. To
calculate the product of these matrices we calculate the scalar product of each row
of the left matrix with each column of the right matrix. The position of the scalar
product in the resulting matrix is given by the row position of the first matrix and the
column position of the second. Hence

$$\begin{pmatrix} 8 & 3 & 2 \\ 2 & 5 & 7 \end{pmatrix}$$

$$\begin{pmatrix} 12 & 0 \\ 1 & 13 \end{pmatrix} \quad \vdots \quad = \begin{pmatrix} 96 & 36 & 24 \\ 34 & \mathbf{68} & 93 \end{pmatrix}. \tag{2.16}$$

The placement of the numbers is shown in detail in Fig. 2.5.

Matrices can be used to change vectors. Consider the following example:

$$\begin{pmatrix} 3 & 1 \\ 2 & 5 \end{pmatrix} \begin{pmatrix} 1 \\ 2 \end{pmatrix} = \begin{pmatrix} 5 \\ 12 \end{pmatrix}. \tag{2.17}$$

Verify this using the rules of matrix multiplication set out above. In this case, the
2×2 matrix $\begin{pmatrix} 3 & 1 \\ 2 & 5 \end{pmatrix}$ "acts" on the vector $\begin{pmatrix} 1 \\ 2 \end{pmatrix}$, and returns the vector $\begin{pmatrix} 5 \\ 12 \end{pmatrix}$. In
other words, the matrix transforms $\begin{pmatrix} 1 \\ 2 \end{pmatrix}$ into $\begin{pmatrix} 5 \\ 12 \end{pmatrix}$. If the vectors represent quantum
states, the matrices represent the changes in the quantum states. This is important
when we try to change the state vector from a photon entering the beam splitter from
the left into a state vector of the photon in a superposition of exiting to the right and
down. We will explore this in detail in the next section.

2.4   The Beam Splitter as a Matrix

We have identified the state of the photon before the beam splitter with a vector of length 1 in a two-dimensional plane spanned by the basis vectors $|\text{left}\rangle_1$ and $|\text{top}\rangle_1$, and the state of the photon after the beam splitter with a vector spanned by the basis vectors $|\text{right}\rangle_2$ and $|\text{down}\rangle_2$. The next question is how we can describe the action of the beam splitter.

Clearly, the beam splitter should turn the state of the photon before the beam splitter into the state of the photon after the beam splitter. This is called a *transformation*, and we can write it symbolically as

$$U_{BS}|\text{in}\rangle_1 = |\text{out}\rangle_2 , \qquad (2.18)$$

where we expressed the output state after the beam splitter $|\text{out}\rangle_2$ in terms of the input state before the beam splitter $|\text{in}\rangle_1$, and an *operator* U_{BS} that denotes the action of the beam splitter, which transforms vectors from state space 1 to state space 2.

We can't do much yet with (2.18) as it is written, but we can use the vector language developed previously to turn (2.18) into a form that we can do calculations with. The beam splitter operator U_{BS} turns one vector into another vector, and it should therefore be described by a matrix. Since the row length of U_{BS} should equal the column length of the vector it acts upon (from the matrix multiplication rule), the matrix representing U_{BS} must have row length 2 (i.e., two columns). Second, after matrix multiplication we want again a state vector of dimension 2, so the matrix representing U_{BS} have two rows. Therefore, U_{BS} must be a 2×2 matrix. We can use the behaviour of the photon in the various experiments to figure out the numerical values of the matrix elements.

First, consider again the example of the beam splitter of experiment 2 in the previous chapter. Using (2.10) and (2.11) in vector notation, we have

$$U_{BS}\begin{pmatrix} 1 \\ 0 \end{pmatrix} = \frac{1}{\sqrt{2}} \begin{pmatrix} 1 \\ 1 \end{pmatrix}, \qquad (2.19)$$

and we can write the 2×2 matrix form of U_{BS} as

$$U_{BS} = \begin{pmatrix} u_{11} & u_{12} \\ u_{21} & u_{22} \end{pmatrix}, \qquad (2.20)$$

where u_{11}, u_{12}, u_{21}, and u_{22} are the numbers we need to determine. We use matrix multiplication to find the first column of the matrix U_{BS}. Equation (2.19) becomes

$$U_{BS}|\text{left}\rangle_1 = \begin{pmatrix} u_{11} & u_{12} \\ u_{21} & u_{22} \end{pmatrix} \begin{pmatrix} 1 \\ 0 \end{pmatrix} = \begin{pmatrix} u_{11} \\ u_{21} \end{pmatrix} = \frac{1}{\sqrt{2}} \begin{pmatrix} 1 \\ 1 \end{pmatrix}. \qquad (2.21)$$

From this, we see directly that

$$u_{11} = u_{21} = \frac{1}{\sqrt{2}} \, . \tag{2.22}$$

This gives us the first column of the matrix U_{BS}. Note that the first column of the matrix is identical to the required output vector because we used the basis vector $\begin{pmatrix} 1 \\ 0 \end{pmatrix}$. In general, the matrix columns are given by the destinations of the basis vectors. We will see this next when we try to determine the second column of U_{BS}.

To obtain the second column of the matrix U_{BS} we must know how U_{BS} acts on the basis state

$$|\text{top}\rangle_1 = \begin{pmatrix} 0 \\ 1 \end{pmatrix} . \tag{2.23}$$

We have not considered this situation before, and we should return momentarily to a beam splitter with a variable reflectivity. Suppose that the beam splitter transmits every photon. Then $|\text{left}\rangle_1$ will turn into $|\text{right}\rangle_2$ and $|\text{top}\rangle_1$ will turn into $|\text{bottom}\rangle_2$. In our vector notation it is clear that the orthogonal input vectors turn into orthogonal output vectors. The same is true when the beam splitter is completely reflective and $|\text{left}\rangle_1$ will turn into $|\text{bottom}\rangle_2$ and $|\text{top}\rangle_1$ will turn into $|\text{right}\rangle_2$. We therefore suppose that *the beam splitter transforms orthogonal input states into orthogonal output states.*

In the case of the beam splitter of experiment 2, the state $|\text{top}\rangle_1$ will be transformed into a state orthogonal to the right-hand side of (2.19). If we determine

$$\frac{1}{\sqrt{2}} \begin{pmatrix} 1 \\ -1 \end{pmatrix} \quad \text{perpendicular to} \quad \frac{1}{\sqrt{2}} \begin{pmatrix} 1 \\ 1 \end{pmatrix} \tag{2.24}$$

then we obtain

$$U_{BS} \begin{pmatrix} 0 \\ 1 \end{pmatrix} = \frac{1}{\sqrt{2}} \begin{pmatrix} 1 \\ -1 \end{pmatrix} . \tag{2.25}$$

The elements u_{12} and u_{22} can then be found in the same way as (2.21), and the matrix U_{BS} is determined to be

$$U_{BS} = \frac{1}{\sqrt{2}} \begin{pmatrix} 1 & 1 \\ 1 & -1 \end{pmatrix} , \tag{2.26}$$

and this is the full mathematical description of a beam splitter with a reflectivity of 50% operating on photons that can enter from the left or the top.

To obtain (2.26) we have assumed that a set of orthogonal vectors are turned into another set of orthogonal vectors, and we need to test wether this is a correct assumption via carefully designed experiments. In the next sections we will see that our choice indeed predicts the right experimental outcomes in the Mach–Zehnder interferometer. However, there is also another argument that may make this rule more plausible. Imagine running the experiment backwards, for example by placing mirrors at the positions of detectors D_1 and D_2. A photon in the state

$$|\text{photon}\rangle_2 = \frac{1}{\sqrt{2}}|\text{right}\rangle_2 + \frac{1}{\sqrt{2}}|\text{down}\rangle_2$$

travelling back through the beam splitter should end up where it started, namely in the state $|\text{left}\rangle_1$. Imaging what would happen if a photon entering from the top would also transform into the state $|\text{photon}\rangle_2$. Then running this process backwards would create an ambiguity: should $|\text{photon}\rangle_2$ return to $|\text{left}\rangle_1$ or $|\text{top}\rangle_1$? Remember that the state description of a physical system uniquely determines everything there is to know about the system. So if we require this reversibility in quantum mechanics, then we *have to* let $|\text{top}\rangle_1$ transform into a state different from $|\text{photon}\rangle_2$. This is why we require the minus sign in (2.24).

Next, we can apply the operation U_{BS} not just to the input states $|\text{left}\rangle_1$ and $|\text{top}\rangle_1$, but also to any superposition of these basis states. For an arbitrary input state $|\text{in}\rangle = a|\text{left}\rangle + b|\text{top}\rangle$ we calculate the output state $|\text{out}\rangle$ as

$$|\text{out}\rangle = U_{\text{BS}}|\text{in}\rangle = \frac{1}{\sqrt{2}}\begin{pmatrix} 1 & 1 \\ 1 & -1 \end{pmatrix}\begin{pmatrix} a \\ b \end{pmatrix} = \frac{1}{\sqrt{2}}\begin{pmatrix} a+b \\ a-b \end{pmatrix}. \qquad (2.27)$$

This is a general property of matrix operators: they apply to any input state vector. If we look at a general 2×2 matrix,

$$U = \begin{pmatrix} u_{11} & u_{12} \\ u_{21} & u_{22} \end{pmatrix}, \qquad (2.28)$$

we can see that the element u_{11} determines how much of the left input ends up in the right output. Similarly, u_{22} determines how much of the top input ends up in the down output. The off-diagonal term u_{21} determines how much the left output ends up in the down output, and u_{12} determines how much the top output ends up in the right output. We could label the matrix elements accordingly:

$$U = \begin{pmatrix} u_{\text{left}\to\text{right}} & u_{\text{top}\to\text{right}} \\ u_{\text{left}\to\text{down}} & u_{\text{top}\to\text{down}} \end{pmatrix}. \qquad (2.29)$$

This is a cumbersome notation, so we will not use it further in this book, but you should appreciate that this is the fundamental meaning of the matrix elements. The numerical values of the matrix elements are deeply connected to the basis we use to write the vectors.

The matrix U transforms two orthogonal basis vectors—in the case above they are $|\text{left}\rangle$ and $|\text{top}\rangle$—into two new orthogonal basis vectors

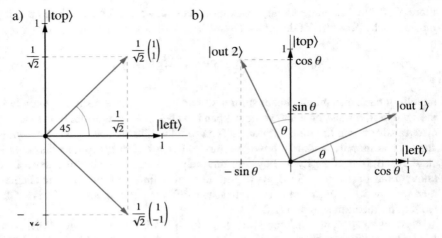

Fig. 2.6 Rotating the basis |left⟩ and |top⟩ to **a** the beam splitter output states, and **b** to a new basis |out 1⟩ and |out 2⟩. In **a** the basis vectors |left⟩ and |top⟩ each rotate differently, while in **b** both rotate in the same way over an angle θ

$$\frac{1}{\sqrt{2}}\begin{pmatrix} 1 \\ 1 \end{pmatrix} \quad \text{and} \quad \frac{1}{\sqrt{2}}\begin{pmatrix} 1 \\ -1 \end{pmatrix}.$$

This is a rotation of the initial basis vectors, as shown in Fig. 2.6a. Notice how the state vector |left⟩ is rotated *counter-clockwise* by 45°, while the state vector |top⟩ is rotated *clockwise* by 135°. The angle between the two vectors after the matrix transformation is still 90°, as it was before the transformation.

Instead of rotating the vectors |left⟩ and |top⟩ over different angles (e.g., +45 and −135° in the case above), it will be of interest to consider the case where both basis vectors rotate in the same way, over the same angle θ. This is shown in Fig. 2.6b. The resulting states are

$$|\text{out } 1\rangle = \cos\theta \, |\text{left}\rangle + \sin\theta \, |\text{top}\rangle \,,$$
$$|\text{out } 2\rangle = -\sin\theta \, |\text{left}\rangle + \cos\theta \, |\text{top}\rangle \,. \tag{2.30}$$

When there is no rotation, i.e., when $\theta = 0$, we see that |out 1⟩ coincides with |left⟩, and |out 2⟩ coincides with |top⟩. The transformation therefore maps the basis vectors to their destinations as follows:

$$\begin{pmatrix} 1 \\ 0 \end{pmatrix} \to \begin{pmatrix} \cos\theta \\ \sin\theta \end{pmatrix} \quad \text{and} \quad \begin{pmatrix} 0 \\ 1 \end{pmatrix} \to \begin{pmatrix} -\sin\theta \\ \cos\theta \end{pmatrix}. \tag{2.31}$$

Using the same construction technique from the start of this section, we find that the resulting vectors form the columns of the matrix:

$$U(\theta) = \begin{pmatrix} \cos\theta & -\sin\theta \\ \sin\theta & \cos\theta \end{pmatrix}. \tag{2.32}$$

This is a so-called *rotation* matrix, because no matter what vector you act upon with this matrix, the result is a vector of the same length rotated counter-clockwise over an angle θ. To summarise, if you know how the basis vectors transform, you can construct the corresponding matrix.

2.5 *Mathematical Intermezzo*: **Complex Numbers**

Quantum mechanics uses complex numbers, so it is important that you are familiar with the basics (that's all we need for now). The idea of a complex number is that, contrary to what you may have learned previously, $\sqrt{-1}$ is not nonsense. However, it is not a normal number either. We can call $\sqrt{-1} = i$, and treat it as a new kind of number, called an imaginary number. All the normal rules of arithmetic apply, so you can multiply i with any other number, or add any number to it. The question is then: what does $z = 4i + 3$ mean? Since the imaginary part ($4i$) is a different kind of number than 3, we can put the two kinds of numbers along perpendicular axes: The real numbers are plotted along the horizontal axis, and the imaginary numbers are plotted along the vertical axis. The combination of a real and an imaginary number is called a *complex number*, and can be represented by a point in a two-dimensional complex plane, as shown in Fig. 2.7.

When we add two complex numbers, we add the real part and the imaginary parts separately:

$$(x + iy) + (u + iv) = (x + u) + i(y + v), \tag{2.33}$$

where x, y, u, and v are real numbers. Notice that the real and imaginary parts really do add separately, like they are the horizontal and vertical components of a two-

Fig. 2.7 The plane of complex numbers (see supplementary material 7)

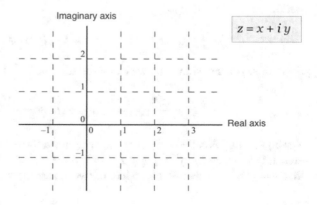

dimensional space (a plane). Do not confuse this space with the two-dimensional state space for the quantum states |photon⟩, though! These are completely different things.

Multiplying two complex numbers is again the same as multiplying two normal numbers, where we use $i \times i = -1$. So we have

$$(x + iy)(u + iv) = xu + ixv + iyu - yv$$
$$= (xu - yv) + i(xv + yu). \quad (2.34)$$

There is again a real part $xu - yv$ and an imaginary part $xv + yu$ to the resulting complex number.

An important procedure is to take the *complex conjugate* of a complex number a, denoted by a^* :

$$(x + iy)^* = x - iy. \quad (2.35)$$

In other words, complex conjugation will change an i into a $-i$ and vice versa, wherever it is in your formula. The product of a complex number $z = x + iy$ with its complex conjugate $z^* = x - iy$ is always a positive real number $|z|^2$:

$$|z|^2 = z^* z = (x + iy)^*(x + iy) = (x - iy)(x + iy) = x^2 + y^2. \quad (2.36)$$

This is the length of the vector from the origin $(0, 0)$ to the point $z = (x, y)$ in the complex plane, and that is why the complex conjugate will show up again and again. If you expect a real number from your calculation (for example a probability), and instead you get a complex number, chances are you have forgotten to take the complex conjugate somewhere along the way.

Numbers in the complex plane can also be represented in polar coordinates, which is particularly convenient for describing phases (which we will need a lot in quantum mechanics). Imagine a line from the point in the complex plane to the origin. The length of that line is called r (we just saw that $r = \sqrt{|z|^2}$) and the angle of that line with the positive horizontal axis is ϕ. Every point in the complex plane can then be written as

$$z = r e^{i\phi} = r \cos \phi + ir \sin \phi. \quad (2.37)$$

From this, you can show that we can write the sine and the cosine as

$$\sin \phi = \frac{e^{i\phi} - e^{-i\phi}}{2i} \quad \text{and} \quad \cos \phi = \frac{e^{i\phi} + e^{-i\phi}}{2}. \quad (2.38)$$

The argument ϕ is called the phase (since it returns the number to the starting position after it has increased by 2π), and r is called the magnitude. You can prove yourself that $r = \sqrt{|z|^2} = |z|$ and the product of two complex numbers becomes

Fig. 2.8 The polar decomposition of complex numbers with radius $r = 1$ (see supplementary material 8)

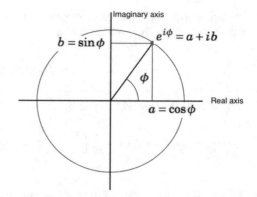

$$z_1 \times z_2 = r_1 \, e^{i\phi_1} \times r_2 \, e^{i\phi_2} = r_1 r_2 \, e^{i(\phi_1 + \phi_2)} . \tag{2.39}$$

The polar decomposition of complex numbers is shown in Fig. 2.8. You should use the polar decomposition form of complex numbers when you can, because it tends to make calculations a lot easier.

2.6 The Phase in an Interferometer

We have to rectify an oversimplification. You may have assumed that the numbers a and b in (2.7) are always real numbers. However, in quantum mechanics we must allow a and b to be complex numbers. Why do we need complex numbers? To answer this question, we have to look at (2.25). We constructed the output state of the beam splitter when the photon enters from the top by requiring that it is orthogonal to the output state when the photon entered from the left. But what does the minus sign mean? We know that in experiment 3 we obtain interference in the Mach–Zehnder interferometer, which prevents one detector from registering photons. The minus sign in (2.25) is the origin of the opposite phase of the two waves entering detector D_1 (and leaving it dark due to destructive interference). Therefore, the minus sign is related to a $\phi = \pi$ phase shift in the wave, and this means that at some point we must also be able to consider phases other than $\phi = 0$ or $\phi = \pi$. Using the polar form of complex numbers $e^{i\phi}$ you see that these complex numbers will occur naturally. In this section we will explore exactly how.

Let us perform a short calculation to show how the Mach–Zehnder interferometer and the interference works with our quantum state vectors. The photon enters the Mach–Zehnder from the left in the first beam splitter:

$$|\text{in}\rangle_1 = \begin{pmatrix} 1 \\ 0 \end{pmatrix} . \tag{2.40}$$

We apply the first beam splitter operator to this state in order to obtain the state of the photon inside the interferometer:

$$|\text{inside}\rangle_2 = U_{\text{BS}}|\text{in}\rangle_1 = \frac{1}{\sqrt{2}}\begin{pmatrix} 1 \\ 1 \end{pmatrix}, \tag{2.41}$$

followed by the second beam splitter

$$|\text{out}\rangle_3 = U_{\text{BS}}|\text{inside}\rangle_2 = \frac{1}{\sqrt{2}}\begin{pmatrix} 1 & 1 \\ 1 & -1 \end{pmatrix} \times \frac{1}{\sqrt{2}}\begin{pmatrix} 1 \\ 1 \end{pmatrix} = \frac{1}{2}\begin{pmatrix} 1+1 \\ 1-1 \end{pmatrix} = \begin{pmatrix} 1 \\ 0 \end{pmatrix} = |D_2\rangle_3, \tag{2.42}$$

where $|D_2\rangle_3$ is the state of the photon entering detector D_2 at some time t_3 after the second beam splitter. This is exactly what we wanted to achieve: a photon entering the interferometer from the left will not trigger detector D_1 because each time it will exit the beam splitter on the side that triggers D_2.

We can do the calculation in (2.42) slightly differently: we can first calculate the product of the two beam splitters:

$$U_{\text{BS}}U_{\text{BS}} = \frac{1}{\sqrt{2}}\begin{pmatrix} 1 & 1 \\ 1 & -1 \end{pmatrix} \times \frac{1}{\sqrt{2}}\begin{pmatrix} 1 & 1 \\ 1 & -1 \end{pmatrix} = \begin{pmatrix} 1 & 0 \\ 0 & 1 \end{pmatrix} = \mathbb{I}. \tag{2.43}$$

The matrix $\mathbb{I}$ is very special, because it always transforms a vector into itself. We call it the *identity*. For our calculation, it means that the expression

$$|\text{out}\rangle_3 = U_{\text{BS}}U_{\text{BS}}|\text{in}\rangle_1 \tag{2.44}$$

will take the form

$$\begin{pmatrix} 1 \\ 0 \end{pmatrix} = \begin{pmatrix} 1 & 0 \\ 0 & 1 \end{pmatrix}\begin{pmatrix} 1 \\ 0 \end{pmatrix} \quad \text{and} \quad \begin{pmatrix} 0 \\ 1 \end{pmatrix} = \begin{pmatrix} 1 & 0 \\ 0 & 1 \end{pmatrix}\begin{pmatrix} 0 \\ 1 \end{pmatrix}. \tag{2.45}$$

Physically, we obtain the interference by carefully tuning the lengths of the two paths between the beam splitters. For classical light (a wave) we must make the path lengths equal in order to send all the light into detector D_2.

If we want all the light to end up in detector D_1 instead, the path lengths must differ by half a wavelength. As we saw in experiment 3, the photons follow the same pattern: they go where the classical signal goes, only probabilistically. How do we describe this change in path length mathematically?

First, applying a path length difference is again an operator, and in matrix notation it should be placed *between* the two beam splitter matrices, because the order of the matrices is determined by the order of the physical elements (beam splitters, phase shifts) that the photon encounters. Second, it must not mix the two paths, because photons cannot jump from one path inside the interferometer to the other. This means that the matrix must be diagonal: only the matrix elements along the diagonal line from the top left to the bottom right of the matrix can be nonzero (see (2.29)). Changing the relative path length by half a wavelength then takes the following form:

$$U_{\lambda/2} = \begin{pmatrix} 1 & 0 \\ 0 & -1 \end{pmatrix}, \tag{2.46}$$

where $\lambda/2$ denotes half a wavelength. Putting this operator together with the beam splitter operators, we use matrix multiplication to describe the successive operations of beam splitter, phase shift, and beam splitter again:

$$\begin{aligned} U_{BS} U_{\lambda/2} U_{BS} &= \frac{1}{\sqrt{2}} \begin{pmatrix} 1 & 1 \\ 1 & -1 \end{pmatrix} \times \begin{pmatrix} 1 & 0 \\ 0 & -1 \end{pmatrix} \times \frac{1}{\sqrt{2}} \begin{pmatrix} 1 & 1 \\ 1 & -1 \end{pmatrix} \\ &= \frac{1}{2} \begin{pmatrix} 1 & 1 \\ 1 & -1 \end{pmatrix} \times \begin{pmatrix} 1 & 1 \\ -1 & 1 \end{pmatrix} \\ &= \begin{pmatrix} 0 & 1 \\ 1 & 0 \end{pmatrix}. \end{aligned} \tag{2.47}$$

Now when we calculate the state of the photon leaving the interferometer $|\text{out}\rangle_4 = U_{BS} U_{\lambda/2} U_{BS} |\text{left}\rangle_1$ at time t_4, we find that

$$|\text{out}\rangle_4 = \begin{pmatrix} 0 & 1 \\ 1 & 0 \end{pmatrix} \begin{pmatrix} 1 \\ 0 \end{pmatrix} = \begin{pmatrix} 0 \\ 1 \end{pmatrix} = |D_1\rangle_4. \tag{2.48}$$

In other words, the photon now triggers detector D_1. This is exactly what we expect from classical physics when we change the path length of one arm by half a wavelength. By changing the phase of the wave in one path relative to the other (i.e., adding half a wavelength in one path), we switch the output from $|D_2\rangle$ to $|D_1\rangle$.

Next, consider how the operator $U_{\lambda/2}$ affects the state in (2.10):

$$U_{\lambda/2} \frac{1}{\sqrt{2}} \begin{pmatrix} 1 \\ 1 \end{pmatrix} = \frac{1}{\sqrt{2}} \begin{pmatrix} 1 & 0 \\ 0 & -1 \end{pmatrix} \begin{pmatrix} 1 \\ 1 \end{pmatrix} = \frac{1}{\sqrt{2}} \begin{pmatrix} 1 \\ -1 \end{pmatrix}. \tag{2.49}$$

The mysterious minus sign just appeared in the state vector! So clearly, the sign relates to the relative phase of the two photon states inside the interferometer. The state

$$\frac{1}{\sqrt{2}} \begin{pmatrix} 1 \\ 1 \end{pmatrix} = \frac{1}{\sqrt{2}} |\text{right}\rangle_3 + \frac{1}{\sqrt{2}} |\text{down}\rangle_3 \tag{2.50}$$

is a superposition of two photon paths with the same phase, while

$$\frac{1}{\sqrt{2}} \begin{pmatrix} 1 \\ -1 \end{pmatrix} = \frac{1}{\sqrt{2}} |\text{right}\rangle_3 - \frac{1}{\sqrt{2}} |\text{down}\rangle_3 \tag{2.51}$$

is a superposition of two photon paths with a relative phase difference of half a wavelength.

The path length is of course something that we can vary continuously, and this raises the question how we can include an arbitrary phase difference ϕ in the interferometer. As a first guess, we could say

$$U_\phi = \begin{pmatrix} 1 & 0 \\ 0 & \phi \end{pmatrix}?$$

where ϕ varies continuously between $+1$ and -1. This will not work, however, because an arbitrary $\phi < 1$ will shorten the state vector, and we said that the state vector must have length 1 (the importance of this will become clear in the next section). Also, a phase must take values between 0 and 2π, rather than between $+1$ and -1. What will work is making the phase *complex*, since $|e^{i\phi}| = 1$ for all values of ϕ:

$$U_\phi = \begin{pmatrix} 1 & 0 \\ 0 & e^{i\phi} \end{pmatrix}. \tag{2.52}$$

This will change an arbitrary photon state as follows:

$$U_\phi \begin{pmatrix} a \\ b \end{pmatrix} = \begin{pmatrix} 1 & 0 \\ 0 & e^{i\phi} \end{pmatrix} \begin{pmatrix} a \\ b \end{pmatrix} = \begin{pmatrix} a \\ b\,e^{i\phi} \end{pmatrix}, \tag{2.53}$$

which means that in general the numbers a and b must allow complex values. It also means that the unit length condition $a^2 + b^2 = 1$ must now be replaced by $|a|^2 + |b|^2 = 1$, where $|a|^2 = a^*a$ and $|b|^2 = b^*b$. We call a and b the *amplitudes* of the photon states |right⟩ and |down⟩. We can combine the two beam splitters and the relative phase shift into one overall transformation of the Mach–Zehnder interferometer U_{MZ} on the state of a photon entering it. Since the interferometer is nothing more than a beam splitter, followed by a phase shift, followed by the second beam splitter, we write

$$U_{MZ} = U_{BS}\, U_\phi\, U_{BS} = \frac{1}{2} \begin{pmatrix} 1 & 1 \\ 1 & -1 \end{pmatrix} \begin{pmatrix} 1 & 0 \\ 0 & e^{i\phi} \end{pmatrix} \begin{pmatrix} 1 & 1 \\ 1 & -1 \end{pmatrix}$$

$$= \frac{1}{2} \begin{pmatrix} 1 + e^{i\phi} & 1 - e^{i\phi} \\ 1 - e^{i\phi} & 1 + e^{i\phi} \end{pmatrix}. \tag{2.54}$$

Note that the first beam splitter the photon encounters in the Mach–Zehnder interferometer is represented by the matrix on the far right in the first line of (2.54), since that matrix acts on the state vector first. When we apply U_{MZ} to the input state $|\text{left}\rangle_1$, the result is

$$|\text{out}\rangle = U_{MZ}|\text{left}\rangle_1 = \frac{1}{2} \begin{pmatrix} 1 + e^{i\phi} \\ 1 - e^{i\phi} \end{pmatrix}. \tag{2.55}$$

Clearly, when $\phi = 0$ the photon exits the interferometer into detector D_2, while for $\phi = \pi$ the photon ends up in detector D_1, as required. The matrix U_{MZ} describes the interferometer acting on a photon completely. And yet it is a simple 2×2 matrix.

Finally, let us return momentarily to the rotation matrix $U(\theta)$ from (2.32). We could choose $\theta = 45°$, and obtain a matrix that is *almost* the same as U_{BS}:

$$U(45°) = \frac{1}{\sqrt{2}} \begin{pmatrix} 1 & -1 \\ 1 & 1 \end{pmatrix}. \tag{2.56}$$

It is easy to check using matrix multiplication that this rotation matrix becomes the same as the beam splitter matrix if we include a π phase shift in the "down" output:

$$U_{\mathrm{BS}} = U(45°)U_{\lambda/2} = \frac{1}{\sqrt{2}} \begin{pmatrix} 1 & -1 \\ 1 & 1 \end{pmatrix} \begin{pmatrix} 1 & 0 \\ 0 & -1 \end{pmatrix} = \frac{1}{\sqrt{2}} \begin{pmatrix} 1 & 1 \\ 1 & -1 \end{pmatrix}. \tag{2.57}$$

Could we use the matrix $U(45°)$ in constructing the Mach–Zehnder interferometer, instead of U_{BS}? Indeed, we could, and you should try it! But that in turn raises the question what is the correct matrix for a real beam splitter.

A physical beam splitter can be implemented in different ways, but the most intuitive way is as a piece of glass with one surface acting as a half-mirror (see Fig. 2.9). The other surface of the glass usually has an anti-reflective coating. From classical optics you may know that typically the reflection from a surface with a higher refractive index picks up a π phase shift. Therefore, in the setup in Fig. 2.9, the photon coming in from the top will pick up a π phase shift when it is reflected into the right output. Therefore, in this case the matrix $U(45°)$ is the correct description. If the beam splitter was oriented in differently (rotated $180°$ in the plane of the page), the left input would pick up a minus sign upon reflection into the downwards direction, and the correct matrix would be $U(-45°)$.

Fig. 2.9 A possible implementation of a physical beam splitter. The minus sign (i.e., the π phase shift) is on the matrix element that connects "top" with "right". If the beam splitter was oriented $180°$ rotated, the minus sign is on the matrix element that connects "left" to "down"

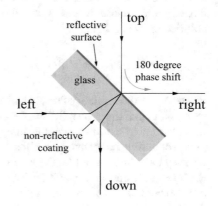

So why did we choose U_{BS}? In practice it does not matter much which beam splitter matrix we use, since they are all related to each other via phase shifts in the incoming and outgoing arms. When we set up experiments like a Mach–Zehnder interferometer, we typically try to make the path lengths of the upper and lower arms equal, and we do this by monitoring the output of the interferometer. Any phase shifts that are due to our beam splitter model will be compensated by changes in the two path lengths.

2.7 *Mathematical Intermezzo*: **Probabilities**

In quantum mechanics, the basis of everything we can say about nature is our ability to calculate probabilities of measurement outcomes. We therefore need to introduce a little bit of probability theory. Luckily, we do not require anything more than a way to calculate averages at this point.

A probability is the likelihood of a certain event happening. If the probability of an event is zero, we know for a fact that it will not happen, while a probability of one means that it certainly will happen. So both 0 and 1 are expressions of certainty. Suppose that we have a coin with two sides, denoted "heads" and "tails". We can give a symbol $\Pr(H)$ to the probability that we find heads in a coin toss, and we give the symbol $\Pr(T)$ to the probability that we find tails. If the coin is sufficiently thin that it never lands on its side, we know that we are going to find either heads or tails. Therefore $\Pr(H) + \Pr(T) = 1$. If the coin is fair we expect no difference between heads and tails, so

$$\Pr(H) = \Pr(T) = \frac{1}{2}. \qquad (2.58)$$

This is the definition of a fair coin. We measure the probabilities by making a very large number of coin tosses N, and the relative frequencies then approach the probabilities $\Pr(H)$ and $\Pr(T)$:

$$\Pr(H) = \lim_{N \to \infty} \frac{N_H}{N} \quad \text{and} \quad \Pr(T) = \lim_{N \to \infty} \frac{N_T}{N}, \qquad (2.59)$$

where N_H is the number of heads and N_T is the number of tails found in the experiment. Note that the probabilities of every possible way in which a situation can play out must always sum to one, because you know with certainty that whatever happens, something happens. Even if that something is "nothing happens". For example, we may calculate (based on the way we throw and the dimensions of the coin) that $\Pr(H) = 0.47$ and $\Pr(T) = 0.48$. Then either we made a mistake in our calculation, or we have not taken into account all possibilities. For example, the probability for the coin to land on its side may be $\Pr(\text{side}) = 0.05$, or we decide not to toss the coin at all with probability $\Pr(\text{not}) = 0.05$, or with probability $\Pr(\text{radiator}) = 0.05$ the coin fell behind the radiator and could be retrieved but not read out; the possibilities are endless. Therefore, probabilities are always numbers between 0 and 1 without any units, and the probabilities of all possible outcomes must sum to 1.

Next, we can use probabilities to calculate averages. Suppose that we use the coin we described earlier to determine how much money you will win: If we throw "heads" you receive £50, and if you throw "tails" you pay £10. In the unlikely event that the coin lands on its side, you win nothing. What is the average amount of money W that you win or lose in this game? To determine this, we multiply each amount of money you can win with the probability of the corresponding outcome of the coin toss:

$$
\begin{aligned}
W &= \Pr(H) \times £50 + \Pr(T) \times (-£10) + p_{\text{side}} \times 0 \\
&= 0.47 \times £50 - 0.48 \times £10 + 0.05 \times 0 = £18.70 .
\end{aligned} \tag{2.60}
$$

So on average you will win money, and this is a game worth playing! (The real question is: who will play this game with you?) In the next section we will determine how we can calculate probabilities in quantum mechanics.

2.8 How to Calculate Probabilities

Quantum mechanics allows us to calculate the probabilities of various measurement outcomes in an experiment. In other words, we can calculate the probability of finding a photon in detector D_1 or D_2 of the Mach–Zehnder interferometer for any value of the phase difference ϕ due to a difference in the two path lengths between the beam splitters. This may seem a calculation of only modest interest, but it will turn out to underpin a very important experiment, detailed in the next section. We will first determine what the probabilities should be in a few special cases, and then extract from this the general formula.

Consider again the beam splitter with a reflectivity of one half and a photon entering from the left. According to (2.10), the state of the photon after the beam splitter can be written as

$$
|\text{photon}\rangle = \frac{1}{\sqrt{2}} |\text{right}\rangle + \frac{1}{\sqrt{2}} |\text{down}\rangle . \tag{2.61}
$$

Furthermore, we know from the previous chapter that the probability of finding the photon in the right output beam is $p_{\text{right}} = 1/2$ and the probability of finding the photon in the down output beam is $p_{\text{down}} = 1/2$, obeying the rule that $p_{\text{right}} + p_{\text{down}} = 1$. When we represent this graphically in Fig. 2.10, we see that the projections of the vector $|\text{photon}\rangle$ onto the axes defined by $|\text{right}\rangle$ and $|\text{down}\rangle$ are equal. This gives us an indication that the components of the vector $|\text{photon}\rangle$ have something to do with the probabilities. A second way of seeing this is to rotate the $|\text{photon}\rangle$ arrow closer to the horizontal $|\text{right}\rangle$ arrow. The component along the horizontal axis will

Fig. 2.10 Probability
amplitudes

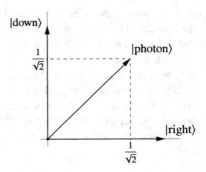

grow from $1/\sqrt{2}$ to a value closer to one, and the component on the vertical axis will
reduce to a value closer to zero. The state $|\text{photon}\rangle$ becomes closer to the state $|\text{right}\rangle$,
and the probability that the photon exits the beam splitter to the right should increase.
In the extreme case where the angle between $|\text{photon}\rangle$ and $|\text{right}\rangle$ becomes zero, the
photon should always come out on the right side, since now $|\text{photon}\rangle = |\text{right}\rangle$.

However, the components do not sum to 1 (instead, they sum to $\sqrt{2} > 1$), so
they cannot be straightforward probabilities. On the other hand, the squares of the
components do sum up to 1 (this was our normalisation condition $|a|^2 + |b|^2 = 1$),
and could therefore be interpreted as probabilities. Indeed, one of the fundamental
rules in quantum mechanics says that the probability of a measurement outcome is
the absolute square of the corresponding amplitude. For example, suppose that the
state of a photon after a beam splitter is given by

$$|\text{photon}\rangle = a|\text{right}\rangle + b|\text{down}\rangle. \tag{2.62}$$

The probability of finding the photon emerging in the beam to the right of the beam
splitter is then given by $|a|^2$. Similarly, the probability of finding the photon emerging
in the beam downwards from the beam splitter is given by $|b|^2$.

This rule is simple enough, but we need a slightly more general way of calculating
it if we are to use it in other situations. First of all, let $\langle \cdot | \cdot \rangle$ denote the scalar product
between two vectors. The scalar product is sometimes also called a "bracket", and
the vector $\langle \cdot |$ is called a "bra" vector. We construct a bra vector from a ket vector by
turning the column into a row vector (the top number in the column vector becomes
the left number in the row vector), and take the *complex conjugate* of all the vector
elements.

Since the states $|\text{right}\rangle$ and $|\text{down}\rangle$ are orthogonal vectors with length 1, we have

$$\langle\text{right}|\text{right}\rangle = \langle\text{down}|\text{down}\rangle = 1$$
$$\langle\text{right}|\text{down}\rangle = \langle\text{down}|\text{right}\rangle = 0. \tag{2.63}$$

This means that we can calculate the following:

$$\langle \text{right}|\text{photon}\rangle = a\langle \text{right}|\text{right}\rangle + b\langle \text{right}|\text{down}\rangle$$
$$= a \times 1 + b \times 0 = a \,. \tag{2.64}$$

The probability of finding the photon on the right of the beam splitter is the absolute square of this, namely

$$p_{\text{right}} = |\langle \text{right}|\text{photon}\rangle|^2 = |a|^2 \,. \tag{2.65}$$

Similarly, the probability of finding the photon propagating downwards from the beam splitter is

$$p_{\text{down}} = |\langle \text{down}|\text{photon}\rangle|^2 = |b|^2 \,. \tag{2.66}$$

The general rule is that we identify a state with each measurement outcome $|\text{outcome}\rangle$ (such as $|\text{right}\rangle$ or $|\text{down}\rangle$). The probability of the measurement outcome is then the absolute square of the scalar product between the state of the system and the state corresponding to the measurement outcome:

$$p(\text{outcome}) = |\langle \text{outcome}|\text{state}\rangle|^2 \tag{2.67}$$

This is called the Born rule, after Max Born,[2] who first stated it. It is the general rule for calculating probabilities of measurement outcomes in quantum mechanics. There is a direct analog in classical physics for this rule that you may already have come across, namely the relation between the amplitude of a wave and its intensity. The intensity of a wave is given by the square of its amplitude. If we consider the intensity as a measure of how much there is of something, you can see how this translates to probabilities. It also gives another glimpse of the close connections quantum mechanics has with wave theory (the first glimpse was when we constructed superpositions of photon states).

When you calculate the scalar product of two complex vectors, you must remember that the entries of the bra vector are the complex conjugate of the ket vector. For example, we can have two vectors $|\psi\rangle$ and $|\phi\rangle$ (remember that the labels are fairly arbitrary; here we use the greek letters "psi": ψ, and "phi": ϕ):

$$|\psi\rangle = \begin{pmatrix} a \\ b \end{pmatrix} \quad \text{and} \quad |\phi\rangle = \begin{pmatrix} c \\ d \end{pmatrix}, \tag{2.68}$$

where a, b, c, and d are complex numbers. Since the bra of $|\psi\rangle$ is

$$\langle \psi| = \begin{pmatrix} a^* & b^* \end{pmatrix}, \tag{2.69}$$

the scalar product between $|\psi\rangle$ and $|\phi\rangle$ becomes

---

[2] M. Born, *Zur Quantenmechanik der Stoßvorgänge*, Zeitschrift für Physik **37** 863, 1926.

$$\langle\psi|\phi\rangle = \begin{pmatrix} a^* & b^* \end{pmatrix}\begin{pmatrix} c \\ d \end{pmatrix} = a^*c + b^*d\,. \tag{2.70}$$

If $a = 2 + 3i$, $b = 1 - 4i$, $c = 8 - 3i$, and $d = 6 + i$ the scalar product $\langle\psi|\phi\rangle$ is

$$\begin{aligned}
\langle\psi|\phi\rangle &= a^*c + b^*d \\
&= (2 - 3i)(8 - 3i) + (1 + 4i)(6 + i) \\
&= 16 - 6i - 24i - 9 + 6 + i + 24i - 4 = 9 - 5i\,.
\end{aligned} \tag{2.71}$$

In polar form this is

$$\langle\psi|\phi\rangle = 9 - 5i = \sqrt{9^2 + (-5)^2}\,e^{i\arctan\frac{-5}{9}} = \sqrt{106}\,e^{-0.507i}\,. \tag{2.72}$$

The vectors $|\psi\rangle$ and $|\phi\rangle$ are not normalised:

$$\begin{aligned}
\langle\psi|\psi\rangle &= |a|^2 + |b|^2 = 30 \\
\langle\phi|\phi\rangle &= |c|^2 + |d|^2 = 110\,,
\end{aligned} \tag{2.73}$$

which means that $|\psi\rangle$ and $|\phi\rangle$ cannot be proper quantum state vectors. We will encounter this situation in later parts of this book. In order to turn $|\psi\rangle$ and $|\phi\rangle$ into proper state vectors with length 1, we have to "normalise" them. This is easy, we just divide the vector by its length to get a new vector of length 1:

$$|\psi_{\mathrm{norm}}\rangle = \frac{|\psi\rangle}{\sqrt{\langle\psi|\psi\rangle}} \quad\text{and}\quad |\phi_{\mathrm{norm}}\rangle = \frac{|\phi\rangle}{\sqrt{\langle\phi|\phi\rangle}}\,, \tag{2.74}$$

where $|\psi_{\mathrm{norm}}\rangle$ and $|\phi_{\mathrm{norm}}\rangle$ are now normalised. You should check that this works. Note that it won't work if you forget to take the complex conjugate when turning a ket into a bra!

There is one more important rule for quantum states that follows from the definition of probabilities of measurement outcomes. Consider again the quantum state $|\mathrm{photon}\rangle$ from (2.62). The probability amplitudes a and b are complex numbers, and they may have a common phase factor $e^{i\phi}$:

$$|\mathrm{photon}\rangle = a'\,e^{i\phi}\,|\mathrm{right}\rangle + b'\,e^{i\phi}\,|\mathrm{down}\rangle = e^{i\phi}\,|\psi\rangle \tag{2.75}$$

where $a' = a\,e^{-i\phi}$ and $b' = b\,e^{-i\phi}$. The two states $|\mathrm{photon}\rangle$ and $|\psi\rangle$ differ only by an overall phase factor $e^{i\phi}$, called a *global* phase. Let's see what this global phase does to the probabilities of any measurement outcome m ("right" or "down"):

$$\Pr(m) = |\langle m|\mathrm{photon}\rangle|^2 = \langle m|\mathrm{photon}\rangle\,\langle\mathrm{photon}|m\rangle\,. \tag{2.76}$$

Similarly, we must have that

$$\Pr(m) = \left| \langle m | \, e^{i\phi} | \psi \rangle \right|^2 = e^{i\phi} \, \langle m | \psi \rangle \; e^{-i\phi} \, \langle \psi | m \rangle$$
$$= e^{i\phi} \, e^{-i\phi} \, \langle m | \psi \rangle \, \langle \psi | m \rangle = |\langle m | \psi \rangle|^2 \, , \tag{2.77}$$

since $e^{i\phi} \, e^{-i\phi} = e^{i\phi - i\phi} = e^0 = 1$ for all values of ϕ. You see that the global phase $e^{i\phi}$ has no influence on the probabilities of the measurement outcomes. And because all observable physical phenomena are ultimately derived from probabilities, the global phase on the quantum state has no observable effects! This means that you may introduce a global phase to your quantum state if that is convenient, and indeed we will do this occasionally to make the maths more elegant. For example, you may wish to make the amplitude a real.

While the global phase of a quantum state has no observable effects, the same is not true for a *relative* phase that is not shared among all the amplitudes. Consider the state

$$a \, |\text{right}\rangle + b \, e^{i\phi} \, |\text{down}\rangle \; .$$

This state has a relative phase $e^{i\phi}$ on the term $|\text{down}\rangle$, and is generally different from

$$a \, |\text{right}\rangle + b \, |\text{down}\rangle \; .$$

You can easily verify this by choosing $a = b = 1/\sqrt{2}$ and $\phi = \pi$. In this case the two states are orthogonal, and will lead to different measurement outcome probabilities if you send the photon into a beam splitter and measure the output (as we have already seen before).

2.9 Gravitational Wave Detection

Let's consider how the Mach–Zehnder interferometer can be employed in a real physics experiment, such as the measurement of gravitational waves. These are waves predicted by Einstein's theory of General Relativity, and occur when two very massive bodies (such as black holes or neutron stars) collide. You can think of them as "ripples" in the fabric of space and time, stretching and compressing space itself. On 14 September 2015, these waves were spotted for the first time by the LIGO collaboration,[3] and it was inferred that two black holes of 29 and 36 solar masses spiralled into each other. In the last moments before the merger (around half a second), the binary system emitted about three solar *masses* worth of energy.[4] When a gravitational wave engulfs Earth, it will stretch space a little in one direction, and compress space in the perpendicular direction. It will therefore stretch one arm of

---

[3] B. P. Abbott et al., *GW170104: Observation of a 50-Solar-Mass Binary Black Hole Coalescence at Redshift 0.2*. Phys. Rev. Lett. **118** 22110, 2015.

[4] You can convert this to Joules by using the famous formula $E = mc^2$, where E is the energy, m the three solar masses in kilograms, and c the speed of light in vacuum in metres per second.

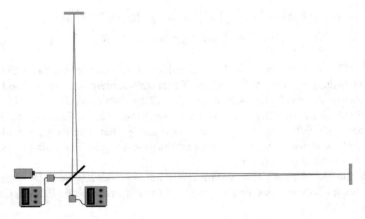

Fig. 2.11 A gravitational wave detector (see supplementary material 11)

an interferometer more than the other, and we can detect this as a phase shift. So the experiment tries to see a tiny phase shift (a thousandth of the width of a proton) by measuring a stream of photons arriving in detectors D_1 and D_2. Let's calculate how this works.

First, we need to modify the Mach–Zehnder interferometer to obtain two long arms. We do this by orienting the mirrors in Fig. 1.4 such that the beams are sent back to the first beam splitter, instead of a second. This is called a Michelson-Morley interferometer, (see Fig. 2.11). The mathematical description of the Michelson-Morley interferometer is exactly the same as the Mach–Zehnder interferometer: the photons don't care whether they encounter one beam splitter or another; a beam splitter does what a beam splitter does.

When we set up the interferometer as before, the output state of the photon is given by

$$|\text{out}\rangle_4 = U_{\text{BS}} U_\phi\, U_{\text{BS}} |\text{left}\rangle_1 = \frac{1}{2}\begin{pmatrix} 1 + e^{i\phi} \\ 1 - e^{i\phi} \end{pmatrix}, \tag{2.78}$$

where ϕ is now the phase shift in the interferometer that results from a gravitational wave passing by.

We can calculate the probability of finding a photon in detector D_1 as

$$p_1 = |\langle D_1 |\text{out}\rangle|^2 = \frac{1}{4}\left| (0\ 1)\begin{pmatrix} 1 + e^{i\phi} \\ 1 - e^{i\phi} \end{pmatrix} \right|^2 = \frac{1}{4}\left| 1 - e^{i\phi} \right|^2$$

$$= \frac{1}{4}(1 - e^{i\phi})(1 - e^{-i\phi}) = \frac{1}{2}(1 - \cos\phi), \tag{2.79}$$

where we used (2.38). Similarly, we can calculate that

$$p_2 = |\langle D_2|\text{out}\rangle|^2 = \frac{1}{2}\left(1 + \cos\phi\right), \tag{2.80}$$

and $p_1 + p_2 = 1$. We can estimate these probabilities by repeating the experiment a large number of times (say N times). If N_1 is the number of detections in D_1 and N_2 is the number of detections in D_2, then the probabilities are estimated as

$$p_1 \simeq \frac{N_1}{N} \quad \text{and} \quad p_2 \simeq \frac{N_2}{N}. \tag{2.81}$$

Using the fact that

$$p_2 - p_1 = \cos\phi, \tag{2.82}$$

the phase ϕ is then estimated as

$$\phi = \arccos(p_2 - p_1) \simeq \arccos\left(\frac{N_2 - N_1}{N}\right). \tag{2.83}$$

In the actual LIGO experiment, the vast majority of work was (and still is) dedicated to eliminating external noise sources, such as thermal fluctuations, electronic noise, the motion of the waves and the tides, and nearby logging operations. LIGO also uses fairly high intensity (classical) laser light that contains lots of photons, instead of sending individual photons through the interferometer. The leaders of the collaboration, Rainer Weiss, Kip Thorne and Barry Barish, received the Nobel prize in physics in 2017 for the detection of gravitational waves. While the Michelson-Morley interferometer can also be described with classical light waves, it is an instructive example how we treat a system like this in quantum mechanics.

2.10   A Quantum Bomb Detector

Finally, we consider a thought experiment that lays bare some of the stranger aspects of quantum mechanics. This is the bomb detection scheme introduced by Elitzur and Vaidman[5] in 1993. It consists of the following setup: A bomb is so sensitive that it is set off even if it is hit by a single photon. The aim is to determine that the bomb is present without setting it off. This seems an impossible task. In order to detect the presence of the bomb, we have to somehow interact with it. For example, we can reflect light off the bomb and measure the scattered light in our retinas. This would allow us to see the bomb, but it would also explode it, since the photons are hitting the bomb. Since reflecting a single photon off an object is one of the most gentle ways of interacting with the bomb, it is not clear how this can be achieved.

---

[5] A. C. Elitzur and L. Vaidman, *Foundations of Physics* **23**, 987–997, 1993.

Fig. 2.12 How to detect the presence of a bomb without setting it off?

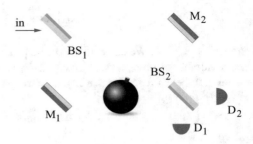

Remarkably, quantum mechanics allows a way to detect the bomb without setting it off. The solution is to construct a Mach–Zehnder interferometer such that the bomb, if it was there, blocks one arm, for example the lower arm (see Fig. 2.12). We balance the interferometer such that *without* a bomb present, an incoming photon will always enter detector D_2. We know how to describe this mathematically:

$$|\text{in}\rangle = \begin{pmatrix} 1 \\ 0 \end{pmatrix}, \tag{2.84}$$

which is transformed into the output state

$$|\text{out}\rangle = U_{\text{BS}_2} U_{\text{BS}_1} |\text{in}\rangle = \frac{1}{\sqrt{2}} \begin{pmatrix} 1 & 1 \\ 1 & -1 \end{pmatrix} \frac{1}{\sqrt{2}} \begin{pmatrix} 1 & 1 \\ 1 & -1 \end{pmatrix} \begin{pmatrix} 1 \\ 0 \end{pmatrix} = \begin{pmatrix} 1 \\ 0 \end{pmatrix} = |D_2\rangle. \tag{2.85}$$

We calculate the probability that the photon will end up in D_2 using the Born rule, and we obtain

$$\Pr(D_2|\text{no bomb}) = |\langle D_2|\text{out}\rangle|^2 = |\langle D_2|D_2\rangle|^2 = 1. \tag{2.86}$$

Therefore, when no bomb is present, detector D_2 always fires. This is because there is destructive interference between the two possible paths for the photon to go to D_1, one from the photon coming into the top of beam splitter 2, and one from the photon coming in from the left.

Next, consider the presence of the bomb in the lower arm of the interferometer. It acts a little bit like the QND detector in Chap. 1, in that it reveals the path of the photon inside the interferometer. The detector signature is rather dramatic, because it is the explosion of the bomb! It also is not a "non-demolition" detector, since the photon is typically lost when it hits the detector.

When the photon takes the upper path, the bomb does not explode and the photon is free to be detected in either detector D_1 or D_2. With the bomb present and the lower path blocked, the photon can end up in detector D_1 since the second beam splitter will split the probability amplitude of the photon coming in from the top over the two detectors. There is no photon coming to the second beam splitter from the left to destructively interfere the pathway to D_1. However, since we now have a possible signal in D_1, that signals the presence of the bomb, since without the bomb we would have only detection events in D_2.

But now something remarkable has happened: we can infer the presence of the bomb *with certainty* from a detection event in D_1. We sent in one photon, and we detect one photon. It must have taken the upper arm, and therefore will not have exploded the bomb. In other words, we detected the presence of the bomb without interacting with it!

Let's work out the probability of this event. We need a matrix that signals the absorption of the photon in the lower arm. It may not be as spectacular as an exploding bomb, but for the maths to work we need a simple 2×2 matrix that we insert between the two beam splitter matrices. Since the upper arm is unaffected and there is no mixing between the upper and lower arm, we know the matrix must have the form

$$U_{\text{abs}} = \begin{pmatrix} 1 & 0 \\ 0 & ? \end{pmatrix}, \tag{2.87}$$

and we need to determine the lower right element. Since we must remove the probability amplitude from that part of the state, we choose the element to be zero. The output state with the bomb present then becomes

$$
\begin{aligned}
|\text{out}\rangle &= U_{\text{BS}_2} U_{\text{abs}} U_{\text{BS}_1} |\text{in}\rangle \\
&= \frac{1}{\sqrt{2}} \begin{pmatrix} 1 & 1 \\ 1 & -1 \end{pmatrix} \begin{pmatrix} 1 & 0 \\ 0 & 0 \end{pmatrix} \frac{1}{\sqrt{2}} \begin{pmatrix} 1 & 1 \\ 1 & -1 \end{pmatrix} \begin{pmatrix} 1 \\ 0 \end{pmatrix} \\
&= \frac{1}{\sqrt{2}} \begin{pmatrix} 1 & 0 \\ 1 & 0 \end{pmatrix} \frac{1}{\sqrt{2}} \begin{pmatrix} 1 & 1 \\ 1 & -1 \end{pmatrix} \begin{pmatrix} 1 \\ 0 \end{pmatrix} = \frac{1}{2} \begin{pmatrix} 1 & 1 \\ 1 & 1 \end{pmatrix} \begin{pmatrix} 1 \\ 0 \end{pmatrix} = \frac{1}{2} \begin{pmatrix} 1 \\ 1 \end{pmatrix}.
\end{aligned} \tag{2.88}
$$

This is not a normalised state, because we suppressed the probability amplitude of the photon going through the lower arm. This makes sense, because it is the state conditioned on the photon making it past the bomb.

To find the probability that the photon ends up in detector D_1 and reveals the presence of the bomb without setting it off, we calculate

$$\Pr(D_1|\text{bomb}) = |\langle D_1|\text{out}\rangle|^2 = \left| \frac{1}{2} \begin{pmatrix} 1 & 0 \end{pmatrix} \begin{pmatrix} 1 \\ 1 \end{pmatrix} \right|^2 = \frac{1}{4}. \tag{2.89}$$

Therefore, we have a probability of $\frac{1}{4}$ of detecting the bomb without setting it off. We can also calculate the probability that the photon ends up in detector D_2:

$$\Pr(D_2|\text{bomb}) = |\langle D_2|\text{out}\rangle|^2 = \left| \frac{1}{2} \begin{pmatrix} 0 & 1 \end{pmatrix} \begin{pmatrix} 1 \\ 1 \end{pmatrix} \right|^2 = \frac{1}{4}. \tag{2.90}$$

In this case we do not learn anything, because the measurement outcome D_2 is also the result when there is no bomb present. The two probabilities for finding the photon in D_1 or D_2 sum up to

$$\Pr(D_1|\text{bomb}) + \Pr(D_2|\text{bomb}) = \frac{1}{4} + \frac{1}{4} = \frac{1}{2},$$

which means that the remaining probability of $\frac{1}{2}$ is for the case where the photon sets off the bomb (since the probabilities must sum to 1). This makes sense, because the first beam splitter creates an equal superposition of the photon taking the upper path (where no bomb is present) and the lower path (where the bomb is present).

Exercises

1. Multiply the following matrices:

$$A = \begin{pmatrix} 12 & -3 \\ 6 & 9 \end{pmatrix} \quad \text{and} \quad B = \begin{pmatrix} 2 & 5 \\ -1 & 0 \end{pmatrix}.$$

Is AB the same as BA?

2. Express the following complex numbers in polar representation $r\,e^{i\phi}$:

$$z_1 = 3 + 4i, \quad z_2 = 12 - 8i, \quad \text{and} \quad z_3 = z_1 + z_2^*.$$

3. Find the complex solutions to the quadratic equation

$$x^2 - 2x + 10 = 0.$$

4. Every matrix has two special properties: the trace and the determinant. The trace is the sum over the diagonal elements (top left to bottom right), while the determinant of a 2×2 matrix is given by

$$\det \begin{pmatrix} a & b \\ c & d \end{pmatrix} = ad - bc.$$

Calculate the trace and the determinant of the following matrices:

$$\begin{pmatrix} 3 & 1 \\ 2 & 6 \end{pmatrix}, \quad \begin{pmatrix} 0 & -i \\ i & 0 \end{pmatrix}, \quad \text{and} \quad \begin{pmatrix} 1 & 1 \\ -1 & -1 \end{pmatrix}.$$

5. Normalise the vector $|\psi\rangle = \begin{pmatrix} 3 \\ 4i \end{pmatrix}$.

6. Calculate the scalar product between

$$|\psi\rangle = \begin{pmatrix} 3i \\ -3 \end{pmatrix} \quad \text{and} \quad |\phi\rangle = \begin{pmatrix} 5 \\ 7 \end{pmatrix}.$$

Show that $\langle\psi|\phi\rangle = \langle\phi|\psi\rangle^*$.

7. Determine the state space of a photon in a Mach–Zehnder interferometer in terms of the values that a and b can take in (2.9).

8. Every month a lottery jackpot is one million pounds. A lottery ticket costs £5, and there are four million people playing the lottery each month. Calculate your average gain at any given month. How long do you have to play the lottery in order to have a 50% chance of winning at least once?

9. A car comes at a fork in the road, and the driver can turn left or right. If we describe the car quantum mechanically (not very realistic!) and the car is twice as likely to turn left as it is to turn right, what is the quantum state of the car after the fork?

10. The state of a photon in an interferometer is given by

$$|\psi\rangle = \frac{3i}{5}|\text{right}\rangle + \frac{4}{5}|\text{down}\rangle.$$

What is the probability of getting the measurement outcome "right"?

11. Construct the matrix for a beam splitter with a 70:30 ratio between reflection and transmission. Show how we can achieve perfect destructive interference in a Mach–Zehnder interferometer using two of these beam splitters (see also Exercise 3 of Chap. 1).

12. A Mach–Zehnder interferometer is described by

$$U_{\text{MZ}} = \frac{1}{4}\begin{pmatrix} 3 + \sqrt{3}\,i & 1 - \sqrt{3}\,i \\ 1 - \sqrt{3}\,i & 3 + \sqrt{3}\,i \end{pmatrix}.$$

Calculate the relative phase difference between the arms in the interferometer. You may use (2.54).

13. A photon in the state

$$|\psi\rangle = \frac{1}{\sqrt{5}}|\text{right}\rangle + \frac{2}{\sqrt{5}}|\text{down}\rangle$$

is sent into a beam splitter. What is the probability of finding the photon in the "down" path?

14. We replace the lower-left mirror of the Mach–Zehnder interferometer in Fig. 1.4 with another 50:50 beam splitter. There are now three input beams and three output beams of the resulting interferometer. Assuming there are no relative path differences, construct the 3×3 transformation matrix describing the interferometer. Calculate the nine probabilities of a photon in any of the three inputs going to any of the three outputs. Compare your results with the 3×3 transformation matrix.

15. Consider again the Mach–Zehnder interferometer in Fig. 1.4. Instead of a single phase shift ϕ in the upper arm, we place two identical phase shifts ϕ, one in the upper arm and one in the lower arm. Calculate the matrix transformation of this interferometer. Sending a photon into the left input, what is the quantum state vector of the photon at the output? Can we measure the phase shift ϕ in both arms based on the probabilities of finding the photon in detectors D_1 and D_2?

16. The input state of a beam splitter is given by

$$|\text{photon}\rangle = a\,|\text{left}\rangle + b\,e^{i\phi}\,|\text{top}\rangle \ .$$

Using (2.27), calculate the probabilities of the measurement outcomes "right" and "down". How do the probabilities depend on the relative phase?

17. Consider a quantum state $|\psi\rangle$ that accumulates a global phase $e^{i\phi}$. Show that any probability calculated using this state does not depend on the phase ϕ. The phase is *unobservable*.

18. We record 1000 photons in a gravity wave detector (see Sect. 2.5), and we find $N_1 = 35$ in detector D_1 and $N_2 = 965$ in detector D_2. Calculate the phase shift ϕ.

19. The beam splitter transformation that we derived in (2.26) is not quite an accurate description of a physical beam splitter used in optics labs. While we constructed the matrix in (2.26) by requiring a π phase shift on the path from the top input to the down output, more realistic beam splitters give the photon a $\frac{\pi}{2}$ phase shift upon reflection, regardless whether it comes from the left or the top. Construct the corresponding beam splitter matrix and show that its columns are orthonormal vectors (i.e., orthogonal and of length one).

Chapter 3
Electrons with Spin

In this chapter, we describe a simple experiment about the spin property of an electron and use it to determine what is a *physical observable* in quantum mechanics. We give a general description of physical systems with two distinct states. As a different observable we introduce the polarisation of a photon, which is described with the same mathematics, but has different physical meaning (but is often referred to as the 'spin' of the photon). We conclude this chapter with a brief description how spin can be used in magnetic resonance imaging, a technique that is used in hospitals to reveal what is going on inside your brain.

3.1 The Stern-Gerlach Experiment

In this section we will follow an approach similar to the previous two chapters: we will describe a simple experiment that will reveal an internal property of electrons, called spin. We will then study various modifications of this experiment and deduce the behaviour of electron spin. This will lead us to describe the spin of the electron in a similar way to the description of a photon in an interferometer, based on states. We use the mathematical theory we developed in the previous chapter to predict measurement outcomes of the electron spin experiment.

In a Stern-Gerlach experiment,[1] shown schematically in Fig. 3.1, electrons are sent from a source to a fluorescent screen. A bright dot will appear at the position

---

[1] W. Gerlach and O. Stern, *Der experimentelle Nachweis der Richtungsquantelung im Magnetfeld.* Zeitschrift für Physik. **9** 349, 1922.

Supplementary Information The online version contains supplementary material available at https://doi.org/10.1007/978-3-031-16165-0_3.

Fig. 3.1 A Stern-Gerlach apparatus. The "up" and "down" regions on the screen where the electron is detected define the electron spin states $|\uparrow\rangle$ and $|\downarrow\rangle$ (see supplementary material 1)

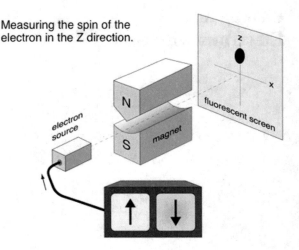

Preparation of the electron spin

where the electron hits the screen. The electrons travel through a region with a strong (non-uniform) magnetic field, generated by two magnets aligned along the vertical axis (we call it the z-axis). In the absence of the magnetic field, the electrons travel along a central axis connecting the source with the screen (we call this the y-axis, which is directed *towards* the source). When the magnetic field is turned on, the dots will appear in two regions: one slightly up from the central axis, and one slightly down from the axis, shown in Fig. 3.1. This means that the electrons have some kind of property that causes a deflection of the path in the presence of a magnetic field. We call this property the *spin* of the electron. One note of caution: electrons are particles with electric charge e, which means that they feel a sideways force $F = evB$ when they move at speed v through a magnetic field B. This is a different effect from the deflection described above, and for simplicity we assume that we compensated for this sideways force.

Since there are only two positions where the electron can hit the screen (the "up" region or the "down" region), we can again describe the state of the electron spin as $|up\rangle$ or $|down\rangle$. Following standard notation in quantum mechanics, we will write this as $|\uparrow\rangle$ and $|\downarrow\rangle$. An arbitrary spin state $|spin\rangle$ can then be written as a superposition of these two states:

$$|spin\rangle = a\,|\uparrow\rangle + b\,|\downarrow\rangle . \tag{3.1}$$

This is similar to the way we constructed the state of a photon inside a Mach-Zehnder interferometer in the previous chapter. In vector notation we can choose

$$|\uparrow\rangle = \begin{pmatrix} 1 \\ 0 \end{pmatrix} \quad \text{and} \quad |\downarrow\rangle = \begin{pmatrix} 0 \\ 1 \end{pmatrix}, \tag{3.2}$$

and the arbitrary spin state in (3.1) can be written as

$$|\text{spin}\rangle = \begin{pmatrix} a \\ b \end{pmatrix}. \tag{3.3}$$

To ensure that the vector associated with $|\text{spin}\rangle$ has length one, we calculate the length of a vector by taking the scalar product with itself:

$$\langle \text{spin}|\text{spin}\rangle = \begin{pmatrix} a^* & b^* \end{pmatrix} \begin{pmatrix} a \\ b \end{pmatrix} = |a|^2 + |b|^2 = 1. \tag{3.4}$$

This leads to the normalisation $|a|^2 + |b|^2 = 1$, and is again similar to the results we derived in the previous chapter, but now with electron spin instead of photon path. In Sect. 3.6 we will construct another of such a quantum state, namely for the polarisation of a photon, and in the next chapter we construct the quantum state for the energy levels of an atom. For every physical system the procedure is the same: identify the distinct states of a system in a *particular* experiment, give them labels inside a ket, and associate with each distinct state an orthogonal vector.

The same mathematics we developed in the previous chapter can now be used to describe the spin of an electron. The fluorescent screen plays the role of the detectors D_1 and D_2, and the probability of finding a dot at position "up" is given by

$$p_\uparrow = |\langle \uparrow |\text{spin}\rangle|^2 = |a|^2, \tag{3.5}$$

and the probability of finding a dot at position "down" is given by

$$p_\downarrow = |\langle \downarrow |\text{spin}\rangle|^2 = |b|^2. \tag{3.6}$$

So we see that $p_\uparrow + p_\downarrow = 1$ is again automatically satisfied via the normalisation condition $|a|^2 + |b|^2 = 1$.

When the source produces electrons in the state $|\uparrow\rangle$, we have $a = 1$ and $b = 0$, and therefore $p_\uparrow = 1$ and $p_\downarrow = 0$. The electron always hits the screen at the position "up". Similarly, if the source produces electrons with spin state $|\downarrow\rangle$ the electron will always hit the screen at position "down". This is how we defined the states $|\uparrow\rangle$ and $|\downarrow\rangle$ in the first place. When we create electrons in the spin state

$$|+\rangle = \frac{1}{\sqrt{2}} |\uparrow\rangle + \frac{1}{\sqrt{2}} |\downarrow\rangle, \tag{3.7}$$

or

$$|-\rangle = \frac{1}{\sqrt{2}} |\uparrow\rangle - \frac{1}{\sqrt{2}} |\downarrow\rangle, \tag{3.8}$$

Fig. 3.2 Creating electrons in spin states $|\pm\rangle$ will give different probabilities of finding the electron in the "up" and "down" regions (see supplementary material 2)

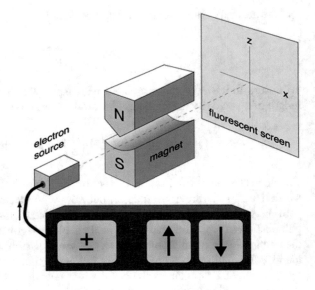

the electrons have a 50:50 chance of hitting positions "up" and "down" on the screen. We can verify this statement by calculating the probabilities:

$$p_\uparrow = |\langle\uparrow|+\rangle|^2 = \left|\frac{\langle\uparrow|\uparrow\rangle}{\sqrt{2}} + \frac{\langle\uparrow|\downarrow\rangle}{\sqrt{2}}\right|^2 = \frac{1}{2}, \tag{3.9}$$

and

$$p_\downarrow = |\langle\downarrow|+\rangle|^2 = \left|\frac{\langle\downarrow|\uparrow\rangle}{\sqrt{2}} + \frac{\langle\downarrow|\downarrow\rangle}{\sqrt{2}}\right|^2 = \frac{1}{2}, \tag{3.10}$$

where we used $\langle\uparrow|\uparrow\rangle = \langle\downarrow|\downarrow\rangle = 1$ and $\langle\uparrow|\downarrow\rangle = \langle\downarrow|\uparrow\rangle = 0$. You should check these probabilities yourself using the vector form for $|\uparrow\rangle$ and $|\downarrow\rangle$. Similarly, you can calculate the probabilities $p_\uparrow$ and $p_\downarrow$ given the electron state $|-\rangle$. The behaviour of electrons with spin $|+\rangle$ and $|-\rangle$ is shown in Fig. 3.2.

At this point, you may have noticed a striking similarity with the state of a photon after the beam splitter in the previous chapter. The state $|\text{spin}\rangle$ in (3.1) is mathematically identical to the state $|\text{photon}\rangle$ in (2.7). The physical meaning is different since $|\text{spin}\rangle$ and $|\text{photon}\rangle$ describe different physical systems, but all the mathematical procedures we described in the previous chapter work just as well with spin states. We just used the Born rule in (3.9) and (3.10) to calculate probabilities of measurement outcomes. Similarly, we can construct unitary matrices that evolve spins from $|\uparrow\rangle$ to $|+\rangle$, etc., just like we constructed a matrix for the action of a beam splitter on incoming photon states. In the case of the spin we will wait with the construction of

Fig. 3.3 A Stern-Gerlach apparatus with input states $|\pm\rangle$ (see supplementary material 3)

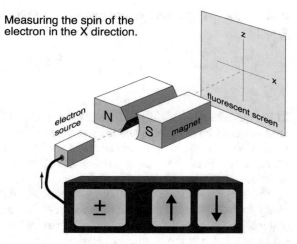

Measuring the spin of the electron in the X direction.

Preparation of the electron spin

this matrix until the next chapter, since there are a few subtleties with the physical meaning of spin that we should address first.

Looking again at the Stern-Gerlach experiment, you see that we can rotate the two magnets around the axis that connects the source and the screen (the dashed line). Let us investigate what happens when we rotate the two magnets over 90° (counterclockwise when facing the screen), so the magnets are aligned along the horizontal axis, called the x-axis. When we do the experiment, we find the results shown in Fig. 3.3 (the behaviour is shown in the interactive figure).

If we prepare the electron in the spin state $|\uparrow\rangle$ we will find the fluorescent spots randomly on the left or on the right of the central vertical axis on the screen. Each electron has a 50:50 chance of going to the left and going to the right. The same behaviour occurs when we create the electron in the spin state $|\downarrow\rangle$. These are experimental facts that form the foundation of conclusions we draw in our subsequent discussion.

Similarly, we find experimentally that when we create the electron in the state $|+\rangle$ and send it through the magnets oriented in the x-direction, we find that the electron *always* ends up on the left. And when we create the electron in the spin state $|-\rangle$, it will always create a fluorescent spot on the right. In other words, the spin states $|+\rangle$ and $|-\rangle$ act the same in the x-direction as the states $|\uparrow\rangle$ and $|\downarrow\rangle$ do in the z-direction. We can say that $|\uparrow\rangle$ is the state of an electron spin in the positive z-direction, and $|+\rangle$ is the state of an electron spin in the positive x-direction.

The behaviour of spin measurements in different directions strongly suggests that spin must be a quantity with a directional character, and in the Stern-Gerlach experiment we measure the *component* of the spin along the axis of the magnets. In other words, when we orient the magnets along the z-axis we measure the z-component of the spin, S_z, and when we orient the magnets along the x-axis we measure the x-component of the spin, S_x. The physical quantity "spin" is then a *vector*:

$$\mathbf{S} = \begin{pmatrix} S_x \\ S_y \\ S_z \end{pmatrix}. \tag{3.11}$$

This kind of vector is different from the vectors $|\uparrow\rangle$ or $|+\rangle$. The quantum states are vectors in an abstract space that we created for the convenience of calculating probabilities, while the spin vector $\mathbf{S}$ is a vector in our own real three-dimensional space. It is crucial that you remember the difference between these two types of vectors.

You may be wondering about the y-direction of the electron spin, S_y, which lies along the axis from the magnets to the screen. We cannot orient the magnets in the y-direction because they would block the path of the electron. We would have to drill small holes in the magnets to let the electrons through. Still, we can create an electron with a spin state oriented in the y-direction—let's not worry about how we achieve this right now; we will return to this in the next chapter. The action of the magnets is to deflect the spins along the axis of the magnets. Electrons travelling along the y-axis through a hole in the magnets will be accelerated or decelerated, and the arrival times of the electrons will again fall into two bins, "early" for accelerated electrons and "late" for decelerated electrons.

Next, imagine that we send such "early" or "late" electrons through the Stern-Gerlach apparatus with the magnets oriented in the x- or z-direction. We expect that these electron states will give 50:50 distributions of fluorescent spots in the up and down region, as well as in the left and right region. How can we achieve this with spin states of the form of (3.1)? We have to make use of the freedom to choose complex numbers for a and b. Indeed, when we create electrons in the spin state

$$|\circlearrowleft\rangle = \frac{1}{\sqrt{2}}\,|\uparrow\rangle + \frac{i}{\sqrt{2}}\,|\downarrow\rangle , \tag{3.12}$$

and

$$|\circlearrowright\rangle = \frac{1}{\sqrt{2}}\,|\uparrow\rangle - \frac{i}{\sqrt{2}}\,|\downarrow\rangle , \tag{3.13}$$

we find that the probabilities of measuring spin $|\uparrow\rangle$ and $|\downarrow\rangle$ are

$$p_\uparrow = |\langle\uparrow|\circlearrowleft\rangle|^2 = \frac{1}{2} \quad \text{and} \quad p_\downarrow = |\langle\downarrow|\circlearrowleft\rangle|^2 = \frac{1}{2}$$
$$p_\uparrow = |\langle\uparrow|\circlearrowright\rangle|^2 = \frac{1}{2} \quad \text{and} \quad p_\downarrow = |\langle\downarrow|\circlearrowright\rangle|^2 = \frac{1}{2} , \tag{3.14}$$

The probabilities of measuring spin $|+\rangle$ and $|-\rangle$ are

$$p_+ = |\langle +|\circlearrowleft\rangle|^2 = \frac{1}{2} \quad \text{and} \quad p_- = |\langle -|\circlearrowleft\rangle|^2 = \frac{1}{2}$$

$$p_+ = |\langle +|\circlearrowright\rangle|^2 = \frac{1}{2} \quad \text{and} \quad p_- = |\langle -|\circlearrowright\rangle|^2 = \frac{1}{2}, \tag{3.15}$$

You should check these results both in Dirac notation and the vector notation. We will generalise these results to spins in any spatial direction in Sect. 3.3.

Finally, we should ask the question: What exactly is it that we are measuring in the Stern-Gerlach experiment? We have seen that the orientation of the magnets is linked to the components of the spin vector. When the magnets are oriented in the z-direction we are measuring the z-component of $\mathbf{S}$, or S_z. Similarly, when the magnets are oriented in the x-direction we are measuring S_x. We can also set the magnets at any angle θ from the z-axis. However, *we can measure only one spin component at a time*, because there is only one orientation of the magnets we can choose at any time. We cannot measure the entire vector $\mathbf{S}$ in a single measurement. This will have some interesting consequences, which we will return to later.

3.2 The Spin Observable

We have seen that the spin of an electron is described by a vector. We should also note that spin is a physical property with units, just like velocity (e.g., metres per second or miles per hour), or energy (Joules, ergs, or electron volts). In the case of spin the units are that of angular momentum: J s, or kg m^2 s^{-1}. This is because the spin of an electron is the quantum mechanical analog of the rotation around its own axis, which was discovered by Samuel Goudsmit and George Uhlenbeck[2] in 1925. As a consequence of the electron charge, the spin of the electron is also closely related to the magnetic moment, which is why the Stern-Gerlach experiment uses magnets. Just like a spinning top, we have the rotation velocity (the magnitude) and the direction of the rotation axis. This is why spin is a vector: it has a magnitude and a direction.

For electrons it is convenient to express the magnitude of the spin in terms of Dirac's constant

$$\hbar = \frac{h}{2\pi} = 1.054572 \times 10^{-34} \, \text{J s}. \tag{3.16}$$

You may recognise h as Planck's constant, and $\hbar$ is also called the reduced Planck constant. The value of the spin in the z-direction S_z of an electron in the state $|\uparrow\rangle$ is $+\hbar/2$, and the value of spin in the z-direction of an electron in the state $|\downarrow\rangle$ is $-\hbar/2$.

---

[2] G. E. Uhlenbeck and S. Goudsmit, *Ersetzung der Hypothese vom unmechanischen Zwang durch eine Forderung bezüglich des inneren Verhaltens jedes einzelnen Elektrons*, Naturwissenschaften **13** 953, 1925.

These values are determined from the experimental data. Every time you see h or $\hbar$, you are are dealing with a quantum mechanical situation.

Classically, one would expect that the spin can take any magnitude (the rotation velocity of the electron). After all, it is easy to imagine that the electron can rotate slower or faster around its own axis. But the Stern-Gerlach experiment tells us that this is not the case. If we measure the spin in the z-direction we find only two possible values, "up" or "down". This tells us that the electron spin is a fundamentally quantum mechanical quantity: its spin does not take a continuous range of values, but is restricted to two distinct values. We say that the spin of the electron is quantised. This happens not only for the spin, but for the energy as well (which we will explore in the next chapter), and it is the origin of the term "quantum" mechanics. The quantisation of spin is an experimental fact of nature that we have to accept, just like we had to accept in the previous chapters that light is quantised (i.e., the quantum of light is the photon).

Next, we will give a more complete mathematical description of the so-called spin observable based on the findings of the Stern-Gerlach experiment. We know how an electron with spin state $|+\rangle$ behaves in a Stern-Gerlach apparatus oriented in the x-direction: the electron will create a fluorescent dot in the "+" region with probability $p_+ = 1$. This is the definition of the state $|+\rangle$. The numerical value of the spin in the x-direction S_x for the electron in state $|+\rangle$ is $+\hbar/2$. What would be the numerical value of S_z for the state $|+\rangle$? When we actually measure it we find that the electron ends up in the "up" or "down" region with equal probability of $1/2$. Since the magnitude associated with $|\uparrow\rangle$ is also $+\hbar/2$ (and the magnitude associated with $|\downarrow\rangle$ is $-\hbar/2$), on average, the value of S_z is given by

$$\langle S_z \rangle = p_\uparrow \left(+\frac{\hbar}{2} \right) + p_\downarrow \left(-\frac{\hbar}{2} \right) = \frac{1}{2} \left(+\frac{\hbar}{2} \right) + \frac{1}{2} \left(-\frac{\hbar}{2} \right) = \frac{\hbar}{4} - \frac{\hbar}{4} = 0 , \quad (3.17)$$

where we used the notation $\langle S_z \rangle$ to denote the average of S_z. So while the individual measurement outcomes can be only $\pm\hbar/2$, the *average* spin in the z-direction over many measurements can take *any* value between $+\hbar/2$ and $-\hbar/2$, depending on the probabilities $p_\uparrow$ and $p_\downarrow$.

For an electron with some arbitrary spin state $|\psi\rangle$, the average can be written as

$$\langle S_z \rangle = p_\uparrow \left(+\frac{\hbar}{2} \right) + p_\downarrow \left(-\frac{\hbar}{2} \right) . \quad (3.18)$$

This is called the *expectation value* of S_z. We will now derive the mathematical form of the physical observable S_z from (3.18).

Using the Born rule for the measurement outcomes of the spin $p_\uparrow = |\langle\uparrow|\psi\rangle|^2$ and $p_\downarrow = |\langle\downarrow|\psi\rangle|^2$ for an electron in spin state $|\psi\rangle$, we can manipulate (3.18) as follows:

$$\langle S_z \rangle = \frac{\hbar}{2} \left(p_\uparrow - p_\downarrow \right)$$

$$= \frac{\hbar}{2} \left(|\langle \uparrow | \psi \rangle|^2 - |\langle \downarrow | \psi \rangle|^2 \right)$$

$$= \frac{\hbar}{2} \langle \psi | \uparrow \rangle \langle \uparrow | \psi \rangle - \frac{\hbar}{2} \langle \psi | \downarrow \rangle \langle \downarrow | \psi \rangle, \tag{3.19}$$

where we expanded the modulus-squared in the second line. Note that both terms on the right-hand side of the last line have a $\langle \psi |$ on the left and a $| \psi \rangle$ on the right. We can take these out as common factors and write

$$\langle S_z \rangle = \langle \psi | \left(\frac{\hbar}{2} | \uparrow \rangle \langle \uparrow | - \frac{\hbar}{2} | \downarrow \rangle \langle \downarrow | \right) | \psi \rangle = \langle \psi | S_z | \psi \rangle, \tag{3.20}$$

where we defined

$$S_z = \frac{\hbar}{2} | \uparrow \rangle \langle \uparrow | - \frac{\hbar}{2} | \downarrow \rangle \langle \downarrow |. \tag{3.21}$$

You may wonder what the expression for S_z in (3.21) means. We can figure it out by using the vector notation for $| \uparrow \rangle$ and $\langle \uparrow |$:

$$| \uparrow \rangle \langle \uparrow | = \begin{pmatrix} 1 \\ 0 \end{pmatrix} \begin{pmatrix} 1 & 0 \end{pmatrix} = \begin{pmatrix} 1 & 0 \\ 0 & 0 \end{pmatrix},$$

$$| \downarrow \rangle \langle \downarrow | = \begin{pmatrix} 0 \\ 1 \end{pmatrix} \begin{pmatrix} 0 & 1 \end{pmatrix} = \begin{pmatrix} 0 & 0 \\ 0 & 1 \end{pmatrix}. \tag{3.22}$$

These "ket-bra" expressions are matrices! By substituting the matrices back into the expression for S_z, this leads to the matrix representation

$$S_z = \frac{\hbar}{2} \begin{pmatrix} 1 & 0 \\ 0 & -1 \end{pmatrix}. \tag{3.23}$$

Note that $| \uparrow \rangle \langle \uparrow |$ is different from $\langle \uparrow | \uparrow \rangle$. The first is a 2×2 matrix, while the second is a single number (namely 1 in this case). If you are careful about the order of the vectors and remember that $| \cdot \rangle$ is a column vector and $\langle \cdot |$ is a row vector (and complex conjugate), matrix multiplication of the two will automatically give the correct result.

We can also derive the matrix representation for S_x. The expectation value $\langle S_x \rangle$ can be written as

$$\begin{aligned}
\langle S_x \rangle &= \frac{\hbar}{2}(p_+ - p_-) \\
&= \frac{\hbar}{2}\left(|\langle +|\psi \rangle|^2 - |\langle -|\psi \rangle|^2\right) \\
&= \frac{\hbar}{2}\langle \psi|+\rangle \langle +|\psi\rangle - \frac{\hbar}{2}\langle \psi|-\rangle \langle -|\psi\rangle \\
&= \langle \psi| \left(\frac{\hbar}{2}|+\rangle \langle +| - \frac{\hbar}{2}|-\rangle \langle -|\right) |\psi\rangle,
\end{aligned} \tag{3.24}$$

and we therefore have

$$S_x = \frac{\hbar}{2}|+\rangle \langle +| - \frac{\hbar}{2}|-\rangle \langle -|. \tag{3.25}$$

This has the same form as S_z, but now with $|+\rangle$ and $|-\rangle$ instead of $|\uparrow\rangle$ and $|\downarrow\rangle$. When we use the vector notation of $|+\rangle$ and $|-\rangle$ we obtain

$$|+\rangle \langle +| = \frac{1}{2}\begin{pmatrix} 1 \\ 1 \end{pmatrix}\begin{pmatrix} 1 & 1 \end{pmatrix} = \frac{1}{2}\begin{pmatrix} 1 & 1 \\ 1 & 1 \end{pmatrix},$$

$$|-\rangle \langle -| = \frac{1}{2}\begin{pmatrix} 1 \\ -1 \end{pmatrix}\begin{pmatrix} 1 & -1 \end{pmatrix} = \frac{1}{2}\begin{pmatrix} 1 & -1 \\ -1 & 1 \end{pmatrix}. \tag{3.26}$$

This leads to the matrix representation of S_x:

$$S_x = \frac{\hbar}{2}\begin{pmatrix} 0 & 1 \\ 1 & 0 \end{pmatrix}. \tag{3.27}$$

Following the exact same procedure again with the states $|\circlearrowleft\rangle$ and $|\circlearrowright\rangle$ we find the matrix representation of S_y:

$$S_y = \frac{\hbar}{2}\begin{pmatrix} 0 & -i \\ i & 0 \end{pmatrix}. \tag{3.28}$$

The average value of the spin in the y-direction for an electron in the state $|\psi\rangle$ is then given by the expectation value $\langle S_y \rangle = \langle \psi|S_y|\psi\rangle$.

The matrices in S_x, S_y, and S_z are used so much in quantum mechanics that they have their own name. They are the Pauli matrices:

$$\sigma_x = \begin{pmatrix} 0 & 1 \\ 1 & 0 \end{pmatrix}, \quad \sigma_y = \begin{pmatrix} 0 & -i \\ i & 0 \end{pmatrix}, \quad \sigma_z = \begin{pmatrix} 1 & 0 \\ 0 & -1 \end{pmatrix}. \tag{3.29}$$

We have now found a remarkable aspect of quantum mechanics: A physical property such as the spin of an electron in the z-direction does not just have a simple value, but must be represented by a matrix, or operator. Operators that represent physical properties are observables. We can determine the value of a physical property only when we combine the state and the observable together. The measurement we perform determines the observable (for example, we choose S_x or S_z by setting the orientation of the magnets in the Stern-Gerlach experiment), while the spin state is determined by the preparation procedure of the electron spin before the measurement (we have not explored this preparation procedure here, but we assume that there is a way to achieve this). We calculate the probabilities of individual measurement outcomes, as well as the average value of physical observables using the state $|\psi\rangle$.

Previously, we encountered operators as a means to describe the transformation of a state. In particular, we have seen in Chap. 2 that the beam splitter is described by an operator that transforms the input state of a photon into the output state. However, the spin operators introduced here are fundamentally different, since they have physical units (for spin it is J s), while transformation matrices—and the quantum states themselves—must be dimensionless. In general, operators such as the beam splitter have mathematical properties that are different from the properties of observables. We will return to this distinction in Chap. 5. For now, it is important that the matrices associated with observables are of a different kind than matrices associated with state transformations.

We will continue our discussion of quantum mechanical spin observables in Sect. 3.4, but before that we will consider all the possible quantum states for the electron spin.

3.3 The Bloch Sphere

We return now to the question how we can describe the spin of an electron in an arbitrary direction. This leads to the full *state space* of the electron spin, in which each point represents a valid spin state. Classically, the spin state of the electron is the three-dimensional vector **S** in (3.11), so the state space is three-dimensional real space: each point in space uniquely defines a vector as an arrow from the origin to that point, where the direction of the arrow is the spin direction, and the length of the arrow is the size of the spin.

How does this translate to the quantum mechanical state space? We immediately notice a problem: the state vectors of electron spin are two-dimensional (the column vectors have two entries, corresponding to two distinguishable measurement outcomes), while the classical state space for spin is three-dimensional. However, we have also seen that the state vectors can have complex numbers (which are composed of two real numbers), and it turns out that you can describe three-dimensional real spin vectors in terms of two-dimensional complex state vectors. We will now see how this is done.

We first write $|\uparrow\rangle$ and $|+\rangle$ in vector notation:

$$|\uparrow\rangle = \begin{pmatrix} 1 \\ 0 \end{pmatrix} \quad \text{and} \quad |+\rangle = \frac{1}{\sqrt{2}} \begin{pmatrix} 1 \\ 1 \end{pmatrix}. \tag{3.30}$$

Notice that the vector $|+\rangle$ is rotated over $45°$ with respect to the vector $|\uparrow\rangle$. Therefore, a $90°$ rotation in real space (rotating the spin direction from the z-axis to the x-axis) corresponds to a $45°$ rotation in the state space of the spin. The angle in state space is half the angle in real space. If we rotate the spin source over an angle θ in real space (around the central axis in the Stern-Gerlach experiment), the spin state will be given by a vector that is rotated over an angle $\theta/2$:

$$|\psi(\theta)\rangle = \begin{pmatrix} \cos\left(\frac{\theta}{2}\right) \\ \sin\left(\frac{\theta}{2}\right) \end{pmatrix} = \cos\left(\frac{\theta}{2}\right)|\uparrow\rangle + \sin\left(\frac{\theta}{2}\right)|\downarrow\rangle. \tag{3.31}$$

We can check that this is true by rotating the magnets over an angle θ in real space, so that we measure the spin component

$$S_\theta = \cos\theta\, S_z + \sin\theta\, S_x = \frac{\hbar}{2}\begin{pmatrix} \cos\theta & \sin\theta \\ \sin\theta & -\cos\theta \end{pmatrix}. \tag{3.32}$$

We should find that the spin in that direction is $+\hbar/2$. We can calculate the expectation value (average) of the rotated spin operator, and if it is $+\hbar/2$ we know that all the spins must contribute $+\hbar/2$ to this average, and no spins contribute $-\hbar/2$ (otherwise the expectation value would be less than $\hbar/2$). Therefore the probability of getting the $+\hbar/2$ measurement result is 1, and the spin state in (3.31) is indeed the spin along the direction θ.

We calculate $\langle S_\theta \rangle$ using (3.31) and (3.32) in vector and matrix form using matrix multiplication:

$$\begin{aligned}
\langle S_\theta \rangle &= \frac{\hbar}{2}\left(\cos\left(\frac{\theta}{2}\right) \ \sin\left(\frac{\theta}{2}\right)\right)\begin{pmatrix} \cos\theta & \sin\theta \\ \sin\theta & -\cos\theta \end{pmatrix}\begin{pmatrix} \cos\left(\frac{\theta}{2}\right) \\ \sin\left(\frac{\theta}{2}\right) \end{pmatrix} \\
&= \frac{\hbar}{2}\left(\cos\left(\frac{\theta}{2}\right) \ \sin\left(\frac{\theta}{2}\right)\right)\begin{pmatrix} \cos\theta\cos\left(\frac{\theta}{2}\right) + \sin\theta\sin\left(\frac{\theta}{2}\right) \\ \sin\theta\cos\left(\frac{\theta}{2}\right) - \cos\theta\sin\left(\frac{\theta}{2}\right) \end{pmatrix} \\
&= \frac{\hbar}{2}\left(\cos\left(\frac{\theta}{2}\right) \ \sin\left(\frac{\theta}{2}\right)\right)\begin{pmatrix} \cos\left(\frac{\theta}{2}\right) \\ \sin\left(\frac{\theta}{2}\right) \end{pmatrix} \\
&= \frac{\hbar}{2}\left(\cos^2\frac{\theta}{2} + \sin^2\frac{\theta}{2}\right) = \frac{\hbar}{2},
\end{aligned} \tag{3.33}$$

where we used the trigonometric identities

$$\cos\theta\cos\left(\frac{\theta}{2}\right) + \sin\theta\sin\left(\frac{\theta}{2}\right) = \cos\left(\frac{\theta}{2}\right),$$

$$\sin\theta\cos\left(\frac{\theta}{2}\right) - \cos\theta\sin\left(\frac{\theta}{2}\right) = \sin\left(\frac{\theta}{2}\right). \tag{3.34}$$

Since we find that $\langle S_\theta \rangle = \hbar/2$, the state $|\psi(\theta)\rangle$ is indeed the spin state in the direction θ, just like $|\uparrow\rangle$ is the spin state in the positive z-direction ($\theta = 0$) and $|+\rangle$ is the spin state in the positive x-direction ($\theta = 90°$). When $\theta = 180°$, we have rotated the spin vector upside-down, and the spin state $|\uparrow\rangle$ has become the orthogonal state $|\downarrow\rangle$.

What about the y-direction? Since all possible values of θ, from 0° to 360°, gives us all possible directions in the xz-plane, we need another angle ϕ to measure the rotation around, say, the vertical (z) axis. We choose ϕ such that $\phi = 0$ puts the spin vector in the xz-plane. A convenient way to work out the ϕ-dependence of a spin state is to consider the spin in the positive x- and y-directions in vector notation:

$$|+\rangle = \frac{1}{\sqrt{2}} \begin{pmatrix} 1 \\ 1 \end{pmatrix} \quad \text{and} \quad |\circlearrowleft\rangle = \frac{1}{\sqrt{2}} \begin{pmatrix} 1 \\ i \end{pmatrix}, \tag{3.35}$$

where now the complex numbers come into play. When we trace out the path from $|+\rangle$ to $|\circlearrowleft\rangle$ to $|-\rangle$ to $|\circlearrowright\rangle$, the lower vector component in (3.35) goes from $+1$ to i to -1 to $-i$, and back to $+1$. In the previous chapter we saw that this is the typical behaviour of a phase factor $e^{i\phi}$. We therefore expect that the spin state in the xy-plane takes the form

$$|\psi(\phi)\rangle = \frac{1}{\sqrt{2}} \begin{pmatrix} 1 \\ e^{i\phi} \end{pmatrix} = \frac{1}{\sqrt{2}} |\uparrow\rangle + \frac{e^{i\phi}}{\sqrt{2}} |\downarrow\rangle. \tag{3.36}$$

Indeed, we can calculate the expectation value with respect to $|\psi(\phi)\rangle$ of the spin component

$$S_\phi = \cos\phi \, S_x + \sin\phi \, S_y, \tag{3.37}$$

and we find that $\langle S_\phi \rangle = \hbar/2$ for the state in (3.36). This proves that $|\psi(\phi)\rangle$ is the state of an electron spin in the direction ϕ.

We combine the xz-plane and the xy-plane to obtain a general expression for a spin state

$$|\psi(\theta, \phi)\rangle = \cos\left(\frac{\theta}{2}\right) |\uparrow\rangle + e^{i\phi} \sin\left(\frac{\theta}{2}\right) |\downarrow\rangle. \tag{3.38}$$

This expression contains two angles, and each value for (θ, ϕ) can therefore be identified with a point on the surface of a sphere. This is the so-called Bloch sphere, shown in Fig. 3.4.

The Bloch sphere is a convenient way of collecting all the possible spin states into a single picture. However, there are two important things to keep in mind:

1. Orthogonal state vectors correspond to vectors at an angle of 180° in the Bloch sphere (antipodal points on the sphere);
2. the identification of the spin direction with the direction in the Bloch sphere is a happy accident.

We need to elaborate on this second point. Every quantum system that has two distinct states, such as the photon states |upper⟩ and |lower⟩ inside the arms of a

Fig. 3.4 The Bloch sphere

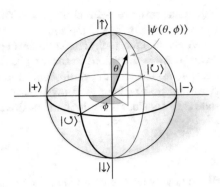

Mach-Zehnder interferometer, can be described by a collection of vectors that fit perfectly on the Bloch sphere. To see this, any normalised complex superposition of |upper⟩ and |lower⟩ is a valid quantum state:

$$|\text{photon}\rangle = a|\text{upper}\rangle + b|\text{lower}\rangle, \qquad (3.39)$$

with $|a|^2 + |b|^2 = 1$. We can put a phase shift $e^{i\phi}$ in the lower arm of the interferometer after a beam splitter with reflectivity $R = \sin^2(\theta/2)$, so we can have the state

$$|\text{photon}(\theta, \phi)\rangle = \cos\left(\frac{\theta}{2}\right)|\text{upper}\rangle + e^{i\phi}\sin\left(\frac{\theta}{2}\right)|\text{lower}\rangle. \qquad (3.40)$$

Since |upper⟩ and |lower⟩ are not states of a vector observable like S_z, but just two possibilities for the photon to go through the interferometer, there is no intrinsic meaning to the angles θ and ϕ in our real three-dimensional space, and the interpretation of the direction of the state vector as the spin direction in space is accidental: It does not hold in general. For spins, this (un)happy accident is a source of much confusion, so be careful not to confuse the state vector in the Bloch sphere with the spin vector in real space.

In real space, the most general spin direction can be constructed from the spin in the z-direction as a rotation over θ around the y-axis followed by a rotation over ϕ around the z-axis. The classical spin vector in three dimensions is then given by

$$\mathbf{S}(\theta, \phi) = S \begin{pmatrix} \cos\phi \, \sin\theta \\ \sin\phi \, \sin\theta \\ \cos\theta \end{pmatrix}, \qquad (3.41)$$

where S is the classical spin magnitude. We can imagine that the angles θ and ϕ depend on time, and that the state of the electron spin evolves over time, tracing out a path on the Bloch sphere. We will consider this in more detail in the next chapter.

Finally, we make the connection between spin observables and points in the Bloch sphere. We have seen that we can measure the spin component in a particular spatial direction. If the spin vector is pointing in the direction given by (3.41), then we obtain the spin in the opposite direction by making the substitution $\theta \to \theta + \pi$ and leaving ϕ unchanged. This will take $\mathbf{S}(\theta, \phi)$ to $-\mathbf{S}(\theta, \phi)$. Since these are opposite spin directions, a measurement in a Stern-Gerlach experiment along this direction can perfectly distinguish between these two spins, and the corresponding quantum states should be orthogonal (since perfectly distinguishable states correspond to orthogonal states). We define $|\psi_1\rangle = |\psi(\theta, \phi)\rangle$ and $|\psi_2\rangle = |\psi(\theta + \pi, \phi)\rangle$, and the scalar product $\langle\psi_1|\psi_2\rangle$ should be equal to zero. First, we calculate $\langle\psi_1|$:

$$\langle\psi_1| = \cos\left(\frac{\theta}{2}\right)\langle\uparrow| + e^{-i\phi}\sin\left(\frac{\theta}{2}\right)\langle\downarrow|. \tag{3.42}$$

Notice the sign change in $i\phi$ due to the complex conjugate. The state $|\psi_2\rangle$ with $\theta \to \theta + \pi$ becomes

$$|\psi_2\rangle = -\sin\left(\frac{\theta}{2}\right)|\uparrow\rangle + e^{i\phi}\cos\left(\frac{\theta}{2}\right)|\downarrow\rangle. \tag{3.43}$$

The scalar product then becomes

$$\langle\psi_1|\psi_2\rangle = -\cos\left(\frac{\theta}{2}\right)\sin\left(\frac{\theta}{2}\right) + e^{-i\phi}e^{i\phi}\cos\left(\frac{\theta}{2}\right)\sin\left(\frac{\theta}{2}\right) = 0, \tag{3.44}$$

as expected (we used $e^{-i\phi}e^{i\phi} = e^{-i\phi+i\phi} = e^0 = 1$). This means that antipodal states in the Bloch sphere are orthogonal. Moreover, each axis passing through the origin in the Bloch sphere connects two antipodal points, and therefore each axis through the origin can be seen as a spin component observable because it connects an "up" and "down" state in that particular direction.

Having defined two angles, θ and ϕ, to cover the entire Bloch sphere, we must be careful not to cover the sphere more than once. If we let both θ and ϕ take values between zero and 2π, we end up covering the Bloch sphere twice (you should verify this!). The standard solution is to adjust the domain of θ, so we have

$$0 \le \theta \le \pi \quad \text{and} \quad 0 \le \phi < 2\pi, \tag{3.45}$$

where we made sure to exclude $\phi = 2\pi$, since it is the same as $\phi = 0$.

Now suppose that we prepare the electron spin in a state described by a vector from the origin to a point on the surface of the Bloch sphere. The spin observable we wish to measure is given by an axis through the origin of the Bloch sphere. The probabilities of the two measurement outcomes are now determined entirely by the *projection of the state vector onto the observable axis*.

As an example, consider the general state

$$|\psi(\theta, \phi)\rangle = \cos\left(\frac{\theta}{2}\right) |\uparrow\rangle + e^{i\phi} \sin\left(\frac{\theta}{2}\right) |\downarrow\rangle ,\qquad(3.46)$$

and a measurement of the S_z observable. The projection K of the state vector onto the z-axis is given by the cosine of the angle between them (since the length of the state vector is 1). Since this angle is just θ, we have

$$K = \cos\theta .\qquad(3.47)$$

Compare this with the classical spin value in the z-direction in (3.41)! At the same time, we can calculate the probabilities for finding measurement outcomes $\uparrow$ and $\downarrow$:

$$p_\uparrow = \cos^2\left(\frac{\theta}{2}\right) \quad \text{and} \quad p_\downarrow = \sin^2\left(\frac{\theta}{2}\right) .\qquad(3.48)$$

Using the double angle formula

$$\cos^2\left(\frac{\theta}{2}\right) - \sin^2\left(\frac{\theta}{2}\right) = \cos\theta ,\qquad(3.49)$$

we relate the outcome probabilities to K via

$$K = p_\uparrow - p_\downarrow .\qquad(3.50)$$

By using $p_\uparrow + p_\downarrow = 1$ we can derive the values of $p_\uparrow$ and $p_\downarrow$ from K alone:

$$p_\uparrow = \frac{1 + K}{2} \quad \text{and} \quad p_\downarrow = \frac{1 - K}{2} .\qquad(3.51)$$

We can repeat this procedure for projections along any other direction as well (there is nothing intrinsically special about the z-direction). A few more examples of projections in the Bloch sphere are shown in Fig. 3.5.

3.4 Uncertainty

Quantum mechanics is all about calculating probabilities, but there is a bit more to this than just calculating the probabilities of measurement outcomes and the average value of physical observables using the expectation value. Sometimes we are interested in the uncertainty we have about certain measurement outcomes. For example, when we prepare an electron in the spin state $|\uparrow\rangle$ and send it through a Stern-Gerlach apparatus aligned in the z-direction, the probability that we will find a fluorescent dot in the "up" region is $p_\uparrow = 1$. In other words, there is no uncertainty about the spin in the z-direction. Similarly, a measurement along the x-direction will give $p_+ = p_- = 1/2$.

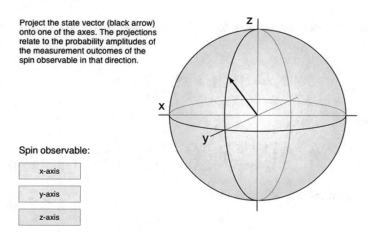

Project the state vector (black arrow) onto one of the axes. The projections relate to the probability amplitudes of the measurement outcomes of the spin observable in that direction.

Spin observable:

x-axis

y-axis

z-axis

Fig. 3.5 Projections in the Bloch sphere (see supplementary material 5)

This means that we have *maximum* uncertainty because no measurement outcome is more likely than the other. When we measure at a small angle θ from the z-axis we have only a little bit of uncertainty. We want to quantify this.

We see that the uncertainty of the spin in the direction θ is related to the spin observable S_θ and the spin state $|\psi\rangle$ of the electron. Instead of the expectation value of the spin we can calculate how much the measurement outcomes deviate from the mean. We can accomplish this by taking the difference between S_θ and its mean using the operator $S_\theta - \langle S_\theta \rangle \mathbb{I}$. However, if we calculate the expectation value of this new operator we find that it is zero:

$$\langle (S_\theta - \langle S_\theta \rangle \mathbb{I}) \rangle = \langle S_\theta \rangle - \langle S_\theta \rangle \langle \psi | \psi \rangle = \langle S_\theta \rangle - \langle S_\theta \rangle = 0 \,, \qquad (3.52)$$

because the deviation can be both positive and negative, and averages out. We need to force it to be positive, which we can do by taking the square. The uncertainty ΔS_θ can then be defined as

$$\Delta S_\theta = \sqrt{\langle \psi | (S_\theta - \langle S_\theta \rangle)^2 | \psi \rangle} \,, \qquad (3.53)$$

where we take the square root in order to preserve the units of S_θ. Another way to calculate this is

$$(\Delta S_\theta)^2 = \langle \psi | S_\theta^2 | \psi \rangle - \langle \psi | S_\theta | \psi \rangle^2 \,. \qquad (3.54)$$

You should check this by expanding the square in (3.53). The quantity $(\Delta S_\theta)^2$ is called the variance of the operator S_θ given the spin state $|\psi\rangle$, and ΔS_θ is the standard deviation.

Suppose that an electron is created in the spin state $|\uparrow\rangle$. The expectation value of S_z is then

$$\langle S_z \rangle = \langle \uparrow | S_z | \uparrow \rangle = \langle \uparrow | \left(\frac{\hbar}{2} | \uparrow \rangle \right) = \frac{\hbar}{2} , \tag{3.55}$$

and using

$$S_z^2 = \frac{\hbar}{2} \begin{pmatrix} 1 & 0 \\ 0 & -1 \end{pmatrix} \cdot \frac{\hbar}{2} \begin{pmatrix} 1 & 0 \\ 0 & -1 \end{pmatrix} = \frac{\hbar^2}{4} \begin{pmatrix} 1 & 0 \\ 0 & 1 \end{pmatrix} , \tag{3.56}$$

the expectation value of S_z^2 is calculated as

$$\langle S_z^2 \rangle = \langle \uparrow | S_z^2 | \uparrow \rangle = \frac{\hbar^2}{4} . \tag{3.57}$$

The variance is therefore

$$(\Delta S_z)^2 = \frac{\hbar^2}{4} - \left(\frac{\hbar}{2} \right)^2 = 0 . \tag{3.58}$$

In other words, there is no uncertainty about the outcome of a spin S_z measurement of an electron in state $|\uparrow\rangle$, as we determined earlier.

However, if we measure the spin of the same electron in the x-direction we obtain entirely different behaviour. The expectation value of S_x is zero:

$$\langle S_x \rangle = \langle \uparrow | S_x | \uparrow \rangle = \frac{\hbar}{2} \begin{pmatrix} 1 & 0 \end{pmatrix} \begin{pmatrix} 0 & 1 \\ 1 & 0 \end{pmatrix} \begin{pmatrix} 1 \\ 0 \end{pmatrix} = 0 , \tag{3.59}$$

but the variance of S_x will not be zero. The matrix form of S_x^2 is

$$S_x^2 = \frac{\hbar}{2} \begin{pmatrix} 0 & 1 \\ 1 & 0 \end{pmatrix} \cdot \frac{\hbar}{2} \begin{pmatrix} 0 & 1 \\ 1 & 0 \end{pmatrix} = \frac{\hbar^2}{4} \begin{pmatrix} 1 & 0 \\ 0 & 1 \end{pmatrix} . \tag{3.60}$$

We therefore calculate

$$\langle S_x^2 \rangle = \langle \uparrow | S_x^2 | \uparrow \rangle = \frac{\hbar^2}{4} , \tag{3.61}$$

so

$$(\Delta S_x)^2 = \langle S_x^2 \rangle - \langle S_x \rangle^2 = \frac{\hbar^2}{4} - 0 = \frac{\hbar^2}{4} . \tag{3.62}$$

There is therefore some uncertainty about the spin components in the x-direction: $\Delta S_x = \hbar/2$. Can you show that $\Delta S_y = \hbar/2$ for the state $|\uparrow\rangle$?

We can measure only one component of the electron spin at any given time in a Stern-Gerlach experiment, due to our choice of a particular orientation of the magnets. As the above example shows, this leaves us uncertain about the other spin components of the electron. Classically, measurements reveal the values of the spin components, and we would expect that we can just perform another spin measurement on the same electron, this time in the y-direction. So instead of recording the electron as a fluorescent dot on the screen, we can send it into a second Stern-Gerlach experiment and measure another spin component of our choice. If we choose to measure the spin in the x-direction we find that the probabilities of measuring "+" or "−" are $p_+ = p_- = 1/2$.

Suppose that our second measurement reveals that the x-component of the spin is +. Can we say that the spin of the electron before it entered the first Stern-Gerlach experiment had spin $+\hbar/2$ in both the z- and x-direction? If this were true, then a third Stern-Gerlach experiment measuring the z component again should give the value $+\hbar/2$. But this is not what we find experimentally. We find instead that the probabilities of the measurement outcomes ↑ and ↓ are $p_\uparrow = p_\downarrow = 1/2$. So we conclude that the measurement of S_x must have *disturbed* the spin state of the electron.

Measuring a particular spin component of an electron *creates* maximum uncertainty of the spin components in the orthogonal directions. Repeating this experiment for another spin component will reduce that uncertainty for the measured component but again create uncertainty in the original spin component. Therefore, we cannot measure the three orthogonal spin components of the original spin state with arbitrary precision. This is a consequence of the fact that we can measure only one of the spin components of **S** at a time. It is also closely related to the example of the QND measurement in Chap. 2, which destroyed the interference in the output beams of a Mach-Zehnder interferometer. We will return to quantum uncertainty in more detail in Chap. 9.

3.5 The Magnitude of the Electron Spin

If we can measure only one component of the spin of an electron at a time, how can we say that the electron is a particle with spin $\frac{1}{2}$? Isn't the amount of spin determined by *all* the components of the spin vector? To answer this, we first have to determine a rule that tells us whether two observables can have simultaneous "sharp" values, i.e., expectation values with zero variance. We will look at the mathematical underpinning of this in Chap. 5, but right now we can state the general rule: if A and B are two observables with corresponding matrices, then they can have sharp expectation values if

$$AB - BA = 0. \tag{3.63}$$

Such matrices are said to *commute*, since the order of the matrices does not matter: $AB = BA$ from (3.63). We often abbreviate the expression $AB - BA$ as

$$[A, B] \equiv AB - BA, \tag{3.64}$$

which is called the *commutator* of A and B. Consider the commutator of S_z and S_x:

$$\begin{aligned}
[S_z, S_x] &= S_z S_x - S_x S_z \\
&= \frac{\hbar}{2} \begin{pmatrix} 1 & 0 \\ 0 & -1 \end{pmatrix} \frac{\hbar}{2} \begin{pmatrix} 0 & 1 \\ 1 & 0 \end{pmatrix} - \begin{pmatrix} 0 & 1 \\ 1 & 0 \end{pmatrix} \frac{\hbar}{2} \begin{pmatrix} 1 & 0 \\ 0 & -1 \end{pmatrix} \\
&= \frac{\hbar^2}{4} \left[\begin{pmatrix} 0 & 1 \\ -1 & 0 \end{pmatrix} - \begin{pmatrix} 0 & -1 \\ 1 & 0 \end{pmatrix} \right] = \frac{\hbar^2}{2} \begin{pmatrix} 0 & 1 \\ -1 & 0 \end{pmatrix} \neq 0.
\end{aligned} \tag{3.65}$$

Since the commutator $[S_z, S_x]$ is not equal to zero, the two observables S_z and S_x do not admit sharp values simultaneously. When the variance of S_z is zero, then the variance of S_x cannot be zero, as we saw in the previous section. However, there we used specific quantum states, whereas the commutator does not involve states, just the matrices. It is therefore a more general test of whether two observables can have sharp values simultaneously. Similarly, you can show that S_y does not commute with either S_x or S_z, and this implies that only one component of the spin can be sharp at any time.

However, there is an observable that commutes with S_x, S_y, and S_z, perhaps somewhat surprisingly. Classical spin is a vector quantity, **S**. To find the magnitude of the classical spin, we have to calculate the magnitude of the vector, which we obtain by taking the dot product with itself:

$$S^2 = \mathbf{S} \cdot \mathbf{S} = S_x^2 + S_y^2 + S_z^2. \tag{3.66}$$

This is a classical expression, but we can construct the corresponding quantum version by replacing the spin components by their quantum mechanical observables. We then have to calculate the sum of the squares of the matrices for S_x, S_y, and S_z. However, the square of each of the spin components is the identity matrix multiplied by $\frac{1}{4}\hbar^2$, and therefore

$$S^2 = \frac{\hbar^2}{4} \begin{pmatrix} 1 & 0 \\ 0 & 1 \end{pmatrix} + \frac{\hbar^2}{4} \begin{pmatrix} 1 & 0 \\ 0 & 1 \end{pmatrix} + \frac{\hbar^2}{4} \begin{pmatrix} 1 & 0 \\ 0 & 1 \end{pmatrix} = \frac{3\hbar^2}{4} \begin{pmatrix} 1 & 0 \\ 0 & 1 \end{pmatrix}. \tag{3.67}$$

This is the magnitude (squared) of a spin-$\frac{1}{2}$ particle. We typically keep the square for convenience. The important property is that the quantum mechanical observable S^2 commutes with every possible component observable $S(\theta, \phi)$, with

$$S(\theta, \phi) = \sin\theta \cos\phi \, S_x + \sin\theta \sin\phi \, S_y + \cos\theta \, S_z. \tag{3.68}$$

This is because S^2 is proportional to the identity matrix, which commutes with all matrices. But this means that regardless which spin component we measure, the overall magnitude of the spin has a sharp value, i.e., no uncertainty. The expectation value of S^2 for *any* normalised spin-$\frac{1}{2}$ state is

$$\langle S^2 \rangle = \frac{3}{4}\hbar^2 . \tag{3.69}$$

Hence, we can talk about the magnitude of the spin in a meaningful way, even if we can only ever determine one component at a time.

In the previous chapter we encountered the matrices describing the beam splitter and phase shifts in an interferometer, and we found that the order of the matrices matters there, as well. In that case, the order of the matrices was determined by the order in which we apply beam splitters and phase shifters (accumulating from right to left in the product order). Here, the order of the matrices matters also, but not as an order of operation. Instead, the non-commutativity of two observables implies here that they cannot have simultaneously sharp values. While the order of operation of the beam splitter and phase shift matrices makes sense also at a classical level (the order of putting on your socks and putting on your shoes matter!), the non-commutativity of observables is a fundamental quantum feature that lies at the heart of why quantum mechanics produces at times such counterintuitive results.

3.6 Photon Polarisation

Earlier in this chapter we identified the mathematical likeness between the quantum mechanical photon states and spin states. Here, we will complete the story and show how all possible states of a single photon also map onto the same Bloch sphere. It will help you understand what is part of the mathematical formalism, and what is the physical content of the quantum theory we developed so far. We will then apply what we have learned to photon polarisation.

Consider the state of a photon entering a beam splitter from the left. The output, if it is a 50:50 beam splitter, can be written as

$$|\text{photon}\rangle = \frac{1}{\sqrt{2}} |\text{right}\rangle + \frac{1}{\sqrt{2}} |\text{down}\rangle . \tag{3.70}$$

However, a beam splitter does not have to be exactly 50:50. In fact, when you buy a beam splitter off the shelf it will never be exactly 50:50. Also, there are plenty of situations where we would want to have a different transmission coefficient, for example in the bomb detection experiment of the previous chapter. If the photon is not absorbed by the beam splitter we require that the transmission T and reflection R obey

$$T + R = 1 .$$

These are fractions of intensities of the light. If we phrase the argument instead in terms of amplitudes, such that $T = t^2$ and $R = r^2$, we obtain[3]

$$t^2 + r^2 = 1 .$$

The state of the photon coming out of this beam splitter is then

$$|\text{photon}\rangle = t \,|\text{right}\rangle + r \,|\text{down}\rangle . \qquad (3.71)$$

There is a very elegant way in which we can automatically ensure that $t^2 + r^2 = 1$, and that is to write $t = \cos\theta$ and $r = \sin\theta$, where θ takes values in the interval $[0, \pi/2]$. This means that for whatever transmission coefficient T you require, you choose the θ that matches its numerical value via $T = \cos^2\theta$. The reflection coefficient is then automatically $R = 1 - T$. With this parametrisation the photon state can be written as

$$|\text{photon}\rangle = \cos\theta \,|\text{right}\rangle + \sin\theta \,|\text{down}\rangle . \qquad (3.72)$$

Next, we can apply a phase shift ϕ in the "down" output. We have seen in Chap. 2 that this leads to a factor $e^{i\phi}$ in the state:

$$|\text{photon}\rangle = \cos\theta \,|\text{right}\rangle + e^{i\phi} \sin\theta \,|\text{down}\rangle . \qquad (3.73)$$

But this is *exactly* the same state space as the spin-$\frac{1}{2}$ particle, where θ took values in the interval $[0, \pi]$ but it appeared in the argument of the sine and cosine as $\theta/2$. This is very convenient, because it suggests that the state space of any object with two distinct outcomes in the measurement of an observable has a state space that looks like the Bloch sphere.

Next, we consider the polarisation of a photon. The polarisation is a property of light that you may already be familiar with from classical physics. Many sunglasses are based on filtering light of one polarisation and letting through the orthogonally polarised light. This removes half of the light that goes into your eyes. When light reflects off surfaces like water it can become polarised, and the light from your laptop screen is also strongly polarised. If we treat classical light as a wave of electric and magnetic fields, its polarisation is given by the direction of the electric field. This is shown in Fig. 3.6. Like spin, polarisation is a vector, since it has a magnitude and a direction. However, we usually do not talk about the components of the polarisation, but rather phrase the discussion in terms of polarising filters or polarising beam splitters.

Consider a photon with some polarisation entering a polarising beam splitter, made in such a way that horizontally polarised light passes through, but vertically polarised light is reflected. Such devices are commonplace in optics labs. We place two detectors, D_H and D_V, in the output beams, one for each polarisation, where H

---

[3] Warning: the symbol t here has nothing to do with time!

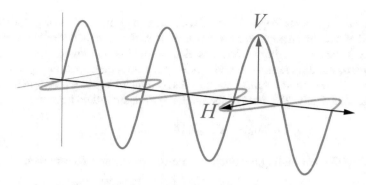

Fig. 3.6 The polarisation of light is determined by the direction of the electric field. The field oscillating in the vertical direction (blue) is called "vertical polarisation" (V), and the field oscillating in the horizontal direction (orange) is called "horizontal polarisation" (H)

and V denote horizontal and vertical polarisation, respectively. We can express the photon state as

$$|\text{photon}\rangle = a\,|H\rangle + b\,|V\rangle \ . \tag{3.74}$$

When this photon enters the polarising beam splitter and the subsequent detectors, we are in effect measuring the polarisation of the photon. We can write the states $|H\rangle$ and $|V\rangle$ in vector form, as before:

$$|H\rangle = \begin{pmatrix} 1 \\ 0 \end{pmatrix} \quad \text{and} \quad |V\rangle = \begin{pmatrix} 0 \\ 1 \end{pmatrix} . \tag{3.75}$$

The probability of finding a horizontally polarised photon is then

$$\Pr(H) = |\langle H|\text{photon}\rangle|^2 = |a\,\langle H|H\rangle + b\,\langle H|V\rangle|^2 = |a|^2 \ . \tag{3.76}$$

Similarly, we find that $\Pr(V) = |b|^2$. I hope this is starting to look familiar!

The polarisation of light is directly determined by the direction of the electric field in space, so if we rotate the polarising beam splitter by $90°$ (or $\pi/2$ radians) we expect that the horizontal polarisation now looks like vertical polarisation to the polarising beam splitter, and instead of transmitting the horizontally polarised light, the polarising beam splitter now reflects it towards D_V (we assume that the detectors rotate along with the polarising beam splitter). This is indeed what is found in experiments. We would get the same behaviour if we were to rotate the polarisation by $90°$ in the opposite direction, since we assume that only the photon and the polarising beam splitter with its detectors are relevant to the measurement outcomes. In that case, a horizontally polarised photon is rotated into a vertically polarised photon. It all checks out.

Next, consider what happens when we rotate the polarisation of a horizontally polarised photon by an angle θ. This can be done with a so-called *birefringent* crystal that we will be considering below. The electric field is no longer aligned perfectly in the horizontal direction, and it will pick up a component $\sin \theta$ along the vertical direction. The horizontal component shrinks by a factor $\cos \theta$, which you can derive purely geometrically. The state of the photon polarisation then becomes

$$|\text{photon}\rangle = \cos \theta \, |H\rangle + \sin \theta \, |V\rangle \, . \tag{3.77}$$

The probabilities of finding the photon in detectors D_H and D_V are then

$$\Pr(H) = |\langle H|\text{photon}\rangle|^2 = |\cos \theta \, \langle H|H\rangle + \sin \theta \, \langle H|V\rangle|^2 = \cos^2 \theta \, ,$$
$$\Pr(V) = |\langle V|\text{photon}\rangle|^2 = |\cos \theta \, \langle V|H\rangle + \sin \theta \, \langle V|V\rangle|^2 = \sin^2 \theta \, . \tag{3.78}$$

As required, these probabilities again sum to one.

Next, we consider how we can change the relative phase of the two polarisation terms in (3.77). When light propagates through a medium, its velocity v is reduced by a factor equal to the index of refraction n compared to the speed of light in vacuum c: $v = c/n$. The wavelength λ_n in the medium is shortened by a factor n as well: $\lambda_n = \lambda/n$, where λ is the wavelength of the light in vacuum. When light travels through a slab of thickness d, the accumulated phase of the light wave is given by

$$\phi = \frac{2\pi n d}{\lambda} \, . \tag{3.79}$$

This is just the phase shift used in optical beams. In a birefringent crystal the speed of light is different for horizontally and for vertically polarised light. This has the effect that the phase of the wave of one polarisation starts lagging relative to the other polarisation. If we have a material of thickness d with two indices of refraction n_H and n_V for the two polarisations H and V, the phases accumulated by the two polarisations are

$$\phi_H = \frac{2\pi n_H d}{\lambda} \quad \text{and} \quad \phi_V = \frac{2\pi n_V d}{\lambda} \, . \tag{3.80}$$

The relative phase between the two polarised waves is thus the difference between these two:

$$\phi = \phi_H - \phi_V = \frac{2\pi d}{\lambda}(n_H - n_V) \, . \tag{3.81}$$

For a single photon, this means that the vertical polarisation term gains a relative phase shift to the horizontal polarisation term:

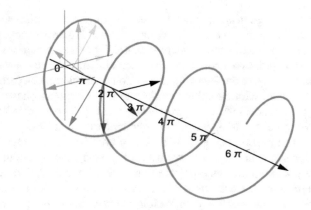

Fig. 3.7 In light with circular polarisation the electric field vector rotates around the propagation axis (the shaded arrows). There are two possibilities, namely right-handed polarisation, shown here, and left-handed polarisation. The orientation is determined by aligning your thumb to the propagation direction, and letting your fingers curl in the rotation direction of the electric field vector

$$|\text{photon}\rangle = \frac{1}{\sqrt{2}}|H\rangle + \frac{e^{-i\phi}}{\sqrt{2}}|V\rangle \ . \tag{3.82}$$

So there is a physical mechanism (a birefringent crystal) that can change the polarisation such that we can create any relative phase between the two polarisation terms. In the special case where $\phi = \pm\pi/2$, we obtain the states of *circular* polarisation

$$|L\rangle = \frac{1}{\sqrt{2}}|H\rangle + \frac{i}{\sqrt{2}}|V\rangle \qquad \text{and} \qquad |R\rangle = \frac{1}{\sqrt{2}}|H\rangle - \frac{i}{\sqrt{2}}|V\rangle \ , \tag{3.83}$$

where L and R refer to left- and right-handed circular polarisation. The electric field vector for such polarisation states rotates as shown in Fig. 3.7. The polarisation rotation in (3.77) can also be achieved using a birefringent crystal, and we will explore this in Exercise 17.

The most general polarisation state of a photon can now be written as

$$|\text{photon}\rangle = \cos\theta |H\rangle + e^{-i\phi} \sin\theta |V\rangle \ , \tag{3.84}$$

and it is clear that the quantum state space of the polarised photon is the same as that for the electron spin and the photon in two optical beams, namely the Bloch sphere. Historically, when referring to polarisation, the Bloch sphere is often called the Poincaré sphere, but it really means the same thing.

It is no coincidence that the polarisation of a photon has exactly the same mathematical description as the spin of an electron. Physically, the polarisation is the direct counterpart of the electron spin, since it too can carry angular momentum (remember that $\hbar$ has units of angular momentum): given the right polarisation state, a photon can impart a torque on the object it hits. However, there is an even more fundamental reason that the mathematics of the photon polarisation and the electron spin are the same. *Any* system with two distinct measurement outcomes can be described using a state space that looks like the Bloch sphere or Poincaré sphere, since that is the state space of unit vectors with complex coefficients up to a global phase. We will see this again in the next chapter, when we consider a simplified atom with two energy levels, and in chapter six when we consider general quantum bits.

Finally, we can create observables for the polarisation of a photon, just like we did for the electron spin. However, we need to think a little bit about the possible values of the polarisation. Since polarisation is really the *direction* of the electric field, any single number we assign it acts mainly as a label. So for example when we measure horizontal and vertical polarisation, we can choose the polarisation value '$+1$' for horizontal polarisation and '-1' for vertical polarisation. Following the same procedure as for the spin S_z observable, we consider the expectation value of the polarisation in the horizontal and vertical direction P_{HV} for an arbitrary polarisation state $|\psi\rangle$:

$$\langle P_{HV} \rangle = p_H \cdot (+1) + p_V \cdot (-1) = p_H - p_V = |\langle H|\psi\rangle|^2 - |\langle V|\psi\rangle|^2$$
$$= \langle \psi | (|H\rangle \langle H| - |V\rangle \langle V|) |\psi\rangle = \langle \psi | P_{HV} |\psi\rangle . \tag{3.85}$$

From this we see that the polarisation observable P_{HV} is given by

$$P_{HV} = |H\rangle \langle H| - |V\rangle \langle V| . \tag{3.86}$$

Similarly, you can show that the circular polarisation observable can be written as

$$P_{LR} = |L\rangle \langle L| - |R\rangle \langle R| = i |H\rangle \langle V| - i |V\rangle \langle H| . \tag{3.87}$$

You should verify this, and also construct the observable for the diagonal polarisation states $(|H\rangle \pm |V\rangle)/\sqrt{2}$. All the mathematical results we have derived for the spin observables also hold for the polarisation observables, including the uncertainties.

3.7 ❧ Magnetic Resonance Imaging

Electrons are not the only particles with spin. For example, protons also have spin with maximum value $\hbar/2$, and we can describe proton spins in the same way as electron spins. In this section we will see how proton spins can be used in an important practical application called magnetic resonance imaging, which among other things allows doctors to take a look inside the brain of a living patient.

Spin is essentially a magnetic moment. In other words, a particle with spin behaves as a little magnet that wants to align itself along an external magnetic field. If you have ever played with little magnets you are familiar with this effect: try to push two magnets together north pole to north pole, and you feel the magnets push back. If you do not hold the magnets tight, they will slip out of your fingers and flip so they are lined up north pole to south pole. It's a fun game until you get a piece of skin caught between the magnets!

We can describe this in terms of the potential energy of the magnets: the potential energy for the north-north alignment is much higher that the potential energy of the north-south alignment of the two magnets. If the magnets are not held in place firmly they will reconfigure themselves in the state of lowest potential energy, just like a ball that is released at the top of a tower will move towards the point of lowest potential energy, i.e., the ground. The potential energy of a proton with spin in a magnetic field is given by

$$E = -\frac{e}{m}\mathbf{S} \cdot \mathbf{B},$$ (3.88)

where $\mathbf{B}$ is the vector that indicates the magnetic field (with its magnitude measured in units of Tesla), $\mathbf{S}$ is the spin vector, e is the charge of the proton, m is the mass of the proton, and c is the speed of light in vacuum. The proton has the highest potential energy when its spin is anti-aligned with the magnetic field (due to the minus sign), and the lowest potential energy when the spin is aligned with the magnetic field. The spin will therefore want to align itself with the magnetic field. Let's suppose that the magnetic field points in the positive z-direction. The proton spin will then relax into the state $|\uparrow\rangle$. Note that $\uparrow$ and $\downarrow$ denote spin directions, and *not* energy. The energy of $|\uparrow\rangle$ is *lower* than that of $|\downarrow\rangle$.

Just like a ball that gets kicked into the air, we can give the proton spin a kick with a small burst of radiation (a photon). If we want to kick the spin into the higher energy state $|\downarrow\rangle$ that is anti-aligned with the magnetic field, the energy of the radiation must overcome the potential energy difference of the two states:

$$\Delta E = \frac{e\hbar B}{m}.$$ (3.89)

Suppose that the magnetic field has a strength of 3.0 T. The frequency f of the photon must then be at least

$$f = \frac{\Delta E}{h} = 46\ \text{MHz}.$$ (3.90)

This is a Very High Frequency (VHF) radio wave, similar to that used in old-fashioned analog FM radio and television broadcasting. We call the pulse of radiation that flips the spin of the proton an "RF" pulse since it has Radio Frequency.

Once the proton spin is in a state of higher potential energy $|\downarrow\rangle$ it will not stay there, just like a kicked ball will not remain hovering in the air. The proton will spontaneously emit a photon of frequency f and relax back to the ground state $|\uparrow\rangle$. We can describe the RF pulse as a transformation U_{RF} on the spin of the proton as

Fig. 3.8 Spins aligned to a magnetic field for magnetic resonance imaging (see supplementary material 8)

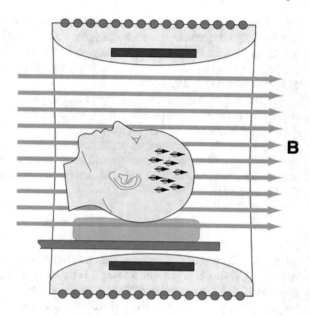

$$U_{\mathrm{RF}} \, |\!\uparrow\rangle = \begin{pmatrix} 0 & 1 \\ 1 & 0 \end{pmatrix} \begin{pmatrix} 1 \\ 0 \end{pmatrix} = \begin{pmatrix} 0 \\ 1 \end{pmatrix} = |\!\downarrow\rangle \; . \tag{3.91}$$

Similarly, the spontaneous emission transformation can also be described by a transformation U_{RF}, with $U_{\mathrm{RF}} |\!\downarrow\rangle = |\!\uparrow\rangle$. This is using the matrix U_{RF} as a transformation (instead of an observable), much like the beam splitter transformation in the previous chapter. If the RF pulse has exactly the right frequency, the pulse is said to be "on resonance" with the transition $|\!\uparrow\rangle \leftrightarrow |\!\downarrow\rangle$. Since a proton is the nucleus of a hydrogen atom, we call the above process Nuclear Magnetic Resonance, or NMR for short. It is used in Magnetic Resonance Imaging (MRI) as shown in Fig. 3.8.

Let's consider how MRI works for a brain scan. The brain consists of different tissues that have a varying amount of hydrogen. In an NMR process, regions with a large amount of hydrogen will return a brighter RF pulse than regions with lower amounts of hydrogen. The MRI scanner measures how much RF radiation is coming from a certain direction, but it cannot measure the depth of the source in the brain. To find this out, MRI scanners use a mathematical technique called "tomography". By looking at the RF radiation coming from many different directions we can work out in great detail where the bright and dark spots are.

Tomography is like solving a puzzle. For a very simple example, consider the interactive tomogram in Fig. 3.9. We can send in light from two perpendicular directions onto a two-dimensional shape, which casts two different shadows. The puzzle is to deduce the 2D shape from the shadows. We record the shadows, and use them as input in a computer programme, which deduces the image from the shadows. The image extraction in MRI works in a similar way.

Fig. 3.9 The principle of tomography (see supplementary material 9)

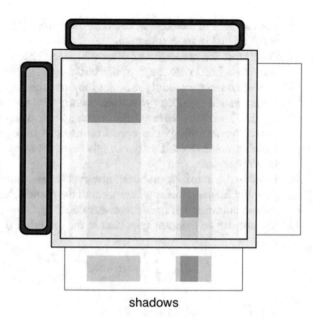

shadows

Exercises

1. The state of an electron spin is given by

$$|\psi\rangle = \frac{1}{\sqrt{3}}|\uparrow\rangle + \sqrt{\frac{2}{3}}|\downarrow\rangle .$$

   What is the probability of finding spin $\downarrow$ in a measurement outcome? What is the probability of finding measurement outcome "+", with $|+\rangle = (|\uparrow\rangle + |\downarrow\rangle)/\sqrt{2}$?
2. Normalise the state $2|\uparrow\rangle + 4|\downarrow\rangle$.
3. An electron is prepared in the spin state $2|\uparrow\rangle - 3i|\downarrow\rangle$. Normalise this state and calculate the probability of finding spin "up" and spin "+", corresponding to $|\uparrow\rangle$ and $|+\rangle = (|\uparrow\rangle + |\downarrow\rangle)/\sqrt{2}$, respectively. What is the expectation value of the z-component of the spin?
4. Construct the matrix form of S_y similar to (3.23).
5. Give the matrix representation of the observable $S_\phi = \cos\phi\, S_x + \sin\phi\, S_y$. Calculate the expectation value of S_ϕ given the spin state

$$|\psi\rangle = \frac{1}{\sqrt{3}}|\uparrow\rangle + \frac{\sqrt{2}\, e^{i\pi/6}}{\sqrt{3}}|\downarrow\rangle ,$$

   with $|\uparrow\rangle$ and $|\downarrow\rangle$ the spin up and down states in the z-direction. What is the uncertainty ΔS_ϕ?

6. For the state in Exercise 5, calculate the probabilities of finding spin "up" and spin "down" in the z-direction.

7. Determine the angles θ and ϕ for which the state in Exercise 5 is the quantum state associated with the "up" direction.

8. Calculate the uncertainty ΔS_θ for the operator $S_\theta = \cos\theta\, S_z + \sin\theta\, S_x$ given the spin state $|\uparrow\rangle$. Does your result conform to your expectation?

9. In (3.45) we have covered the Bloch sphere using two angles. However, we have covered two special points on the sphere too many times. Can you tell which points? Does it matter in this case? Hint: remember that a global phase is unobservable.

10. Construct a matrix form of the observable associated with the path in a Mach-Zehnder interferometer where we find the photon. You will need to take special care in choosing the measurement values.

11. We prepare an electron spin state in the direction (θ, ϕ), which can be written as

$$\frac{\hbar}{2}\begin{pmatrix} \sin\theta\cos\phi \\ \sin\theta\sin\phi \\ \cos\theta \end{pmatrix} = \frac{\hbar}{2}\begin{pmatrix} 0.433 \\ 0.750 \\ 0.500 \end{pmatrix}.$$

What is the quantum state of the electron?

12. For the electron spin state of the previous question, relate the probabilities of finding outcomes $+$ and $-$ in the x-direction to the projection onto the x-axis.

13. An electron with spin state

$$|\psi\rangle = \frac{3}{5}|\uparrow\rangle + \frac{4}{5}|\downarrow\rangle$$

has its spin measured in the x-direction. What is the expectation value $\langle S_x\rangle$? What is the uncertainty ΔS_x?

14. An electron with spin state

$$|\psi\rangle = \frac{3}{5}|\uparrow\rangle + \frac{4}{5}|\downarrow\rangle$$

has its spin measured in the y-direction. What is the expectation value $\langle S_y\rangle$? What is the uncertainty ΔS_y?

15. An electron with spin state

$$|\psi\rangle = \frac{3}{5}|\uparrow\rangle + \frac{4}{5}|\downarrow\rangle$$

has its spin measured in the z-direction. What is the expectation value $\langle S_z\rangle$? What is the uncertainty ΔS_z?

16. The state of an electron spin is given by

$$|\psi\rangle = \frac{2}{\sqrt{13}} |\uparrow\rangle + \frac{3i}{\sqrt{13}} |\downarrow\rangle .$$

Calculate $\langle S_\theta \rangle$ and ΔS_θ.

17. A birefringent crystal is a crystal in which light of different polarisation propagates at different speeds. Let us suppose that the polarisation direction $|F\rangle$ that travels fastest through the crystal is expressed in terms of $|H\rangle$ and $|V\rangle$ is given by

$$|F\rangle = \cos\theta \, |H\rangle - \sin\theta \, |V\rangle ,$$

and suppose that the polarisation direction $|S\rangle$ that travels slowest through the crystal is expressed in terms of $|H\rangle$ and $|V\rangle$ is given by

$$|S\rangle = \sin\theta \, |H\rangle + \cos\theta \, |V\rangle .$$

A so-called *half-wave plate* will slow down the photon with polarisation $|S\rangle$ such that it will pick up a π phase shift relative to $|F\rangle$:

$$|F\rangle \rightarrow |F\rangle \qquad \text{and} \qquad |S\rangle \rightarrow -|S\rangle .$$

By writing $|H\rangle$ in terms of $|F\rangle$ and $|S\rangle$, show that the half-wave plate oriented at angle θ induces a polarisation rotation 2θ.

Chapter 4
Atoms and Energy

In this chapter, we look at the energy of electrons in an atom and determine how the quantum state of the atom changes over time. This leads us to introduce the energy operator, or Hamiltonian, and the Schrödinger equation that governs the behaviour of quantum systems. We show how electric and magnetic fields can interact with an atom, and as a demonstration of the quantum theory developed so far we explain how atomic clocks work.

4.1 The Energy Spectrum of Atoms

Atoms consist of electrons in orbits around a nucleus made of protons and neutrons. The negatively charged electrons are attracted to the positively charged protons in the nucleus by the Coulomb force, and this is how atoms are held together. It is very tempting to think of an atom as a mini solar system, with the nucleus in place of the sun and the electrons as planets, but this is a misleading picture. As in the previous chapters, we again consider a simple experiment to guide our investigations.

We fill a transparent container with hydrogen gas, and heat it in order to break up the hydrogen molecules into hydrogen atoms. White light is sent first through the container and second through a diffraction grating that separates the different frequencies of the white light into a spectrum. This is a technique called *spectroscopy* (see Fig. 4.1). When we inspect the spectrum, we will see the familiar rainbow pattern. In addition, we observe dark lines at specific colours in the spectrum. These are absorption lines, shown in Fig. 4.2. Somehow, the hydrogen interacts strongly

Supplementary Information The online version contains supplementary material available at https://doi.org/10.1007/978-3-031-16165-0_4.

Fig. 4.1 Spectroscopy

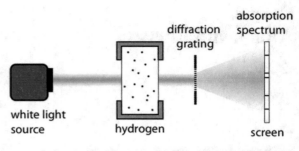

Fig. 4.2 Spectral lines (see supplementary material 2)

with light of these particular wavelengths (the absorption lines are not there if there isn't any hydrogen in the container).

The explanation of this phenomenon, first given by Niels Bohr[1] in 1913 and based on Einstein's explanation of the photoelectric effect[2] from 1905, is simple: photons of a particular wavelength λ carry an energy $E = hc/\lambda$, with h Planck's constant and c the speed of light. We often write this in terms of the (angular) frequency ω:

$$E = \frac{hc}{\lambda} = \hbar\omega \,. \tag{4.1}$$

The angular frequency is measured in radians per second (rad s^{-1}), and is related to the frequency f via $\omega = 2\pi f$. The photon is a packet of electromagnetic energy that can be absorbed by the electron in a hydrogen atom. This leads to a more energetic electron, and the disappearance of the photon. When lots of hydrogen atoms are placed in the path of a beam of light, many photons are absorbed, leading to a reduced intensity of the light. The experimental fact that the reduced intensity happens for very specific frequencies (the absorption lines) indicates that electrons can absorb only photons of these frequencies. Bohr argued that the electrons in the hydrogen atom can exist only in very specific energy states, and the frequency lines in the absorption spectrum correspond to the energy it takes to go from one energy state to another. Moreover, the electrons have a lowest possible energy state called the ground state, which is why the electrons do not crash into the nucleus (see Fig. 4.3). This is related to the uncertainty in position and momentum of the electron, and we will explore this further in Chap. 9.

---

[1] N. Bohr, *On the Constitution of Atoms and Molecules*, Philosophical Magazine **26** 1, 1913.

[2] A. Einstein, Einstein, *Über einen die Erzeugung und Verwandlung des Lichtes betreffenden heuristischen Gesichtspunkt*, Annalen der Physik **17** 132, 1905.

Fig. 4.3 The origin of spectral lines due to energy levels of the hydrogen atom (see supplementary material 3)

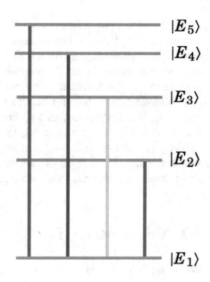

We can now identify the energy states of an electron in a hydrogen atom by carefully studying the absorption lines. An electron with energy E will be in the energy state $|E\rangle$. We can enumerate the energy states $|E_n\rangle$ with $n = 1, 2, 3, \ldots$, where E_1 is the energy of the ground state, E_2 is the energy of the first excited state, and so on. From the absorption spectrum of hydrogen we find that

$$E_n = \frac{E_0}{n^2}, \tag{4.2}$$

with $E_0 = -13.6\,\text{eV}$ a constant energy that is specific to hydrogen. Since only energy *differences* are recorded in the absorption spectrum, we can choose the energy of the ground state as we like by adding a constant to E_n. Here, $E_0 = -13.6\,\text{eV}$ is chosen such that if an electron in the ground state absorbs more than 13.6 eV it will no longer be bound to the nucleus. This number will be different for other types of atoms, and also the exact dependence on n will change.

In the previous chapters, we considered systems with two distinct states, such as a photon in two arms of an interferometer, and an electron with two spin states. Here, an electron bound to a proton in a hydrogen atom has infinitely many energy states, one for each value of n. Clearly, we cannot write this as a column vector, since it has infinitely many elements. In the remainder of this chapter, we will consider simplified atoms with only a few energy levels in order to keep the mathematics simple. However, the full theory of quantum mechanics has no problem dealing with these infinite vectors.

We would like to have a device that can perform a measurement of the energy of the electron in an atom. However, this is impossible since only energy *differences* are physically meaningful. We will have to settle for something slightly more modest.

We will create a procedure that will tell us if the electron is in a specific energy state or not, via the physical phenomenon of *resonance fluorescence*. In this process, we send light into the atom with a frequency that corresponds to the energy difference between the energy level of interest E_n and a higher energy level E_m. If the electron is in the state $|E_n\rangle$, it will absorb a photon and jump to energy level $|E_m\rangle$. After some time, the electron will jump back[3] to the state $|E_n\rangle$ and emit a photon with energy $E_m - E_n$. We can measure this photon if it is emitted in a direction that differs from the incoming light. Repeating this procedure will produce a lot of photons that are unlikely to be all missed, and detecting some of these photons means that the electron was in the state $|E_n\rangle$. If we do not measure any photons it is overwhelmingly likely that the electron was not in the state $|E_n\rangle$.

4.2 Changes Over Time

So far we have described how to calculate probabilities of measurement outcomes and averages of physical quantities (such as the expectation value of the electron spin). We have considered simple transformations in the form of a beam splitter, and in this chapter we will study in more detail how quantum states change over time. We can gain some insight into how this works by considering the parallels between photon and electron energy states.

Classically, light is an electromagnetic wave, and the time behaviour of such a wave can be written as $\mathbf{E}(t) = \mathbf{E}_0 \cos(\omega t)$, where $\mathbf{E}$ is the electric field and ω is the angular frequency of the wave. We can also write this as $\mathbf{E}(t) = \mathbf{E}_0\, e^{-i\omega t}$ and take the real part of $\mathbf{E}(t)$ as the physical value (see the complex numbers intermezzo in Chap. 2). The quantity ωt is the phase of the light (but we will also call $e^{-i\omega t}$ a phase, or sometimes a phase factor). Since a photon is a quantum of light, the natural time evolution of a photon with frequency ω is given by

$$|\text{photon}\rangle \rightarrow e^{-i\omega t}|\text{photon}\rangle, \tag{4.3}$$

exactly the same as for the electric field $\mathbf{E}(t)$. The photon with frequency ω is a quantum of energy of the electromagnetic field wave of the same frequency, and its energy is $E = \hbar\omega$. The phase can therefore be rewritten as

$$e^{-i\omega t} = e^{-iEt/\hbar} \tag{4.4}$$

In other words, the energy state $|\text{photon}\rangle$ accumulates a phase $e^{-iEt/\hbar}$ over time.

Energy is a very general concept that applies to all systems. It is therefore tempting to assume that the energy state of an electron in an atom behaves in a similar way:

$$|E_n\rangle \rightarrow e^{-iE_n t/\hbar}|E_n\rangle. \tag{4.5}$$

---

[3] This is a bit of an over-simplification, but it is good enough for our purpose right now.

Indeed, this assumption leads to behaviour that is observed in experiments, as we will now show.

Consider a (simplified) atom with a ground state $|g\rangle = |E_g\rangle$ and an excited state $|e\rangle = |E_e\rangle$. Furthermore, we will assume that we can measure whether the atom is in the state $|g\rangle$ or $|e\rangle$ via resonance fluorescence with some higher energy level that we will otherwise ignore. First, at time $t = 0$ we prepare the atom in an equal superposition of the ground state and the excited state

$$|\psi(0)\rangle = \frac{1}{\sqrt{2}}|g\rangle + \frac{1}{\sqrt{2}}|e\rangle . \tag{4.6}$$

After a time t, the ground state $|g\rangle$ and the excited state $|e\rangle$ will each have accumulated a phase that is proportional to their energies:

$$|\psi(0)\rangle \rightarrow |\psi(t)\rangle = \frac{e^{-iE_g t/\hbar}}{\sqrt{2}}|g\rangle + \frac{e^{-iE_e t/\hbar}}{\sqrt{2}}|e\rangle . \tag{4.7}$$

We can take out $e^{-iE_g t/\hbar}$ as a global unobservable phase (see Exercise 17 in Chap. 2), and obtain

$$|\psi(t)\rangle = \frac{1}{\sqrt{2}}|g\rangle + \frac{e^{-i(E_e - E_g)t/\hbar}}{\sqrt{2}}|e\rangle . \tag{4.8}$$

In vector notation, this becomes

$$|\psi(t)\rangle = \frac{1}{\sqrt{2}} \begin{pmatrix} 1 \\ e^{-i(E_e - E_g)t/\hbar} \end{pmatrix} , \tag{4.9}$$

with

$$|g\rangle = \begin{pmatrix} 1 \\ 0 \end{pmatrix} \quad \text{and} \quad |e\rangle = \begin{pmatrix} 0 \\ 1 \end{pmatrix} . \tag{4.10}$$

The relative phase is proportional to the energy difference. We could have guessed this beforehand, because the absolute value of the ground state energy is a matter of convention, and if the relative phase depended on individual energies we would be able to measure the value of the ground state energy.

Compare (4.8) with (3.36) and identify $|\uparrow\rangle$ with $|g\rangle$ and $|\downarrow\rangle$ with $|e\rangle$. The state of the two-level atom (consisting of $|g\rangle$ and $|e\rangle$) we consider here can then also be described by a vector in the Bloch sphere, and the time evolution is equivalent to a counter-clockwise rotation of the state vector around the vertical (z) axis. Similarly, if $E_e - E_g = \hbar\omega$, that is, the transition frequency from energy state $|g\rangle$ to $|e\rangle$ is ω, the time evolution of a general state of the atom is given by

Fig. 4.4 Time evolution as a
rotation of the state vector in
the Bloch sphere (see
supplementary material 4)

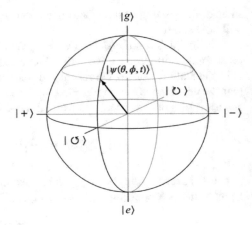

$$|\psi(\theta, \phi, t)\rangle = \cos\left(\frac{\theta}{2}\right)|g\rangle + e^{i\phi - i\omega t}\sin\left(\frac{\theta}{2}\right)|e\rangle \,. \qquad (4.11)$$

This corresponds to circular paths around the vertical axis in the Bloch sphere as
shown in Fig. 4.4. In other words, the state precesses around the z-axis. When there
are more than two energy levels to consider, the Bloch sphere is no longer the correct
state space, and the time evolution is not quite as simple.

4.3   The Hamiltonian

Let's consider a general quantum state of our two-level atom $|\psi\rangle = a|g\rangle + b|e\rangle$ with
$|a|^2 + |b|^2 = 1$. As usual, we can write this in vector form:

$$|\psi\rangle = a\begin{pmatrix}1\\0\end{pmatrix} + b\begin{pmatrix}0\\1\end{pmatrix} = \begin{pmatrix}a\\b\end{pmatrix} \,. \qquad (4.12)$$

The probability of finding the atom in the ground state is given by $p_g = |\langle g|\psi\rangle|^2$,
and the probability of finding the atom in the excited state is given by $p_e = |\langle e|\psi\rangle|^2$.
The average energy in the atom is then given by

$$\begin{aligned}\langle E\rangle &= p_g E_g + p_e E_e \\ &= |\langle g|\psi\rangle|^2 E_g + |\langle e|\psi\rangle|^2 E_e \,.\end{aligned} \qquad (4.13)$$

We can expand the modulus-squared to find the energy observable, just like we did
with the spin and polarisation observables in the previous chapter:

$$\langle E \rangle = E_g \langle \psi | g \rangle \langle g | \psi \rangle + E_e \langle \psi | e \rangle \langle e | \psi \rangle$$
$$= \langle \psi | \left(E_g | g \rangle \langle g | + E_e | e \rangle \langle e | \right) | \psi \rangle . \tag{4.14}$$

We can then define an energy operator

$$H = E_g | g \rangle \langle g | + E_e | e \rangle \langle e | . \tag{4.15}$$

In matrix form, this is given by

$$H = \begin{pmatrix} E_g & 0 \\ 0 & E_e \end{pmatrix} . \tag{4.16}$$

This is such an important operator in quantum mechanics that it has a special name: the Hamiltonian (hence the symbol H instead of E). We thus found that the average energy $\langle E \rangle$ is given by the expectation value of the Hamiltonian $\langle \psi | H | \psi \rangle$.

In Chap. 2 we saw that the beam splitter operator could be applied to the state of a photon before it entered the beam splitter. This gave us the state of the photon after the beam splitter. The Hamiltonian works differently because it is an observable, but since it is an operator, we can also apply it to a state. If we apply H to an energy state we obtain the following:

$$H | g \rangle = \begin{pmatrix} E_g & 0 \\ 0 & E_e \end{pmatrix} \begin{pmatrix} 1 \\ 0 \end{pmatrix} = \begin{pmatrix} E_g \\ 0 \end{pmatrix} = E_g | g \rangle . \tag{4.17}$$

Similarly,

$$H | e \rangle = \begin{pmatrix} E_g & 0 \\ 0 & E_e \end{pmatrix} \begin{pmatrix} 0 \\ 1 \end{pmatrix} = \begin{pmatrix} 0 \\ E_e \end{pmatrix} = E_e | e \rangle . \tag{4.18}$$

In other words, the Hamiltonian H does not change the energy states but multiplies it with the energy value. This is not true in general. Consider the general state $| \psi \rangle$:

$$H | \psi \rangle = \begin{pmatrix} E_g & 0 \\ 0 & E_e \end{pmatrix} \begin{pmatrix} a \\ b \end{pmatrix} = \begin{pmatrix} a E_g \\ b E_e \end{pmatrix}$$
$$= a E_g | g \rangle + b E_e | e \rangle . \tag{4.19}$$

Since generally $E_e \neq E_g$ they do not factor out and $H | \psi \rangle$ is *not* the same as the state $| \psi \rangle$ multiplied by some number. In general, H will change the state it operates on. The energy states are special because they do not change when acted on by the Hamiltonian. They are called *eigenstates* (or eigenvectors) of the Hamiltonian matrix. The energies E_g and E_e are called the *eigenvalues* of the states $| g \rangle$ and $| e \rangle$, respectively. Equations (4.17) and (4.18) are eigenvalue equations for the Hamiltonian.

Since the Hamiltonian is the energy operator of a system, it is often fairly straightforward to find from general principles. The problem is then to find the eigenvalues (the allowed energy values of the system, or spectrum) and the eigenstates. There is a whole area of mathematics devoted to finding the eigenvalues and eigenvectors of matrices, and we will briefly touch upon this in Sect. 4.5.

The Hamiltonian is so important because it is related in a very fundamental way to the time evolution of a quantum system. Consider again the quantum state of the atom $|\psi(0)\rangle = a|g\rangle + b|e\rangle$ at time $t = 0$. According to (4.7), the time-evolved state can be written as

$$|\psi(t)\rangle = a\,e^{-iE_g t/\hbar}|g\rangle + b\,e^{-iE_e t/\hbar}|e\rangle . \qquad (4.20)$$

You should check that this state is still properly normalised. But what if we know only the Hamiltonian, and not the energy eigenstates and eigenvalues? We need some procedure to find $|\psi(t)\rangle$ based only on the Hamiltonian.

The change of a state over time is related to the time derivative. You are already familiar with this: if we are interested in the change of position of a particle, then we take the time derivative of the position, which gives the velocity, or change in position over time. Similarly, we can take the time derivative of a general quantum state $|\psi(t)\rangle$ for a two-level atom:

$$\frac{d}{dt}|\psi(t)\rangle = -\frac{i}{\hbar}E_g a\,e^{-iE_g t/\hbar}|g\rangle - \frac{i}{\hbar}E_e b\,e^{-iE_e t/\hbar}|e\rangle . \qquad (4.21)$$

We can re-write this using (4.17) and (4.18) as

$$\frac{d}{dt}|\psi(t)\rangle = -\frac{i}{\hbar}H a\,e^{-iE_g t/\hbar}|g\rangle - \frac{i}{\hbar}H b\,e^{-iE_e t/\hbar}|e\rangle , \qquad (4.22)$$

or more compactly,

$$i\hbar\frac{d}{dt}|\psi(t)\rangle = H|\psi(t)\rangle . \qquad (4.23)$$

This is the famous *Schrödinger equation*,[4] which describes how the state of a system evolves over time. It is a differential equation that must be solved for $|\psi(t)\rangle$ given our knowledge of H and initial condition $|\psi(0)\rangle$. A large part of mastering quantum mechanics is to find solutions of the Schrödinger equation for various forms of H.

Sometimes it is convenient to look at a different way to find the time evolution of a quantum state. Consider again the time evolution of the energy states

$$|g\rangle \to e^{-iE_g t/\hbar}|g\rangle \quad\text{and}\quad |e\rangle \to e^{-iE_e t/\hbar}|e\rangle . \qquad (4.24)$$

Each time-dependent phase has the form of an exponential, which we can write as

---

[4] E. Schrödinger, *An Undulatory Theory of the Mechanics of Atoms and Molecules*, Phys. Rev. **28** 1049, 1926.

$$e^x = 1 + \frac{x}{1!} + \frac{x^2}{2!} + \frac{x^3}{3!} + \cdots, \qquad (4.25)$$

with $x = -itE_g/\hbar$ or $x = -itE_e/\hbar$. Let's consider E_g. The phase factor can then be written as

$$e^{-itE_g/\hbar}|g\rangle = \left[1 + \frac{-it}{\hbar}E_g + \frac{(-it)^2}{2\hbar^2}E_g^2 + \cdots \right]|g\rangle. \qquad (4.26)$$

We can use (4.17) to replace $E_g|g\rangle$ with $H|g\rangle$, $E_g^2|g\rangle$ with $H^2|g\rangle$, and so on. When we collapse the series again into an exponential, we obtain

$$|g\rangle \rightarrow e^{-iHt/\hbar}|g\rangle, \qquad (4.27)$$

where the phase factor is now an operator. The crucial observation is that we can do exactly the same for the energy eigenstate $|e\rangle$:

$$|e\rangle \rightarrow e^{-iHt/\hbar}|e\rangle. \qquad (4.28)$$

The phase operators in (4.27) and (4.28) are the same, so we can take *any* state $|\psi\rangle = a|g\rangle + b|e\rangle$ and find the time evolution by applying the phase operator $\exp(-iHt/\hbar)$:

$$|\psi(t)\rangle = e^{-iHt/\hbar}|\psi(0)\rangle. \qquad (4.29)$$

Note that because H is an operator, this is not the same as a global phase. It is a so-called *formal* solution of the Schrödinger equation (4.23): we have removed the derivative from the equation, but we don't quite have an explicit form of $|\psi(t)\rangle$, because the exponential may be hard to calculate. However, for the two-level atom we consider here the exponential is often quite easy to find.

We have essentially already solved the Schrödinger equation in (4.23) for the case where H is given by (4.16). The solution is given by (4.11). This represents the physical situation where the atom is prepared in some superposition of $|g\rangle$ and $|e\rangle$ and left isolated. The mere passing of time will change the state. This is called the *free* evolution of the atom, since we are not doing anything to it, such as shining lasers on it.

4.4 Interactions

The Hamiltonian in the previous section gives rise to the simple time evolution when the atom is left to its own. But often we are interested in situations where the atom interacts with other physical systems. For example, we can shine a laser onto the atom with a frequency ω that matches the energy transition between $|g\rangle$ and $|e\rangle$. How do we describe such interactions?

The effect of the laser will be to drive the atom from the ground state $|g\rangle$ into the excited state $|e\rangle$. In addition, it will drive the atom from the excited state $|e\rangle$ into the ground state $|g\rangle$ (this may surprise you, but it is true. It is called *stimulated emission*[5] and lies at the heart of how lasers operate: Light Amplification by Stimulated Emission of Radiation). This interaction of the atom with the laser is also part of the time evolution, and there is indeed a Hamiltonian that describes this mechanism. It is called an *interaction Hamiltonian*, and it can be constructed as

$$H_{\text{int}} = \gamma |e\rangle\langle g| + \gamma^* |g\rangle\langle e|, \tag{4.30}$$

with an interaction strength γ that may be a complex number. For reasons that we will explore in the next chapter, we must include both γ and its complex conjugate γ^*. In matrix form, this is given by

$$H_{\text{int}} = \begin{pmatrix} 0 & \gamma^* \\ \gamma & 0 \end{pmatrix}. \tag{4.31}$$

Typically, the total Hamiltonian for the atom is the Hamiltonian of the free evolution H plus the interaction Hamiltonian H_{int}, which can be expressed as

$$H_{\text{total}} = H + H_{\text{int}} = \begin{pmatrix} E_g & \gamma^* \\ \gamma & E_e \end{pmatrix}. \tag{4.32}$$

Let's break the interaction Hamiltonian down into its components and see what it does. First of all, the factor γ is the strength of the interaction. It can be complex, since it is directly related to the amplitude of the laser light (which has a complex phase factor). When the strength of the interaction goes to zero ($\gamma \to 0$) we retrieve the free evolution, as we should.

The term $|e\rangle\langle g|$ represents the physical transition from the ground state to the excited state, since

$$|e\rangle\langle g| \times |g\rangle = |e\rangle\langle g|g\rangle = |e\rangle, \tag{4.33}$$

and $\langle g|g\rangle = 1$. Similarly, applying $|e\rangle\langle g|$ to $|e\rangle$ will be zero, since $\langle g|e\rangle = 0$. The last term in the interaction Hamiltonian does exactly the opposite, it gives the transition from the excited state to the ground state:

$$|g\rangle\langle e| \times |e\rangle = |g\rangle\langle e|e\rangle = |g\rangle. \tag{4.34}$$

This is the time reversal of the first term in H_{int}. Every properly constructed Hamiltonian must have terms for both the physical process and its time-reversed process. This is a consequence of the fact that the physical laws at the microscopic level have

---

[5] A. Einstein, *Strahlungs-emission und -absorption nach der Quantentheorie*, Verhandlungen der Deutschen Physikalischen Gesellschaft **18** 318, 1916.

no preferred direction in time. It is only when we consider larger systems that the arrow of time becomes apparent.

We can study the effect of the interaction Hamiltonian H_{int} separately from H. In particular, we want to calculate the matrix form of the phase operator in (4.31). We start by calculating the square of H_{int}:

$$H_{int}^2 = \begin{pmatrix} 0 & \gamma^* \\ \gamma & 0 \end{pmatrix} \begin{pmatrix} 0 & \gamma^* \\ \gamma & 0 \end{pmatrix} = |\gamma|^2 \begin{pmatrix} 1 & 0 \\ 0 & 1 \end{pmatrix}. \tag{4.35}$$

In other words, H_{int}^2 is proportional to the identity, which means that H_{int}^3 will be proportional to H_{int}, and so on. At this point it is convenient to choose γ as a real number (i.e., the imaginary component is zero) and independent of time in order to keep the formulas reasonably short. We will also use the following abbreviations:

$$\sigma_x = \begin{pmatrix} 0 & 1 \\ 1 & 0 \end{pmatrix} \quad \text{and} \quad \mathbb{I} = \begin{pmatrix} 1 & 0 \\ 0 & 1 \end{pmatrix}. \tag{4.36}$$

We can write the series expansion of $\exp(-i H_{int} t/\hbar)$ as

$$e^{-i H_{int} t/\hbar} = \mathbb{I} + \frac{-i\gamma t}{\hbar}\sigma_x + \frac{(-i\gamma t)^2}{2\hbar^2}\mathbb{I} + \frac{(-i\gamma t)^3}{6\hbar^3}\sigma_x + \cdots \tag{4.37}$$

Next, we collect the even and odd powers of $-i\gamma t/\hbar$, and note the following mathematical identities:

$$\cos x = 1 - \frac{x^2}{2} + \frac{x^4}{4!} + \cdots \tag{4.38}$$

and

$$\sin x = x - \frac{x^3}{3!} + \frac{x^5}{5!} + \cdots \tag{4.39}$$

This leads to

$$e^{-i H_{int} t/\hbar} = \cos\left(\frac{\gamma t}{\hbar}\right)\mathbb{I} - i \sin\left(\frac{\gamma t}{\hbar}\right)\sigma_x. \tag{4.40}$$

Setting $\Omega = \gamma/\hbar$, this becomes in matrix form

$$e^{-i H_{int} t/\hbar} = \begin{pmatrix} \cos(\Omega t) & -i \sin(\Omega t) \\ -i \sin(\Omega t) & \cos(\Omega t) \end{pmatrix}. \tag{4.41}$$

When we apply this operator to the ground state $|g\rangle$ we obtain

$$|\psi(t)\rangle = e^{-iH_{\text{int}}t/\hbar}|g\rangle = \begin{pmatrix} \cos(\Omega t) & -i\sin(\Omega t) \\ -i\sin(\Omega t) & \cos(\Omega t) \end{pmatrix}\begin{pmatrix} 1 \\ 0 \end{pmatrix}$$

$$= \cos(\Omega t)|g\rangle - i\sin(\Omega t)|e\rangle . \tag{4.42}$$

We can draw the path of the evolving quantum state again in the Bloch sphere and see that this interaction Hamiltonian generates a rotation around the x-axis. In the special case where $\Omega t = \pi/2$, we have

$$|\psi(t)\rangle = -i\begin{pmatrix} 0 & 1 \\ 1 & 0 \end{pmatrix}\begin{pmatrix} 1 \\ 0 \end{pmatrix} = -i\begin{pmatrix} 0 \\ 1 \end{pmatrix} = |e\rangle , \tag{4.43}$$

where we ignored the global phase factor $-i$ in the last term. Clearly the interaction term drives the atom from the ground state $|g\rangle$ to the excited state $|e\rangle$.

However, you also see that it is easy to over- or undershoot: A slightly weaker or stronger interaction or a slightly shorter or longer interaction time will leave the atom not in the excited state, but in a superposition of mostly excited state and a small amount of ground state. The magnitude of γ (or Ω) gives the speed, or frequency, of the rotation. When $\gamma \gg E_e$, the rotation from $|g\rangle$ to $|e\rangle$ occurs much faster than the free evolution. Therefore, with a sufficiently strong and short laser pulse we can excite the atom almost instantly from $|g\rangle$ to $|e\rangle$.

4.5   Interaction Energy and the Zeeman Effect

We constructed the Hamiltonian of the two-level atom using the energy values of the ground state $|g\rangle$ and excited state $|e\rangle$. We assigned them the energy values E_g and E_e, respectively. This led to the matrix

$$H = \begin{pmatrix} E_g & 0 \\ 0 & E_e \end{pmatrix} . \tag{4.44}$$

However, after including an interaction term H_{int}, the total Hamiltonian is

$$H_{\text{total}} = \begin{pmatrix} E_g & \gamma^* \\ \gamma & E_e \end{pmatrix} . \tag{4.45}$$

This raises the question whether this changes the energy of the system, and if so: how?

For our example of the Hamiltonian H, we have the eigenvalue equations

$$H|g\rangle = E_g|g\rangle \quad \text{and} \quad H|e\rangle = E_e|e\rangle , \tag{4.46}$$

where $|g\rangle$ and $|e\rangle$ are the eigenstates of H with corresponding eigenvalues E_g and E_e, respectively. When a system is in an eigenstate of H, its energy value is sharp.

In other words, the uncertainty in the energy is zero. Take for example the excited state $|e\rangle$. The energy uncertainty given the Hamiltonian H can be calculated as

$$(\Delta H)^2 = \langle e|H^2|e\rangle - \langle e|H|e\rangle^2 = \langle e|E_e^2|e\rangle - \langle e|E_e|e\rangle^2$$
$$= E_e^2 \langle e|e\rangle - (E_e \langle e|e\rangle)^2 = E_e^2 - E_e^2 = 0. \tag{4.47}$$

On the other hand, when we calculate the uncertainty in H_{total} given the same state $|e\rangle$, we find that it is not zero. You should verify this.

If $|g\rangle$ and $|e\rangle$ are not the eigenstates of H_{total}, how do we find the eigenstates of H_{total}? For this, we need to solve the eigenvalue equation

$$H_{\text{total}} |E_j\rangle = E_j |E_j\rangle , \tag{4.48}$$

where E_j is an energy eigenvalue of H_{total} and $|E_j\rangle$ its corresponding eigenstate. The index j takes on the values 1 and 2, since there are two eigenvalues and two eigenvectors for a 2×2 matrix. In the next chapter we will present a simple way to calculate the eigenvalues E_j. For now, we solve the eigenvalue equation explicitly for an arbitrary quantum state $|\psi\rangle = a |g\rangle + b |e\rangle$:

$$H_{\text{total}} |\psi\rangle = \begin{pmatrix} E_g & \gamma^* \\ \gamma & E_e \end{pmatrix} \begin{pmatrix} a \\ b \end{pmatrix} = E \begin{pmatrix} a \\ b \end{pmatrix} . \tag{4.49}$$

From matrix multiplication we obtain two equations, one for the top component of the resulting vector, and one for the bottom component:

$$aE_g + b\gamma^* = aE \quad \text{and} \quad a\gamma + bE_e = bE . \tag{4.50}$$

We can express b in terms of a:

$$b = \frac{E - E_g}{\gamma^*}a \quad \text{and} \quad b = -\frac{\gamma}{E_e - E}a . \tag{4.51}$$

Since both equations must hold, we can set the right-hand sides equal to each other (since they are both equal to b) and divide by a. After a little algebra this yields

$$(E - E_g)(E_e - E) = -|\gamma|^2 . \tag{4.52}$$

Solving the resulting quadratic equation in E, we find that the solutions to the two eigenvalues E_1 and E_2 are

$$E_1 = \frac{1}{2}(E_g + E_e) - \frac{1}{2}\sqrt{(E_e - E_g)^2 + 4|\gamma|^2} ,$$
$$E_2 = \frac{1}{2}(E_g + E_e) + \frac{1}{2}\sqrt{(E_e - E_g)^2 + 4|\gamma|^2} . \tag{4.53}$$

Fig. 4.5 Changes in
eigenvalues with increasing
interaction strength $|\gamma|^2$. The
energy units on the axes are
arbitrary

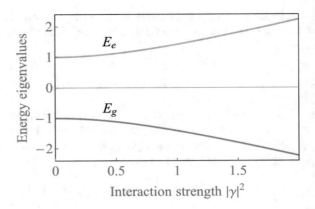

These are the energy eigenvalues of H_{total}. Note that they are different from E_g and
E_e: the interaction γ changes the energy of the system. Also verify that when $\gamma \to 0$
the energy eigenvalues reduce to E_g and E_e, as they should. The energy eigenvalues
as a function of $|\gamma|^2$ are shown in Fig. 4.5.

Next, we can use the same technique to find the eigenstates $|E_1\rangle$ and $|E_2\rangle$. We
start again with (4.49) and substitute for E the values E_1 and E_2 to find the values
of a and b of the corresponding states $|E_1\rangle$ and $|E_2\rangle$. A very quick way to achieve
this is to substitute E_1 and E_2 into (4.51), and set $a = 1$. We have some freedom
in which of the two expressions we choose, and one particular choice leads to the
(unnormalised) eigenvectors

$$\begin{pmatrix} 1 \\ \frac{E_1 - E_g}{\gamma^*} \end{pmatrix} \quad \text{and} \quad \begin{pmatrix} 1 \\ \frac{\gamma}{E_2 - E_e} \end{pmatrix}.$$

We still have to normalise these vectors to obtain $|E_1\rangle$ and $|E_2\rangle$. The result after
some algebra is

$$|E_1\rangle = \frac{1}{N_1} \begin{pmatrix} \gamma^* \\ E_1 - E_g \end{pmatrix} \quad \text{and} \quad |E_2\rangle = \frac{1}{N_2} \begin{pmatrix} E_2 - E_e \\ \gamma \end{pmatrix}, \qquad (4.54)$$

where the normalisation constants are

$$N_1 = \sqrt{|\gamma|^2 + (E_1 - E_g)^2} \quad \text{and} \quad N_2 = \sqrt{|\gamma|^2 + (E_2 - E_e)^2}. \qquad (4.55)$$

This is a very general result that gives the energy eigenvalues and eigenstates for any
two-level system that interacts with its environment via the coupling γ.

As an example, we consider the situation where the electron in our two-level atom
also carries spin. We assume for simplicity that there is only a single ground state
$|g\rangle$, but that the excited state $|e\rangle$ is made up of two spin states, $|e_\uparrow\rangle$ and $|e_\downarrow\rangle$. In the
absence of any magnetic fields, these two states have the same energy relative to the

ground state, and we call such states *degenerate*. Suppose that both these states have energy E relative to the ground state. The Hamiltonian for the two excited states is then

$$H = \begin{pmatrix} E & 0 \\ 0 & E \end{pmatrix}. \tag{4.56}$$

Note that in this description the ground state is not included, which is why H is still a 2×2 matrix. We want to know the energy of the excited states changes as we turn up the magnetic field. First, we need to know how the magnetic field interacts with the spin of the electron. We have already seen in the previous chapter how the electron spin responds to the magnetic field in a Stern-Gerlach experiment. Here, we want to know the energy of an electron spin **S** in a magnetic field **B**, so that we can construct the interaction Hamiltonian:

$$H_{\text{int}} = -\frac{e}{m} \mathbf{B} \cdot \mathbf{S}, \tag{4.57}$$

where e is the electron charge and m the mass of the electron. To find the matrix form of this operator, we expand the dot product $\mathbf{B} \cdot \mathbf{S}$ and substitute the quantum operators for S_x, S_y, and S_z:

$$\begin{aligned}
H_{\text{int}} &= -\frac{\hbar e}{2m} \left[B_x \begin{pmatrix} 0 & 1 \\ 1 & 0 \end{pmatrix} + B_y \begin{pmatrix} 0 & -i \\ i & 0 \end{pmatrix} + B_z \begin{pmatrix} 1 & 0 \\ 0 & -1 \end{pmatrix} \right] \\
&= -\frac{\hbar e}{2m} \begin{pmatrix} B_z & B_x - iB_y \\ B_x + iB_y & -B_z \end{pmatrix}.
\end{aligned} \tag{4.58}$$

For simplicity, we will consider two separate cases. First, when the magnetic field has magnitude B and points in the z direction we have $B_z = B$ and $B_x = B_y = 0$. We substitute these values into the interaction Hamiltonian H_{int} and calculate the total Hamiltonian

$$H_{\text{total}} = H + H_{\text{int}} = \begin{pmatrix} E - \frac{\hbar e B}{2m} & 0 \\ 0 & E + \frac{\hbar e B}{2m} \end{pmatrix}. \tag{4.59}$$

You see that the energies split, with the spin "↑" electron state lowering its energy, and the spin "↓" electron state increasing its energy. This splitting is the so-called *Zeeman effect*, discovered in 1896 by Pieter Zeeman.[6] In 1908, Georges Ellery Hale observed the splitting of spectral lines in light emitted by sun spots,[7] thus proving conclusively that sun spots are regions of intense magnetic field activity.

---

[6] P. Zeeman, *Over de invloed eener magnetisatie op den aard van het door een stof uitgezonden licht*, Royal Academy of Sciences in Amsterdam **5**, 181 and 242 (1896). Technically, the Zeeman effect described here based on the electron spin is known as the *anomalous* Zeeman effect.

[7] G. E. Hale, *Solar Vortices*, Astrophysical Journal **28**, 100 (1908).

Second, we may consider the situation where the magnetic field points in the x direction, again with magnitude B. In this case the total Hamiltonian becomes

$$H_{\text{total}} = H + H_{\text{int}} = \begin{pmatrix} E & -\frac{\hbar e B}{2m} \\ -\frac{\hbar e B}{2m} & E \end{pmatrix}. \tag{4.60}$$

We can identify the variable $\gamma = -\hbar e B/2m$, and the new energy eigenvalues become

$$E_1 = E - \frac{\hbar e B}{2m} \quad \text{and} \quad E_2 = E + \frac{\hbar e B}{2m}. \tag{4.61}$$

The energies are the same as before, but in this case the states $|e_\uparrow\rangle$ and $|e_\downarrow\rangle$ are no longer eigenstates of the Hamiltonian, and the states will evolve over time according to the Schrödinger equation.

4.6 ॐ The Atomic Clock

One of the most fundamental and important measuring devices in physics is the atomic clock.[8] Using the theory we have developed so far we can now explain the basic principle of such devices. To do this, we will need to know a little bit about the background of clocks. First, a clock measures time, so we better have a good definition of the unit of time, the second. According to the *Bureau International des Poids et Mesures*, which maintains the standards of the SI units, the definition of the second is:

> The duration of 9 192 631 770 periods of the radiation corresponding to the transition between the two hyperfine levels of the ground state of the caesium 133 atom.

Here, hyperfine means that the nucleus of the caesium atom has a spin that interacts with the spin of the outer electron. This interaction will cause a splitting of the energy levels, because it is energetically more favourable for the electron and nuclear spin vectors to be parallel rather than anti-parallel. We call the two hyperfine levels of the ground state $|g\rangle$ and $|e\rangle$, which have spin values of $-\hbar/2$ and $+\hbar/2$, respectively. The fact that the two levels have different electron spin values means that we can manipulate the atoms using the Stern-Gerlach apparatus introduced in the previous chapter. The 133 refers to the fact that the particular isotope of caesium has 55 protons +78 neutrons = 133 nucleons.

In its most general description, a clock is a device made of a physical system that naturally oscillates with a very regular frequency, coupled to a mechanism that counts the number of cycles in the oscillation. The higher the frequency, the shorter the counting steps, and the more accurate the clock is. For example, a grandfather clock

---

[8] L. Essen and J. V. L. Parry, *An Atomic Standard of Frequency and Time Interval: A Caesium Resonator*, Nature **176** 280, 1955.

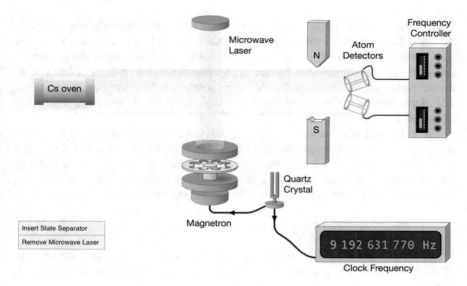

Fig. 4.6 Working principle of the atomic clock (see supplementary material 6)

has a pendulum (the *oscillator*) that swings back and forth with regular intervals, and a mechanism called an *escapement* that pushes the hands of the clock a tiny bit forward with each swing of the pendulum. The escapement also pushes the pendulum a little at each swing, which overcomes the natural drag of the pendulum and prevents it from coming to a halt after a few minutes.

Another example is the quartz clock used in computers and digital wrist watches. In these clocks a quartz crystal is cut to a specific shape (typically the shape of a tuning fork), and made to vibrate at its resonance frequency. Quartz is a piezo-electric material, which means that deformations of the crystal cause an electric signal. This signal is used both to drive the oscillation and to measure the number of periods. The quartz crystal is the oscillator, and the electronics serve as the escapement.

For our most accurate clocks, we need an extremely accurate oscillator. As we have seen in the definition of the second, the radiation between the lowest energy states $|g\rangle$ and $|e\rangle$ of caesium 133 can be used as the oscillator, since it can be described as a wave and therefore has a periodic behaviour. The question is: how do we get a practical clock read-out from this oscillator? In other words, how do we build the escapement? The answer is that we use a feedback mechanism to lock the atomic transition to the frequency of a quartz crystal that we can read out electronically. This mechanism relies critically on the interaction of the caesium atoms with an external laser, as studied in the previous section.

First, as shown in Fig. 4.6, we set up a vacuum tube with a source of caesium 133 on one side and atom detectors on the other side. The source is often called an oven, since it heats the caesium and lets it escape into the tube towards the detectors. Immediately after the oven, the atoms pass through a Stern Gerlach apparatus that deflects the atoms in the state $|e\rangle$. Only caesium atoms in the state $|g\rangle$ pass through the tube. This is called a state separator. At the other end, before the detectors, the caesium atoms pass through another Stern-Gerlach apparatus separating atoms in states $|g\rangle$ and $|e\rangle$ into different beams. Each beam is sent into a detector that counts the number of atoms. Clearly, when the atoms pass through the vacuum tube from the first Stern-Gerlach to the second, their state should not change and all atoms should be detected in the ground state $|g\rangle$.

Second, we aim a microwave laser (or *maser*, generated by a magnetron, the active component in your microwave oven) into the path of the caesium atoms between the two state separators. While the atoms are in the beam of the maser, they experience the interaction described in (4.29). If the maser is on resonance with the transition between $|g\rangle$ and $|e\rangle$ the atoms entering the second state separator are in the state

$$|\psi(t)\rangle = \cos \Omega t |g\rangle - i \sin \Omega t |e\rangle , \qquad (4.62)$$

where t is the time the atoms spent in the maser beam and Ω is the coupling strength of the interaction. The probability that such an atom is detected in the state $|e\rangle$ is given by

$$p_e = |\langle e|\psi(t)\rangle|^2 = \sin^2 \Omega t . \qquad (4.63)$$

When the maser is off resonance, meaning that the frequency of the maser does not match the transition frequency between $|g\rangle$ and $|e\rangle$, the coupling Ω' is much reduced. As a consequence, $\sin \Omega' t$ is much closer to 0, and the probability of finding an atom in the state $|e\rangle$ is suppressed.

Finally, we can control the frequency of the maser using a quartz crystal. This crystal is cut into the shape of a tuning fork such that the frequency is centred around the transition between $|g\rangle$ and $|e\rangle$ at 9 192 631 770 Hz. However, it is not nearly this precise, for if it was we would not need the atomic transition. As a result, the frequency of the maser will drift away from the resonance frequency, and the coupling strength of the interaction will drop. This means that fewer atoms will be detected in the state $|e\rangle$. A frequency counter monitors the relative number of atoms detected in the state $|e\rangle$, and when it drops the controller applies a small voltage to the quartz crystal, nudging it back to the resonance frequency. This feedback mechanism will keep the quartz crystal locked at 9 192 631 770 Hz, and the frequency is much more stable than the quartz crystal can maintain on its own. By counting the oscillations of the—now stabilised—quartz crystal (which can be done electronically due to the piezo-electric effect), we can mark the passage of time with great accuracy. This is how an atomic clock works.

Clocks like this have a typical precision of one part in 10^{13}. Modern atomic clocks are still an area of active research, and can achieve precisions as high as one part in 10^{18}. They are used for all sorts of applications, from GPS to measuring the spectrum of anti-hydrogen.

Exercises

1. Normalise the state $3|g\rangle - i|e\rangle$.
2. An atomic state

$$|\psi\rangle = \frac{1}{\sqrt{3}}|g\rangle + \sqrt{\frac{2}{3}}|e\rangle$$

evolves with an evolution matrix $\begin{pmatrix} e^{-i\omega t/2} & 0 \\ 0 & e^{i\omega t/2} \end{pmatrix}$. Calculate the state $|\psi(t)\rangle$

at time t. Calculate the probability of finding $|+\rangle = (|g\rangle + |e\rangle)/\sqrt{2}$ at time t.
3. A two-level atom with ground state $|g\rangle$ and excited state $|e\rangle$ is prepared in the state

$$|+\rangle = \frac{1}{\sqrt{2}}|g\rangle + \frac{1}{\sqrt{2}}|e\rangle$$

at time $t = 0$. The energy difference between $|g\rangle$ and $|e\rangle$ is $\hbar\omega$.

a. Calculate the probability of finding the atom in the state

$$|R\rangle = \frac{1}{\sqrt{2}}|g\rangle + \frac{i}{\sqrt{2}}|e\rangle .$$

b. The atom evolves according to the Schrödinger equation with the Hamiltonian

$$H = \frac{\hbar\omega}{2} \begin{pmatrix} 1 & 0 \\ 0 & -1 \end{pmatrix}.$$

Calculate the probability of finding the atom in the state $|R\rangle$ at time t.
c. Sketch this probability as a function of time.
d. Sketch the trajectory of the quantum state over time in the Bloch sphere.
4. An electron with spin $1/2$ is prepared in the state $|\uparrow\rangle$ in the z-direction. A magnetic field is pointing $45°$ away from the z-axis and towards the x-axis.

(a) Using equation (3.88), construct the matrix form of the Hamiltonian H in the Schrödinger equation.
(b) Show that the states

$$|+45\rangle = \cos\left(\frac{\pi}{8}\right)|\uparrow\rangle + \sin\left(\frac{\pi}{8}\right)|\downarrow\rangle ,$$

$$|+135\rangle = -\sin\left(\frac{\pi}{8}\right)|\uparrow\rangle + \cos\left(\frac{\pi}{8}\right)|\downarrow\rangle ,$$

are eigenstates of the matrix H. What are the corresponding eigenvalues?
(c) The eigenstates form the poles of the rotation axis determined by H. Sketch the time trajectory of the electron spin in the Bloch sphere.
(d) Does the electron spin ever reach the state $|\downarrow\rangle$?

5. At time $t = 0$ an atom is prepared in the state

$$|\psi\rangle = \frac{|g\rangle - i\sqrt{2}\,|e\rangle}{\sqrt{3}}.$$

   (a) Calculate the probability of finding the measurement outcome "+" and "–", associated with the states $|\pm\rangle = (|g\rangle \pm |e\rangle)/\sqrt{2}$.

   The free evolution of the atom is determined by the Hamiltonian

$$H = -\frac{1}{2}\hbar\omega\left(|g\rangle\langle g| - |e\rangle\langle e|\right).$$

   (b) Calculate the state at time t.
   (c) Sketch the rotation axis of the evolution due to H on the Bloch sphere.

6. (a) A photon in a polarisation state $|\psi\rangle = \alpha|H\rangle + \beta|V\rangle$ with $|\alpha|^2 + |\beta|^2 = 1$ is measured in the polarisation basis $\{|H\rangle, |V\rangle\}$. What are the possible measurement outcomes and their corresponding probabilities?
   (b) If we measure an identically prepared photon in the *circular* polarisation basis $(|H\rangle \pm i|V\rangle)/\sqrt{2}$, what will be the measurement outcomes and their corresponding probabilities?
   (c) The free evolution of the photon is governed by the Hamiltonian

$$H = i\hbar\omega(|H\rangle\langle V| - |V\rangle\langle H|),$$

   with ω the angular frequency of the light. Find the state of the photon at time $t = T$ given that the state at $t = 0$ is given by $|\psi\rangle = \alpha|H\rangle + \beta|V\rangle$.
   (d) Calculate and sketch the probability of finding the outcomes of a measurement in the polarisation basis $\{|H\rangle, |V\rangle\}$ in the time interval $0 \le t \le T$ when $\alpha = \beta = 1/\sqrt{2}$.

7. Show that ΔH_{total}, given the Hamiltonian in (4.45), is not equal to zero for the states $|g\rangle$ and $|e\rangle$.

8. Consider a two-level atom with ground state $|g\rangle$ and excited state $|e\rangle$, and energy eigenvalues E_g and E_e, respectively. We place this atom in an external constant electric field with magnitude E and pointing in the z direction. The additional potential energy due to the interaction with the external field is $U = -eEz$, with e the electron charge, and z the position of the electron along the z-axis.

(a) Construct the interaction Hamiltonian matrix H_{int} from U using the matrix elements $[H_{int}]_{eg} = \langle e|H_{int}|g\rangle$, etc.
   Use the abbreviation $\alpha = \langle e|z|g\rangle$, and $\langle g|z|g\rangle = \langle e|z|e\rangle = 0$.

(b) Calculate the energy eigenvalues of the two-level atom in the external electric field. The shift of the energies due to the electric field is called the Stark effect.

Chapter 5
Operators

In this chapter, we recall the material from the previous chapters and summarise how vectors and operators play different roles in quantum mechanics. After an introduction to so-called eigenvalue problems, we discuss observables and evolution operators, as well as projectors and commutators. They provide us with the essential mathematical techniques for later chapters.

5.1 Eigenvalue Problems

Mathematically, quantum mechanics is a theory about calculating probabilities using the concepts of states and operators. In this chapter we are going to study the most important aspects of operators for quantum mechanics, and show how it fits in with the theory. First, we will address the most fundamental properties of operators, namely their eigenvalues and eigenstates.

In the previous chapter you have encountered the eigenvalue equation of the Hamiltonian

$$H|E_n\rangle = E_n|E_n\rangle, \qquad (5.1)$$

where H is an operator, and $|E_n\rangle$ is an eigenstate of H with eigenvalue E_n. The Hamiltonian is the energy operator and the eigenvalues E_n are the energies that the system can possess. The states (or vectors) $|E_n\rangle$ describe the system when it has the energy E_n. Often we need to find out the energy states $|E_n\rangle$ given a particular Hamiltonian H. Sometimes the energy eigenvalues E_n are known straight away or

Supplementary Information The online version contains supplementary material available at https://doi.org/10.1007/978-3-031-16165-0_5.

can be found from applying well-known physical principles, and other times the eigenvalues must be calculated from a general form of H, for example when there is some interaction present. In this section we will show how you can find both E_n and $|E_n\rangle$ given H.

Let's write the previous equation in matrix form with $|g\rangle$ and $|e\rangle$ the energy eigenstates, E_g and E_e the corresponding energies, and

$$H = \begin{pmatrix} E_g & 0 \\ 0 & E_e \end{pmatrix}. \tag{5.2}$$

It is straightforward to show that (5.1) holds. First, we define

$$|g\rangle = \begin{pmatrix} 1 \\ 0 \end{pmatrix} \quad \text{and} \quad |e\rangle = \begin{pmatrix} 0 \\ 1 \end{pmatrix}. \tag{5.3}$$

Next, we substitute H, $|g\rangle$, and $|e\rangle$ into equation (5.1). We find that

$$H|g\rangle = \begin{pmatrix} E_g & 0 \\ 0 & E_e \end{pmatrix} \begin{pmatrix} 1 \\ 0 \end{pmatrix} = \begin{pmatrix} E_g \\ 0 \end{pmatrix} = E_g \begin{pmatrix} 1 \\ 0 \end{pmatrix} = E_g|g\rangle,$$

$$H|e\rangle = \begin{pmatrix} E_g & 0 \\ 0 & E_e \end{pmatrix} \begin{pmatrix} 0 \\ 1 \end{pmatrix} = \begin{pmatrix} 0 \\ E_e \end{pmatrix} = E_e \begin{pmatrix} 0 \\ 1 \end{pmatrix} = E_e|e\rangle. \tag{5.4}$$

You see that there is a direct link between the numerical values of the vector components of $|g\rangle$, $|e\rangle$, and the fact that H is diagonal. When a matrix is diagonal, the only non-zero matrix elements are along the line from the top left to the bottom right, and the diagonal elements are the eigenvalues (which can also include zero). The eigenvectors are then simply of the form of (5.3). This is true for matrices of all dimensions, so if we define a diagonal matrix M according to

$$M = \begin{pmatrix} m_1 & 0 & 0 & 0 \\ 0 & m_2 & 0 & 0 \\ 0 & 0 & m_3 & 0 \\ 0 & 0 & 0 & m_4 \end{pmatrix}, \tag{5.5}$$

then the four eigenvectors $\mathbf{v}_1$, $\mathbf{v}_2$, $\mathbf{v}_3$, and $\mathbf{v}_4$ are given by

$$\mathbf{v}_1 = \begin{pmatrix} 1 \\ 0 \\ 0 \\ 0 \end{pmatrix}, \quad \mathbf{v}_2 = \begin{pmatrix} 0 \\ 1 \\ 0 \\ 0 \end{pmatrix}, \quad \mathbf{v}_3 = \begin{pmatrix} 0 \\ 0 \\ 1 \\ 0 \end{pmatrix}, \quad \mathbf{v}_4 = \begin{pmatrix} 0 \\ 0 \\ 0 \\ 1 \end{pmatrix}, \tag{5.6}$$

and the eigenvalue equation is given by

$$M\mathbf{v}_j = m_j \mathbf{v}_j, \tag{5.7}$$

where $j = 1, 2, 3, 4$. You see that the generalisation to arbitrary dimension is straightforward.

We often have to deal with matrices that are not diagonal. Consider the situation where the atom interacts with a laser field, also discussed in the previous chapter. The Hamiltonian H_{total} is a combination of the free evolution Hamiltonian H and the interaction Hamiltonian H_{int} from (4.30):

$$H_{total} = \begin{pmatrix} E_g & \gamma^* \\ \gamma & E_e \end{pmatrix}, \tag{5.8}$$

If we use this Hamiltonian in (5.1) with the vectors $|g\rangle$ and $|e\rangle$, we see that it is no longer satisfied, and therefore $|g\rangle$ and $|e\rangle$ are no longer eigenstates of the Hamiltonian. In other words, and very much against our intuition, $|g\rangle$ and $|e\rangle$ are not the states of the system with a well-defined energy when there is an interaction present! The states with a particular value of the total energy change as the interaction strength γ is increased, and the allowed energy values tend to change also.

We will approach the problem of finding the eigenvalues and eigenvectors of a matrix in a slightly more general way than in the previous chapter. Lucky for us, this is a problem that has been solved a long time ago in a branch of mathematics called linear algebra, which deals with matrices and vectors of all kinds. Here we will give the recipe for finding the eigenvalues and eigenvectors without showing explicitly where it comes from, and for a full explanation where this all comes from we refer you to any introductory linear algebra book (see Further Reading).

First, we need to use a property of matrices called the *determinant*, which is a kind of measure for the "size" of a matrix. The determinant of a 2×2 matrix is given by

$$\det \begin{pmatrix} a & b \\ c & d \end{pmatrix} = ad - bc. \tag{5.9}$$

We take the eigenvalue equation and rearrange it so the right-hand side becomes zero:

$$H |\lambda\rangle - \lambda |\lambda\rangle = (H - \lambda \mathbb{I}) |\lambda\rangle = 0, \tag{5.10}$$

where λ is the eigenvalue and $|\lambda\rangle$ the eigenvector of H. We included the identity matrix $\mathbb{I}$ when we factor out $|\lambda\rangle$. Since we don't want $|\lambda\rangle = 0$, somehow the operation of $H - \lambda \mathbb{I}$ on $|\lambda\rangle$ must give zero in order to satisfy (5.10). This can be achieved by requiring that the determinant of $H - \lambda \mathbb{I}$ is zero:

$$\det(H - \lambda \mathbb{I}) = \det \begin{pmatrix} E_g - \lambda & \gamma^* \\ \gamma & E_e - \lambda \end{pmatrix}$$
$$= (E_g - \lambda)(E_e - \lambda) - |\gamma|^2 = 0. \tag{5.11}$$

Note that this gives us immediately the quadratic equation we obtained in (4.52). The eigenvalues of H are the two values of λ that make the determinant zero. Since the equation for λ in (5.11) is a quadratic equation of the form $a\lambda^2 + b\lambda + c = 0$, we can use the standard solution:

$$\lambda_\pm = \frac{-b \pm \sqrt{b^2 - 4ac}}{2a} = \frac{1}{2}\left(E_g + E_e \pm \sqrt{(E_e - E_g)^2 + 4|\gamma|^2} \right). \tag{5.12}$$

When $\gamma \to 0$ the eigenvalues are $\lambda_- = E_g$ and $\lambda_+ = E_e$, as you would expect. The eigenvectors are again found by substituting the eigenvalue λ into the eigenvalue equation and calculating $|\lambda\rangle$ explicitly.

We can apply this mathematical procedure to any matrix, and find the eigenvalues and eigenvectors. The eigenvalues are often most important, since these are the physically allowed values of the energy, and determine the spectral absorption and emission lines in spectroscopy. We also often want to know the ground state of a system, since this is the state the system most likely relaxes into after some time. The ground state is the eigenstate of the lowest energy level. If we want to calculate the probability p_n that a particular state $|\psi\rangle$ is measured in an energy eigenstate $|E_n\rangle$, we must find the eigenstates $|E_n\rangle$ in order to evaluate the Born rule $p_n = |\langle E_n|\psi\rangle|^2$.

How do we evaluate the determinant of higher-dimensional matrices? This is determined recursively by the determinant of 2×2 matrices. For example, given a 3×3 matrix we can calculate the determinant as follows: choose any row or column (preferably with as many zero entries as possible, as you will see shortly). For each element in the row or column we define a new matrix that we obtain by removing the row and column of that element. This is the sub-matrix of that element. If we start with a 3×3 matrix, the sub-matrix of any element will be a 2×2 matrix. The determinant of that sub-matrix (which we can evaluate) is called the *cofactor* of that element.

We now evaluate the determinant of the 3×3 matrix as follows: we multiply each element in our chosen row or column by its cofactor. Depending on the position of the element in the matrix, this product picks up a minus sign:

$$\begin{pmatrix} + & - & + \\ - & + & - \\ + & - & + \end{pmatrix}, \tag{5.13}$$

which for larger matrices extends in a regular checker board pattern. The determinant is the sum of these products.

As an example, consider the determinant of the 3×3 matrix A:

$$A = \begin{pmatrix} 5 & 1 & 2 \\ 1 & -8 & -4 \\ 4 & 6 & -7 \end{pmatrix}. \tag{5.14}$$

Let's pick the top row to evaluate the determinant:

$$\det A = 5 \times \det \begin{pmatrix} -8 & -4 \\ 6 & -7 \end{pmatrix} - 1 \times \det \begin{pmatrix} 1 & -4 \\ 4 & -7 \end{pmatrix} + 2 \times \det \begin{pmatrix} 1 & -8 \\ 4 & 6 \end{pmatrix}$$
$$= 5 \times 80 - (-23) + 2 \times 38 = 499. \tag{5.15}$$

You see that you want to pick the row or column with the most zeros, since it will reduce the amount of work.

We can calculate the determinant of any matrix using the above procedure. The determinant is expressed in terms of determinants of smaller matrices, which in turn can be expressed in terms of determinants of smaller matrices, and so on, until we arrive at 2×2 matrices.

5.2 Observables

We have seen that there are two types of matrices that play an important role in quantum mechanics. The first type we encountered was the matrix describing the beam splitter, and it tells us how the state of the photon changes after we do something to it. In other words, it describes the *evolution* of the quantum state. The second type of matrix we encountered was the matrix describing the spin of an electron, and later we discussed the Hamiltonian matrix describing the energy of a system. This type of matrix generally describes a physical property of the system (such as spin and energy), and we call these matrices *observables*.

When you think about it, it is a little bit strange that observables are matrices that are completely independent of the state of the system. You may have expected that the state is simply a list of physical properties of the system. However, we have seen in Chap. 1 that "a phenomenon is not a phenomenon until it is an observed phenomenon", which means that quantum systems have no fixed properties independent of the measurement of the system. So instead, we have an abstract vector denoting the state of the system, and a separate matrix representing the physical observable. Only using this vector and matrix can we calculate the probabilities of measurement outcomes and the average values of physical observables. In the standard interpretation of quantum mechanics, this is all we can say about the actual physical system (more on that later in Chap. 10).

The key requirement for an observable is that its eigenvalues are the possible physical values of the corresponding physical property. For example, the eigenvalues of the spin operator are the possible values of the electron spin in a given direction (since spin is a vector), and the eigenvalues of the Hamiltonian are the possible energy values that a system can possess. Since these are real-world experimental values, they should be real numbers. After all, no physical measurement device can give you a complex number as a valid result. Our use of complex numbers is simply a mathematical convenience (quantum mechanics without complex numbers is possible, but much more complicated!).

We can view this requirement of real eigenvalues as the defining property of observables: a matrix is an observable if it has real eigenvalues. We would then like to have a quick way of telling whether a matrix has real or complex eigenvalues. Of course, we can always calculate the actual eigenvalues, but this may be a lot of work. Luckily, there is another way.

First, consider the simplest possible matrix, namely a single number a. This is a 1×1 matrix. You know already how we can tell whether it is a real number: just check whether or not $a^* = a$. Next, let's try this same technique with the Pauli matrices:

$$\sigma_z^* = \begin{pmatrix} 1 & 0 \\ 0 & -1 \end{pmatrix}^* = \begin{pmatrix} 1^* & 0 \\ 0 & -1^* \end{pmatrix} = \sigma_z \tag{5.16}$$

and

$$\sigma_x^* = \begin{pmatrix} 0 & 1 \\ 1 & 0 \end{pmatrix}^* = \begin{pmatrix} 0 & 1^* \\ 1^* & 0 \end{pmatrix} = \sigma_x , \tag{5.17}$$

so this clearly works for σ_x and σ_z. However, it does not work for σ_y :

$$\sigma_y^* = \begin{pmatrix} 0 & -i \\ i & 0 \end{pmatrix}^* = \begin{pmatrix} 0 & -i^* \\ i^* & 0 \end{pmatrix} = -\sigma_y . \tag{5.18}$$

Yet, the eigenvalues of σ_y are ± 1 and therefore σ_y is a proper observable since it has real eigenvalues. From a physical perspective you already know that σ_y should be an observable, because it is directly related to the spin in the y-direction (up to a factor $\hbar/2$).

So the complex conjugate is not a good indicator for telling whether the eigenvalues of a matrix are real. However, when we inspect the effect on σ_y, we can make a small modification that may work: In addition to taking the complex conjugate, we can also mirror the matrix in the main diagonal. In other words, the top left element swaps places with the bottom right. The diagonal elements stay where they are. This operation on a matrix is called the *transpose*, and you have already encountered this in Chap. 2, when we turned a column vector (a ket $|\cdot\rangle$) into a row vector (a bra $\langle\cdot|$) in order to calculate the scalar product. This operation also required us to take the complex conjugate in order to arrive at the normalisation condition $|a|^2 + |b|^2 = 1$. The combination of taking the transpose and the complex conjugate is called the adjoint, and is denoted by a † as a superscript.

An important mathematical theorem states that a matrix M has real eigenvalues if $M = M^\dagger$. A matrix with this property is called *Hermitian*. All physical observables are described by Hermitian matrices (or Hermitian operators, since we have a direct correspondence between matrices and operators).

In quantum mechanics, every Hermitian matrix is considered a valid physical observable, even though it may not be obvious how the matrix relates to an experimental setup. For example, in the case of the Stern-Gerlach experiment we can rotate the apparatus around the y-axis from alignment in the z-direction to the x-direction, and we can measure any spin direction in the xz-plane. However, we cannot rotate the magnets so as to measure the y-component, because the magnets would block the path of the electrons. Nevertheless, it makes intuitive sense to talk about the spin in the y-direction, and there are other ways to measure this component, involving more complicated setups.

For the simplest case of a 2×2 matrix we can construct all possible observables in terms of the Pauli matrices and the identity matrix $\mathbb{I}$. Since σ_x, σ_y, σ_z, and $\mathbb{I}$ are Hermitian, a linear combination of them is also Hermitian:

$$(a\sigma_x + b\sigma_y + c\sigma_z + d\mathbb{I})^\dagger = a^*\sigma_x + b^*\sigma_y + c^*\sigma_z + d^*\mathbb{I}, \tag{5.19}$$

so if a, b, c, and d are real the linear combination in equation (5.18) is Hermitian. This can be written as

$$M = \begin{pmatrix} c+d & a-ib \\ a+ib & -c+d \end{pmatrix}. \tag{5.20}$$

All Hermitian 2×2 matrices can be written like this, because $c + d$ and $-c + d$ can take on any real value, and $a + ib$ can take on any complex value and fully determines $a - ib$. The Pauli matrices are thus not just important matrices for describing the spin of an electron (up to a numerical factor of $\hbar/2$), but play a crucial role in the mathematics of 2×2 matrices.

We can take the Hermitian adjoint of a vector and a matrix:

$$M \to M^\dagger$$
$$|\psi\rangle \to (|\psi\rangle)^\dagger = \langle\psi|, \tag{5.21}$$

as well as a product of two matrices:

$$AB \to (AB)^\dagger = B^\dagger A^\dagger, \tag{5.22}$$

so the adjoint applied to a product of two operators changes the order of the operators! This is important, because AB is generally not the same as BA. The Hermitian adjoint of an expectation value is

$$\langle\psi|A|\psi\rangle^\dagger = \langle\psi|A^\dagger|\psi\rangle, \tag{5.23}$$

which is the same as the complex conjugate, since the expectation $\langle\psi|A|\psi\rangle$ is a single complex number. From this you can work out what happens when you take the adjoint twice.

Finally, we give a very convenient mathematical form of observables. Note how we could write the Hamiltonian for a two-level atom in its diagonal basis

$$H = E_g \, |g\rangle \, \langle g| + E_e \, |e\rangle \, \langle g| \,. \tag{5.24}$$

Notice how the eigenstates multiply with the eigenvalues, and are summed. Any observable can be written this way, also if it has more than two eigenstates. In general, when an observable has eigenstates $|a_j\rangle$ and eigenvalues a_j, the observable can be written as

$$A = \sum_j a_j \, |a_j\rangle \, \langle a_j| \,, \tag{5.25}$$

where j runs over all the eigenvalues. The set of values a_j are called the *spectrum* of A, and (5.25) is called the *spectral decomposition* of A. Some of these eigenvalues may have the same numerical value, in which case the spectrum is degenerate.

5.3   Evolution

We have seen that physical observables such as spin and energy are described by Hermitian operators, and that these operators have real eigenvalues. Can we give a similarly straightforward criterion that tells us whether an operator is a valid evolution? It turns out we can, by considering how evolved state vectors must remain normalised.

First, note that the expectation value of a (possibly Hermitian) matrix A is given by

$$\langle A \rangle = \langle \psi | A | \psi \rangle \,. \tag{5.26}$$

Next, let us evolve the state $|\psi\rangle$ to $U|\psi\rangle$, where U is the operator that describes the evolution. The evolved state must again be normalised, which restricts the form U can take. The new expectation value is then

$$\langle A \rangle' = \langle \psi | U^\dagger A U | \psi \rangle \,. \tag{5.27}$$

Now look at the special case where A is the identity operator ($A = \mathbb{I}$). In this case (5.26) becomes $\langle \psi | \mathbb{I} | \psi \rangle = \langle \psi | \psi \rangle = 1$, and (5.27) gives

$$\langle \psi | U^\dagger \, \mathbb{I} \, U | \psi \rangle = \langle \psi | U^\dagger U | \psi \rangle \,. \tag{5.28}$$

However, this is the same as the scalar product of the state $U|\psi\rangle$ with itself:

$$(\langle \psi | U^\dagger)(U | \psi \rangle) = 1 \,, \tag{5.29}$$

which is equal to 1 due to the normalisation condition. More importantly, this must be true for any state $|\psi\rangle$. We can also write $\langle \psi | \psi \rangle = 1$, and we therefore have

$$\langle\psi|U^{\dagger}U|\psi\rangle = \langle\psi|\psi\rangle\,, \tag{5.30}$$

Since this must be true for any $|\psi\rangle$, this means that

$$U^{\dagger}U = \mathbb{I}\,, \tag{5.31}$$

or $U^{\dagger}$ is the inverse of U. Any U that takes any valid quantum state $|\psi\rangle$ to another valid quantum state is bound to obey this rule. These are called *unitary* matrices, since they do not change the length of $|\psi\rangle$. They describe the evolution of the quantum system. The same reasoning, but with U and $U^{\dagger}$ swapped, will give $UU^{\dagger} = \mathbb{I}$.

We now derive an important mathematical property about unitary operators and vectors. If we apply a unitary evolution to two different vectors, the scalar product between the vectors before and after the evolution remains the same:

$$\langle\phi|\psi\rangle \xrightarrow{U} ((\langle\phi|U^{\dagger})(U|\psi\rangle)) = \langle\phi|U^{\dagger}U|\psi\rangle = \langle\phi|\psi\rangle\,. \tag{5.32}$$

In other words, U preserves the angle between $|\psi\rangle$ and $|\phi\rangle$. In particular, this means that orthogonal states evolve into new states that are again orthogonal. This was the assumption we physically motivated when we worked out the matrix form of the beam splitter in Chap. 2.

As we have seen in (4.29), we can write the time evolution as an exponential function of the Hamiltonian:

$$U(t) = \exp\left(-\frac{i}{\hbar}Ht\right)\,. \tag{5.33}$$

This is a function of a matrix, and it is itself a matrix. The inverse of this matrix is given by the Hermitian adjoint

$$U(t)^{-1} = U(t)^{\dagger} = \exp\left(\frac{i}{\hbar}H^{\dagger}t\right) = \exp\left(\frac{i}{\hbar}Ht\right)\,, \tag{5.34}$$

where we used that $H^{\dagger} = H$. Therefore, it is easy to find the inverse of a unitary matrix when it is given in the form of (5.33). We can take this result one step further, though: *every* unitary operator can be written as the exponential of a Hermitian matrix A times i:

$$U = \exp(iA)\,, \tag{5.35}$$

with $A^{\dagger} = A$. The prove this, we may write U in the basis where its matrix is diagonal. All the diagonal elements of U then become the eigenvalues e^{ia_j} with a_j real. For example, a 2×2 unitary matrix can be written as

$$U = \begin{pmatrix} e^{ia_1} & 0 \\ 0 & e^{ia_2} \end{pmatrix} = \exp \begin{pmatrix} ia_1 & 0 \\ 0 & ia_2 \end{pmatrix} = \exp \left[i \begin{pmatrix} a_1 & 0 \\ 0 & a_2 \end{pmatrix} \right]. \qquad (5.36)$$

However, the matrix

$$A = \begin{pmatrix} a_1 & 0 \\ 0 & a_2 \end{pmatrix}$$

is Hermitian, because a_1 and a_2 are real. Therefore, we have explicitly constructed the Hermitian matrix A that gives us U via $U = \exp(iA)$. In turn, this means that any unitary evolution of a quantum state can be described by some form of Hamiltonian H such that $A = Ht/\hbar$.

5.4 The Commutator

We have noted earlier that the product of two matrices AB is generally not the same as BA. As a quick example, we can multiply the Pauli σ_x and σ_y matrices in two ways:

$$\sigma_x \sigma_y = \begin{pmatrix} 0 & 1 \\ 1 & 0 \end{pmatrix} \begin{pmatrix} 0 & -i \\ i & 0 \end{pmatrix} = \begin{pmatrix} i & 0 \\ 0 & -i \end{pmatrix} = i\sigma_z,$$

$$\sigma_y \sigma_x = \begin{pmatrix} 0 & -i \\ i & 0 \end{pmatrix} \begin{pmatrix} 0 & 1 \\ 1 & 0 \end{pmatrix} = \begin{pmatrix} -i & 0 \\ 0 & i \end{pmatrix} = -i\sigma_z, \qquad (5.37)$$

and see that the two products $\sigma_x \sigma_y$ and $\sigma_y \sigma_x$ are not the same. They differ by a minus sign. We can capture this difference by subtracting $\sigma_y \sigma_x$ from $\sigma_x \sigma_y$:

$$\sigma_x \sigma_y - \sigma_y \sigma_x = 2i\sigma_z. \qquad (5.38)$$

This non-commutative property of matrices turns out to be of momentous importance in quantum mechanics. It is what makes quantum mechanics so different from classical physics.

In general, for two operators A and B we define

$$[A, B] = AB - BA. \qquad (5.39)$$

The quantity $[A, B]$ is called the commutator of A and B, and when $[A, B] = 0$, we say that the operators A and B commute: it does not matter in which order they are written down because $AB = BA$. However, when $[A, B] \neq 0$, the order does make a difference. In fact, you can consider $[A, B]$ the *correction* that you need to take into account when you want to swap the order of the product of AB:

$$AB = BA + [A, B]. \qquad (5.40)$$

You should always be careful with the order when you do calculations with operators in quantum mechanics.

For unitary operators it is not so surprising that the order of the operation matters. Suppose that operator U describes "putting on my socks", while operator V describes "putting on my shoes". It makes a difference whether I apply U first and then V, or V first and then U. The difference is between looking normal and looking silly.

However, the interpretation is not so clear for observables. These are also described by matrices and therefore also do not commute in general. As an example, the Pauli matrices above are Hermitian and represent (up to a factor $\hbar/2$) the spin of an electron in the x, y and z-direction. What does it mean that the spin observables in different directions do not commute? To answer this, we first note that two diagonal matrices of the same dimension always commute. For example,

$$\begin{pmatrix} a & 0 \\ 0 & b \end{pmatrix} \begin{pmatrix} c & 0 \\ 0 & d \end{pmatrix} = \begin{pmatrix} c & 0 \\ 0 & d \end{pmatrix} \begin{pmatrix} a & 0 \\ 0 & b \end{pmatrix} = \begin{pmatrix} ac & 0 \\ 0 & bd \end{pmatrix}. \tag{5.41}$$

In general, matrices commute if and only if they have the same eigenvectors. This is a key property of matrices that holds true in general, with a subtle modification when two or more eigenvalues of the operators are the same (but we will not consider those cases here).

The second key property of quantum mechanics is that a measurement of an observable will put the state of the quantum system in the eigenstate corresponding to the measurement outcome. If we measure the spin of an electron in the z-direction (S_z) and find spin ↑, a second measurement of S_z will always give spin ↑ again. That is, the measurement outcome ↑ has probability one, and ↓ has probability zero. So if two observables A and B commute, you can measure A and leave the system in an eigenstate of A. Then when you measure B, the system is already in an eigenstate of B (because A and B share the same eigenvectors) and you will find the corresponding measurement outcome (the eigenvalue) with probability one. Moreover, the state of the system is not disturbed by the measurement of B. Consequently, you can measure A again after B, and because the system is still in the same eigenstate you will again find the same measurement outcome for A as you did the first time.

If A and B do *not* commute, the measurement of A leaves the system in an eigenstate of A, and a subsequent measurement of B will leave the system in an eigenstate of B. A third measurement—of A again—will now no longer be guaranteed to give the same measurement outcome as the first measurement of A. Take, for example, the spin of an electron in the x and the z-direction. These observables do not commute. Suppose that we measure first spin ↑ in the z-direction. The state of the spin after the measurement is then $|\uparrow\rangle$. We can calculate the probabilities of the spin measurement outcomes $+$ and $-$ in the x-direction as

$$p_+ = |\langle +|\uparrow\rangle|^2 = \frac{1}{2} \quad \text{and} \quad p_- = |\langle -|\uparrow\rangle|^2 = \frac{1}{2}. \tag{5.42}$$

So the measurement outcome of the second observable, S_x, is uncertain. Moreover, performing a third measurement, again in the z-direction, will not yield outcome ↑ with probability one, but will rather be $p_\uparrow = p_\downarrow = 1/2$. The measurement of S_x disturbs the quantum system if it is in an eigenstate of S_z.

So when two observables commute, their measurements can be carried out in different orders, or even simultaneously, which makes no difference for the actual measurement outcomes. However, when the observables do not commute, each measurement disturbs the quantum system, and repeated measurements will generally not yield the same measurement outcomes (e.g., the first measurement of S_z gave ↑, but after S_x the second measurement of S_z gives either ↑ or ↓). This means that two observables can be measured together if they commute (they are compatible), but not if they do not commute.

5.5 Projectors

The scalar product of two state vectors $|\psi\rangle$ and $|\phi\rangle$ is given by $\langle\phi|\psi\rangle$. You can also calculate $\langle\psi|\phi\rangle = \langle\phi|\psi\rangle^*$. The easiest way to calculate the scalar products is via the vector form of $|\psi\rangle$ and $|\phi\rangle$. For example, if

$$|\psi\rangle = \begin{pmatrix} a \\ b \end{pmatrix} \quad \text{and} \quad |\phi\rangle = \begin{pmatrix} c \\ d \end{pmatrix}, \tag{5.43}$$

then the scalar product $\langle\phi|\psi\rangle$ is

$$\langle\phi|\psi\rangle = \begin{pmatrix} c^* & d^* \end{pmatrix} \begin{pmatrix} a \\ b \end{pmatrix} = c^*a + d^*b, \tag{5.44}$$

where a, b, c, and d are complex numbers. You should check that $\langle\psi|\phi\rangle$ is indeed the complex conjugate of $\langle\phi|\psi\rangle$.

These scalar products are used to calculate probabilities of measurement outcomes, for example when $|\psi\rangle$ is the spin state of an electron and $|\phi\rangle$ is the state $|\uparrow\rangle$ the probability of finding measurement outcome ↑ in a Stern-Gerlach experiment is given by

$$p_\uparrow = |\langle\uparrow|\psi\rangle|^2. \tag{5.45}$$

From geometry you know that the scalar product measures the overlap of two vectors, or more precisely, how much of one vector lies in the direction of the other. It so happens that it is very convenient to describe this in terms of projection operators, or projectors for short (see Fig. 5.1).

Let's consider a vector **v** in ordinary three-dimensional space. It can be written as a column of three numbers

Fig. 5.1 Projections of vectors (see supplementary material 1)

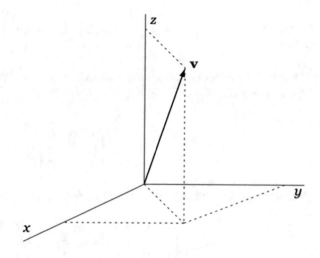

$$\mathbf{v} = \begin{pmatrix} v_x \\ v_y \\ v_z \end{pmatrix}, \tag{5.46}$$

which are the coordinates with respect to some coordinate system. We can project this vector onto the x-axis by setting $v_y = v_z = 0$. This is an operation on the vector, and we therefore expect that we can write this as a matrix. Indeed, it is straightforward to verify that the matrix

$$P_x = \begin{pmatrix} 1 & 0 & 0 \\ 0 & 0 & 0 \\ 0 & 0 & 0 \end{pmatrix} \tag{5.47}$$

does the job for any value of v_x, v_y, and v_z:

$$P_x \mathbf{v} = \begin{pmatrix} 1 & 0 & 0 \\ 0 & 0 & 0 \\ 0 & 0 & 0 \end{pmatrix} \begin{pmatrix} v_x \\ v_y \\ v_z \end{pmatrix} = \begin{pmatrix} v_x \\ 0 \\ 0 \end{pmatrix}. \tag{5.48}$$

Similarly, we can construct projectors onto the y- and z-axis:

$$P_y = \begin{pmatrix} 0 & 0 & 0 \\ 0 & 1 & 0 \\ 0 & 0 & 0 \end{pmatrix} \quad \text{and} \quad P_z = \begin{pmatrix} 0 & 0 & 0 \\ 0 & 0 & 0 \\ 0 & 0 & 1 \end{pmatrix}. \tag{5.49}$$

You should check that

$$P_y \mathbf{v} = \begin{pmatrix} 0 \\ v_y \\ 0 \end{pmatrix} \quad \text{and} \quad P_z \mathbf{v} = \begin{pmatrix} 0 \\ 0 \\ v_z \end{pmatrix}. \tag{5.50}$$

We can also project vectors onto planes instead of axes. Projecting the vector $\mathbf{v}$ onto the xy-plane is achieved by setting $z = 0$. The projector P_{xy} that accomplishes this is

$$P_{xy} = \begin{pmatrix} 1 & 0 & 0 \\ 0 & 1 & 0 \\ 0 & 0 & 0 \end{pmatrix}. \tag{5.51}$$

Similarly, projections onto the xz- and yz-planes are given by

$$P_{xz} = \begin{pmatrix} 1 & 0 & 0 \\ 0 & 0 & 0 \\ 0 & 0 & 1 \end{pmatrix} \quad \text{and} \quad P_{yz} = \begin{pmatrix} 0 & 0 & 0 \\ 0 & 1 & 0 \\ 0 & 0 & 1 \end{pmatrix}. \tag{5.52}$$

Finally, the projection onto the entire three-dimensional space is the identity operator.

One of the key properties of projectors is that the square of the projector is equal to itself: $P^2 = P$. You should try to prove that the eigenvalues of the matrices must therefore be 0 or 1. Indeed, the projectors we constructed above all obey this rule. In addition, every projection operator must be Hermitian: $P^\dagger = P$.

In the quantum mechanical example of finding the measurement outcome ↑ for the spin of an electron in a Stern-Gerlach experiment we calculate the probability by projecting the state vector of the electron onto the eigenstate $|\uparrow\rangle$ and determining the length of the resulting vector. Suppose that the state vector of the electron before the measurement is given by

$$|\psi\rangle = a|\uparrow\rangle + b|\downarrow\rangle = \begin{pmatrix} a \\ b \end{pmatrix}, \tag{5.53}$$

where we used

$$|\uparrow\rangle = \begin{pmatrix} 1 \\ 0 \end{pmatrix} \quad \text{and} \quad |\downarrow\rangle = \begin{pmatrix} 0 \\ 1 \end{pmatrix}. \tag{5.54}$$

Then, as we know, the probability of finding measurement outcome ↑ is calculated via

$$p_\uparrow = |\langle\uparrow|\psi\rangle|^2 = \langle\psi|\uparrow\rangle\langle\uparrow|\psi\rangle = |a|^2. \tag{5.55}$$

Alternatively, we could use our new knowledge about projectors: We project the state $|\psi\rangle$ onto the state vector corresponding to the measurement outcome ↑ using a projector $P_\uparrow$ and calculate the length-squared of the resulting vector $|\phi_\uparrow\rangle = P_\uparrow|\psi\rangle$.

But calculating the length-squared of a vector is nothing more than taking the scalar product with itself:

$$p_\uparrow = \langle \phi_\uparrow | \phi_\uparrow \rangle = \langle \psi | P_\uparrow^\dagger P_\uparrow | \psi \rangle . \tag{5.56}$$

Simplifying this using the rules $P^\dagger = P$ and $P^2 = P$, and comparing to (5.55), we find that

$$p_\uparrow = \langle \psi | P_\uparrow | \psi \rangle = \langle \psi | \uparrow \rangle \langle \uparrow | \psi \rangle = |a|^2 . \tag{5.57}$$

Since this is true for all $|\psi\rangle$, we find that

$$P_\uparrow = |\uparrow\rangle\langle\uparrow| . \tag{5.58}$$

We can verify that this is true by calculating $P_\uparrow$ using the vector form of $|\uparrow\rangle$ and $\langle\uparrow|$:

$$P_\uparrow = \begin{pmatrix} 1 \\ 0 \end{pmatrix} \begin{pmatrix} 1 & 0 \end{pmatrix} = \begin{pmatrix} 1 & 0 \\ 0 & 0 \end{pmatrix} . \tag{5.59}$$

Acting with $P_\uparrow$ on $|\psi\rangle$ gives

$$P_\uparrow |\psi\rangle = \begin{pmatrix} 1 & 0 \\ 0 & 0 \end{pmatrix} \begin{pmatrix} a \\ b \end{pmatrix} = \begin{pmatrix} a \\ 0 \end{pmatrix} , \tag{5.60}$$

projecting the state vector $|\psi\rangle$ onto the axis defined by the vector $|\uparrow\rangle$.

Using (5.58) we now see how we can construct projectors onto an arbitrary axis such as the vector $|\psi\rangle$. We just take the "outer" product $P_\psi = |\psi\rangle\langle\psi|$. You can easily verify that

$$P_\psi^2 = P_\psi \quad \text{and} \quad P_\psi^\dagger = P_\psi , \tag{5.61}$$

and the operator P_ψ is therefore a projector. For a vector $|\psi\rangle = \begin{pmatrix} a \\ b \end{pmatrix}$ with $a = i/\sqrt{3}$ and $b = \sqrt{2/3}$, the projector in matrix form is given by

$$P_\psi = |\phi\rangle\langle\psi| = \begin{pmatrix} \frac{i}{\sqrt{3}} \\ \sqrt{\frac{2}{3}} \end{pmatrix} \begin{pmatrix} -\frac{i}{\sqrt{3}} & \sqrt{\frac{2}{3}} \end{pmatrix} = \begin{pmatrix} \frac{1}{3} & \frac{i\sqrt{2}}{3} \\ -\frac{i\sqrt{2}}{3} & \frac{2}{3} \end{pmatrix} . \tag{5.62}$$

You should explicitly verify for this matrix that $P_\psi^2 = P_\psi$ and $P_\psi^\dagger = P_\psi$.

When we want to calculate the overlap of the state vector $|\psi\rangle$ with the eigenvector $|\uparrow\rangle$, we can project both the ket and the bra in P_ψ onto $|\uparrow\rangle$:

$$P_\uparrow P_\psi \, P_\uparrow^\dagger = \begin{pmatrix} 1 & 0 \\ 0 & 0 \end{pmatrix} \begin{pmatrix} |a|^2 & ab^* \\ a^*b & |b|^2 \end{pmatrix} \begin{pmatrix} 1 & 0 \\ 0 & 0 \end{pmatrix}$$

$$= \begin{pmatrix} 1 & 0 \\ 0 & 0 \end{pmatrix} \begin{pmatrix} |a|^2 & 0 \\ a^*b & 0 \end{pmatrix} = \begin{pmatrix} |a|^2 & 0 \\ 0 & 0 \end{pmatrix}. \tag{5.63}$$

Similarly, we can calculate the projection of $P_\downarrow$ as

$$P_\downarrow P_\psi \, P_\downarrow^\dagger = \begin{pmatrix} 0 & 0 \\ 0 & 1 \end{pmatrix} \begin{pmatrix} |a|^2 & ab^* \\ a^*b & |b|^2 \end{pmatrix} \begin{pmatrix} 0 & 0 \\ 0 & 1 \end{pmatrix}$$

$$= \begin{pmatrix} 0 & 0 \\ 0 & |b|^2 \end{pmatrix}. \tag{5.64}$$

Somehow we want to convert these two matrices to the numbers $|a|^2$ and $|b|^2$, respectively. You see that we can accomplish this by adding the diagonal elements. For $P_\uparrow P_\psi \, P_\uparrow^\dagger$ this gives $|a|^2 + 0 = |a|^2$, and for $P_\downarrow P_\psi \, P_\downarrow^\dagger$ this gives $0 + |b|^2 = |b|^2$.

There is a matrix operation called the trace that does exactly what we want, namely summing over all the diagonal elements. We denote the trace of a matrix M by $\mathrm{Tr}M$, so we have

$$\mathrm{Tr}\,(P_\uparrow P_\psi \, P_\uparrow^\dagger) = |a|^2 \quad \text{and} \quad \mathrm{Tr}(\,P_\downarrow P_\psi \, P_\downarrow^\dagger) = |b|^2. \tag{5.65}$$

A very useful property of the trace is the cyclic property:

$$\mathrm{Tr}\,(ABC) = \mathrm{Tr}\,(CAB). \tag{5.66}$$

We can use this, and the properties of the projectors, to write the probabilities of the measurement outcomes as

$$p_\uparrow = \mathrm{Tr}\,(P_\uparrow \, P_\psi) \quad \text{and} \quad p_\downarrow = \mathrm{Tr}\,(P_\downarrow \, P_\psi). \tag{5.67}$$

You should verify that this is true in the example above.

Finally, let's consider the expectation value of the spin operator S_z. In terms of the eigenvectors $|\uparrow\rangle$ and $|\downarrow\rangle$ this can be written as

$$S_z = \frac{\hbar}{2}|\uparrow\rangle\langle\uparrow| - \frac{\hbar}{2}|\downarrow\rangle\langle\downarrow| = \frac{\hbar}{2}P_\uparrow - \frac{\hbar}{2}P_\downarrow. \tag{5.68}$$

The expectation value of S_z with respect to the state vector $|\psi\rangle$ can be written as

$$\langle S_z \rangle = \langle \psi | S_z | \psi \rangle = \frac{\hbar}{2} \langle \psi | P_\uparrow | \psi \rangle - \frac{\hbar}{2} \langle \psi | P_\downarrow | \psi \rangle$$

$$= \frac{\hbar}{2} \mathrm{Tr} (P_\uparrow \, P_\psi) - \frac{\hbar}{2} \mathrm{Tr} (P_\downarrow \, P_\psi)$$

$$= \mathrm{Tr} \left(\frac{\hbar}{2} P_\uparrow - \frac{\hbar}{2} P_\downarrow \right) P_\psi = \mathrm{Tr} \, S_z \, P_\psi \, . \tag{5.69}$$

This is a general rule: for any operator A the expectation value is given by

$$\langle A \rangle = \mathrm{Tr} (A \, P_\psi) \, . \tag{5.70}$$

You may wonder what's the point of these mathematical exercises. All we gained is a different way to calculate quantities that we already knew how to calculate. However, it turns out that it is very convenient to use projectors when we deal with decoherence in Chap. 7.

Exercises

1. Calculate the determinant of the following matrices:

   (a) $\begin{pmatrix} 3 & 0 \\ 4 & -1 \end{pmatrix}$,   (b) $\begin{pmatrix} 2i & 5 \\ 1 & 7 \end{pmatrix}$,   (c) $\begin{pmatrix} 1 & 0 & -1 \\ 2 & 3 & 8 \\ 2 & 9 & 5 \end{pmatrix}$.

2. Calculate the eigenvalues and eigenvectors of the Pauli matrices σ_x, σ_y and σ_z. Do your results agree with the spin vectors in Chap. 3? Are the Pauli matrices Hermitian and/or unitary?

3. Prove that the eigenvalues of a unitary operator are complex numbers with modulus 1. In other words, they are of the form $e^{i\phi}$ with ϕ in $[0, 2\pi)$.

4. Indicate whether the following matrices are Hermitian, unitary, or both:

$$\frac{1}{\sqrt{2}} \begin{pmatrix} 1 & 1 \\ 1 & -1 \end{pmatrix}, \quad \begin{pmatrix} 3 & 4 - 2i \\ 4 - 2i & 5 \end{pmatrix}, \quad \frac{1}{2} \begin{pmatrix} 1 & e^{-3t} \\ e^{-3t} & e^{-t} \end{pmatrix} .$$

5. For an arbitrary Hamiltonian H whose square is not proportional to the identity matrix, for example the one given in (4.32), it is not so easy to calculate the unitary evolution $U = \exp(-iHt/\hbar)$ directly. We will calculate U in two steps:

   (a) Let V be the unitary matrix that diagonalises H such that $H_D = VHV^{-1}$ is a diagonal matrix. Show that

$$U = V \, e^{-\frac{i}{\hbar} H_D t} V^{-1} .$$

(b) Let $H = \frac{1}{2}\hbar\omega\sigma_z + g\hbar\sigma_x$, with g a real number. Find the matrix V and calculate U.

6. An electron with spin 'up' in the positive z-direction ($|\uparrow\rangle$) is measured using a Stern-Gerlach apparatus oriented in the x-direction.

   (a) Before the electron spin is measured, it evolves for a time T due to a uniform magnetic field in the y-direction. The Hamiltonian can be written as

$$H = -\frac{e\hbar}{2m}\,\sigma \cdot \mathbf{B},$$

   where $\sigma = (\sigma_x, \sigma_y, \sigma_z)$ are the standard Pauli matrices, e is the charge of the electron, m its mass, and $\mathbf{B}$ is the magnetic field. Calculate the new spin state of the electron entering the Stern-Gerlach apparatus.
   (b) Calculate the probabilities of the measurement outcomes of the Stern-Gerlach apparatus for the spin state calculated in part (a).
   (c) Sketch the evolution of the electron spin state in part (b) in the Bloch sphere, and indicate the essential elements.

7. Prove the following properties of the commutator:

   (a) $[A, A] = 0$,
   (b) $[A, B] = -[B, A]$,
   (c) $[A, BC] = B[A, C] + A, BC$,
   (d) $[A, [B, C]] + [B, [C, A]] + [C, [A, B]] = 0$.

8. Calculate P^2 with P given by

$$P = \begin{pmatrix} a \\ b \end{pmatrix}(a^*, b^*) = \begin{pmatrix} |a|^2 & ab^* \\ a^*b & |b|^2 \end{pmatrix} \quad \text{and} \quad |a|^2 + |b|^2 = 1. \tag{5.71}$$

9. Show that two operators commute if they have the same eigenvectors.
10. Show that the projectors on eigenstates of σ_x do not commute wth the projectors on eigenstates of σ_y.
11. Consider three (possibly non-orthogonal) states $|a\rangle$, $|b\rangle$, and $|c\rangle$ for an electron spin.

   (a) Construct the projectors onto the three states, and write down the formulas for the probabilities that the electron has spin a, b, c in terms of these projectors.
   (b) What is the probability that the electron does *not* have spin a, and what is the projector associated with this situation?
   (c) What is the probability that the electron has spin a *or* b, and what is the projector associated with this situation?

The Rules of Quantum Mechanics

We have arrived at the halfway point of this book, and it is time to take stock of what we have learned so far. We introduced the quantum state of a system, which can be seen as a mathematical short-hand for how the system was prepared. After preparation, the system may evolve to some other state, either as a change over time, or as a change in the motion, for example due to a beam splitter. Finally, we choose to measure a physical observable of the quantum system (like spin or energy), and the measurement outcome indicates the value of the observable. We find the probabilities of the measurement outcomes by taking the squared modulus of the scalar product between the state and the eigenvector associated with the measurement outcome. We can schematically represent the structure of quantum mechanics as follows:

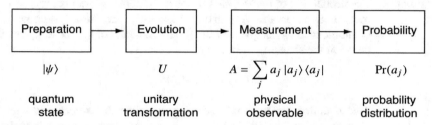

We will capture quantum mechanics in five rules that will completely determine the theory. These are called the *postulates of quantum theory*:

Postulate 1: The state of a quantum system is described by a vector $|\psi\rangle$ of length $\langle\psi|\psi\rangle = 1$. The vector components are complex numbers and carry no units. Two vectors that differ only by a global phase $e^{i\phi}$ are considered the same quantum state; the global phase ϕ is unobservable. The state does not directly describe the physical properties of a system; for that we need observables.

Postulate 2: Physical observables are described by Hermitian operators (matrices) A, which obey the rule $A^\dagger = A$. This relation ensures that the eigenvalues $a_1, \ldots, a_n$ of A are real and have physical dimensions, so they can be interpreted as the possible physical values of the observable. The eigenvalues therefore carry units. Any observable A can be written in the spectral decomposition

$$A = \sum_j a_j |a_j\rangle\langle a_j|,$$

where the $|a_j\rangle$ are the dimensionless eigenstates of a_j and $|a_j\rangle\langle a_j|$ are projectors onto the eigenstates.

Postulate 3: Changes in the state of the quantum system are described by unitary operators U. An operator or matrix is unitary if $U^\dagger = U^{-1}$. We can write $U(t) = \exp(-iHt/\hbar)$, where H is the Hamiltonian, or energy operator of the system. This leads to the Schrödinger equation for the quantum state:

$$i\hbar \frac{d}{dt}|\psi(t)\rangle = H|\psi(t)\rangle.$$

If U is known, we can write the evolved state directly as $|\psi(t)\rangle = U(t)|\psi(0)\rangle$ and solve this as a matrix equation, where $|\psi(t)\rangle$ and $|\psi(0)\rangle$ are vectors, and $U(t)$ is a matrix.

Postulate 4: The probability of finding measurement outcome a_j after the measurement of an observable A with eigenvalues $a_1, \ldots a_n$ for a system in the quantum state $|\psi\rangle$ is given by the square modulus of the scalar product between the state and the eigenvector:

$$\Pr(a_j) = |\langle a_j|\psi\rangle|^2.$$

This is called the *Born rule*. We can calculate the expectation value of A for the system in state $|\psi\rangle$ by taking the weighted sum over the eigenvalues:

$$\langle A \rangle = \sum_j a_j \Pr(a_j) = \sum_j a_j \langle \psi|a_j\rangle \langle a_j|\psi\rangle$$

$$= \langle \psi| \left(\sum_j a_j |a_j\rangle\langle a_j| \right) |\psi\rangle = \langle \psi|A|\psi\rangle.$$

Postulate 5: Immediately after the measurement of an observable, the system is in the eigenstate $|a_j\rangle$ corresponding the eigenvalue a_j found in the measurement. Measuring the observable for a second time right after the first measurement will give the same outcome a_j with probability $\Pr(a_j) = 1$. This is called the "Projection Postulate".

Chapter 6
Entanglement

In this chapter, we will find out how we can apply quantum mechanics to more than one system. In doing so, we encounter what is truly strange in quantum mechanics, namely *entanglement*. We also explore some of the more remarkable applications of quantum mechanics, including teleportation and quantum computing.

6.1 The State of Two Electrons

We have looked at thought experiments with a single photon in a Mach–Zehnder interferometer, a single electron in a Stern-Gerlach apparatus, and the interaction of a two-level atom with an optical pulse. We found that their states are all described by a two-dimensional complex vector, and the observables and evolution operators are described by 2×2 matrices. But what if we have *two* photons, or two electrons, or two atoms? Or three?

Let's consider two electrons, and assume that they are spatially well-separated so we can label "electron 1" and "electron 2" without any ambiguity. It is easy to write the state of electron 1 as

$$|\psi\rangle_1 = a|\uparrow\rangle_1 + b|\downarrow\rangle_1 \quad \text{with} \quad |a|^2 + |b|^2 = 1, \tag{6.1}$$

where we add a subscript "1" to emphasise that we refer to electron 1. Similarly, we can write down the state of electron 2:

$$|\phi\rangle_2 = c|\uparrow\rangle_2 + d|\downarrow\rangle_2 \quad \text{with} \quad |c|^2 + |d|^2 = 1. \tag{6.2}$$

Supplementary Information The online version contains supplementary material available at https://doi.org/10.1007/978-3-031-16165-0_6.

The question is: what is the state of the composite system consisting of these two electrons?

We can construct the states of the composite system from the states of the individual systems. Remember that the symbols and writing inside the ket are nothing more than convenient labels, indicating the measurement outcomes. We therefore have four possible measurement outcomes if we measure the spin of each electron in the z-direction:

$$|\text{electron } 1 = \uparrow, \text{ electron } 2 = \uparrow\rangle,$$
$$|\text{electron } 1 = \uparrow, \text{ electron } 2 = \downarrow\rangle,$$
$$|\text{electron } 1 = \downarrow, \text{ electron } 2 = \uparrow\rangle,$$
$$|\text{electron } 1 = \downarrow, \text{ electron } 2 = \downarrow\rangle. \tag{6.3}$$

This is hardly convenient, so instead we may write

$$|\uparrow_1, \uparrow_2\rangle, \quad |\uparrow_1, \downarrow_2\rangle, \quad |\downarrow_1, \uparrow_2\rangle, \quad |\downarrow_1, \downarrow_2\rangle. \tag{6.4}$$

By convention, the ordering of the $\uparrow$ and $\downarrow$ arrow is fixed, and do we really need the comma? Clearly, we can reduce this further to

$$|\uparrow\uparrow\rangle, \quad |\uparrow\downarrow\rangle, \quad |\downarrow\uparrow\rangle \quad \text{and} \quad |\downarrow\downarrow\rangle. \tag{6.5}$$

These are four quantum states that make perfect sense in the light of measuring S_z on each electron separately. And because this is quantum mechanics, we can take superpositions of these states, such as

$$|\Psi\rangle = \frac{1}{2}|\uparrow\uparrow\rangle + \frac{1}{2}|\uparrow\downarrow\rangle + \frac{1}{2}|\downarrow\uparrow\rangle + \frac{1}{2}|\downarrow\downarrow\rangle. \tag{6.6}$$

Next, suppose that the electrons are in the states given in (6.1) and (6.2). How do we write this in terms of the states of (6.5)? We can look first at the probabilities of the measurement outcomes of S_z on the two electrons. Since the electrons are completely independent, the probabilities multiply:

$$\Pr(\uparrow\uparrow) = \Pr(\uparrow_1) \times \Pr(\uparrow_2) = |a|^2|c|^2$$
$$\Pr(\uparrow\downarrow) = \Pr(\uparrow_1) \times \Pr(\downarrow_2) = |a|^2|d|^2$$
$$\Pr(\downarrow\uparrow) = \Pr(\downarrow_1) \times \Pr(\uparrow_2) = |b|^2|c|^2$$
$$\Pr(\downarrow\downarrow) = \Pr(\downarrow_1) \times \Pr(\downarrow_2) = |b|^2|d|^2 \tag{6.7}$$

It is easy to verify that the probabilities sum to one:

$$\Pr(\uparrow\uparrow) + \Pr(\uparrow\downarrow) + \Pr(\downarrow\uparrow) + \Pr(\downarrow\downarrow) = 1. \tag{6.8}$$

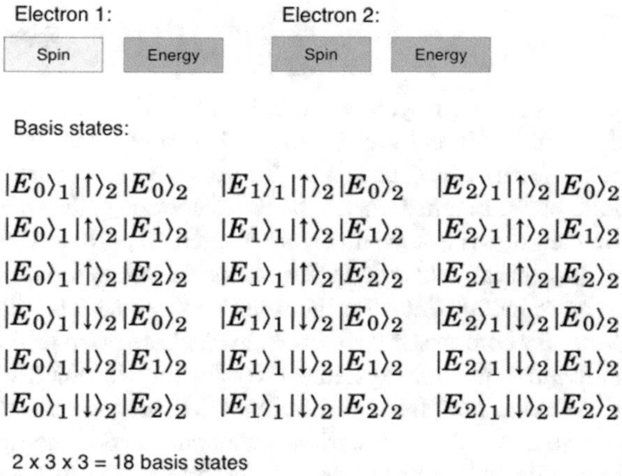

Fig. 6.1 Constructing states for composite systems (see supplementary material 1)

Electron 1:   Electron 2:

| Spin | Energy | | Spin | Energy |

Basis states:

$$|E_0\rangle_1 |\uparrow\rangle_2 |E_0\rangle_2 \quad |E_1\rangle_1 |\uparrow\rangle_2 |E_0\rangle_2 \quad |E_2\rangle_1 |\uparrow\rangle_2 |E_0\rangle_2$$

$$|E_0\rangle_1 |\uparrow\rangle_2 |E_1\rangle_2 \quad |E_1\rangle_1 |\uparrow\rangle_2 |E_1\rangle_2 \quad |E_2\rangle_1 |\uparrow\rangle_2 |E_1\rangle_2$$

$$|E_0\rangle_1 |\uparrow\rangle_2 |E_2\rangle_2 \quad |E_1\rangle_1 |\uparrow\rangle_2 |E_2\rangle_2 \quad |E_2\rangle_1 |\uparrow\rangle_2 |E_2\rangle_2$$

$$|E_0\rangle_1 |\downarrow\rangle_2 |E_0\rangle_2 \quad |E_1\rangle_1 |\downarrow\rangle_2 |E_0\rangle_2 \quad |E_2\rangle_1 |\downarrow\rangle_2 |E_0\rangle_2$$

$$|E_0\rangle_1 |\downarrow\rangle_2 |E_1\rangle_2 \quad |E_1\rangle_1 |\downarrow\rangle_2 |E_1\rangle_2 \quad |E_2\rangle_1 |\downarrow\rangle_2 |E_1\rangle_2$$

$$|E_0\rangle_1 |\downarrow\rangle_2 |E_2\rangle_2 \quad |E_1\rangle_1 |\downarrow\rangle_2 |E_2\rangle_2 \quad |E_2\rangle_1 |\downarrow\rangle_2 |E_2\rangle_2$$

2 x 3 x 3 = 18 basis states

The two-electron spin state that is consistent with these probabilities is

$$|\Psi\rangle = ac|\uparrow\uparrow\rangle + ad|\uparrow\downarrow\rangle + bc|\downarrow\uparrow\rangle + bd|\downarrow\downarrow\rangle. \tag{6.9}$$

But this is nothing more than the product of the two spin states:

$$|\psi\rangle_1 |\phi\rangle_2 = (a|\uparrow\rangle_1 + b|\downarrow\rangle_1)(c|\uparrow\rangle_2 + d|\downarrow\rangle_2)$$
$$\equiv ac|\uparrow\uparrow\rangle + ad|\uparrow\downarrow\rangle + bc|\downarrow\uparrow\rangle + bd|\downarrow\downarrow\rangle. \tag{6.10}$$

So we can combine the states of two quantum system into a composite quantum system by multiplying out the states in the manner above, and keep the order of the symbols in the ket! You can now show that the state in (6.6) is the same as two electrons that are each in the state

$$|\psi\rangle = |\phi\rangle = \frac{1}{\sqrt{2}}|\uparrow\rangle + \frac{1}{\sqrt{2}}|\downarrow\rangle. \tag{6.11}$$

As a convention, in this book we will use lower case Greek letters for simple quantum systems, and upper case Greek letters for composite quantum systems. In Fig. 6.1 we encounter other examples of composite systems.

6.2 Entanglement

The four states in (6.5) form a basis for all quantum states of two electron spins. In other words, we can take any superposition of these four states, and obtain another valid quantum state, as long as the normalisation condition is satisfied. However, this can lead to strange situations. For example, consider the state

$$|\Phi^+\rangle = \frac{1}{\sqrt{2}}|\uparrow\uparrow\rangle + \frac{1}{\sqrt{2}}|\downarrow\downarrow\rangle. \tag{6.12}$$

Try as you might, you will not be able to find values of a, b, c, and d in (6.10) such that you obtain the state in (6.12). This means that there are no single electron states of the form of (6.1) and (6.2) that give the same behaviour as the two electrons in the state $|\Phi^+\rangle$. In other words, the two electrons in the state $|\Phi^+\rangle$ are entangled, because we cannot write down the state of each electron individually.

In general, two quantum systems are entangled if you cannot write the total state as a product of states for the separate systems. The total state is called an *entangled state*. By contrast, if the total state can be written as the product of two states for the individual quantum systems, the total state is called separable.

What is the physical difference between entangled and separable states? Let's construct a thought experiment that can shed some light on this. Consider a source that emits electrons in opposite directions, one to Alice, and the other to Bob. They both measure the spin of the electron they receive in the z-direction. If the state of the electrons is given by $|\Phi^+\rangle$, it is not hard to see that they both find $\uparrow$ and $\downarrow$ measurement outcomes 50% of the time. But crucially, when Alice measures $\uparrow$ so does Bob, and vice versa when Alice measures $\downarrow$, Bob gets $\downarrow$ as well.

So maybe this is what entanglement means. Does it just say that the source creates the state $|\uparrow\uparrow\rangle$ half the time, and $|\downarrow\downarrow\rangle$ the other half, making the choice, for example, via a coin toss (see Fig. 6.2)? If that were true, suppose that in one particular run of the experiment the source sent $|\uparrow\uparrow\rangle$, but now Alice and Bob measure the spin in the x-direction instead. We know that we can write the states $|\uparrow\rangle$ and $|\downarrow\rangle$ in terms of the eigenstates $|+\rangle$ and $|-\rangle$ of the S_x observable:

$$|\uparrow\rangle = \frac{1}{\sqrt{2}}|+\rangle + \frac{1}{\sqrt{2}}|-\rangle, \tag{6.13}$$

and

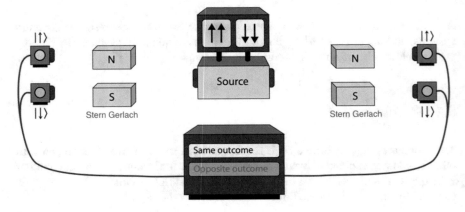

Fig. 6.2 Creating classical correlations (see supplementary material 2)

$$|\downarrow\rangle = \frac{1}{\sqrt{2}}|+\rangle - \frac{1}{\sqrt{2}}|-\rangle . \tag{6.14}$$

The state $|\uparrow\uparrow\rangle$ translates into

$$|\uparrow\uparrow\rangle = \frac{1}{2}|++\rangle + \frac{1}{2}|+-\rangle + \frac{1}{2}|-+\rangle + \frac{1}{2}|--\rangle , \tag{6.15}$$

which means that Alice and Bob obtain all measurement outcomes "++", "+−", "−+", and "−−", all with probability 1/4. In other words, there is no correlation between the measurement outcomes of Alice and Bob, because Alice's outcome tells her nothing about Bob's measurement outcome and vice versa. The same measurement outcomes are obtained when the source creates two electrons with spin $|\downarrow\downarrow\rangle$, since

$$|\downarrow\downarrow\rangle = \frac{1}{2}|++\rangle - \frac{1}{2}|+-\rangle - \frac{1}{2}|-+\rangle + \frac{1}{2}|--\rangle , \tag{6.16}$$

The minus signs in the amplitudes have no effect on the probabilities, since the probabilities are the square of the amplitudes. These measurement statistics are the result of a source that creates two electrons in the state $|\uparrow\uparrow\rangle$ or $|\downarrow\downarrow\rangle$, each with probability 1/2.

However, when the source truly creates the entangled state $|\Phi^+\rangle$, the measurement statistics will be different. Specifically, the minus signs in (6.16) will become important. When we translate the entangled state $|\Phi^+\rangle$ into the eigenvectors for the spin observables in the x-direction, we obtain

$$\begin{aligned}
|\Phi^+\rangle &= \frac{1}{\sqrt{2}}|\uparrow\uparrow\rangle + \frac{1}{\sqrt{2}}|\downarrow\downarrow\rangle \\
&= \frac{1}{2\sqrt{2}}\left(|++\rangle + |+-\rangle + |-+\rangle + |--\rangle\right) + \frac{1}{2\sqrt{2}}\left(|++\rangle - |+-\rangle - |-+\rangle + |--\rangle\right) \\
&= \frac{1}{\sqrt{2}}|++\rangle + \frac{1}{\sqrt{2}}|--\rangle .
\end{aligned} \tag{6.17}$$

You see that the minus signs are responsible for the cancellation of the terms $|+-\rangle$ and $|-+\rangle$. As a result, there is a perfect correlation between the measurement outcomes of Alice and Bob when they measure the spin in the x-direction. Somehow, the correlations in the entangled state $|\Phi^+\rangle$ are stronger than the correlations we get when we choose $|\uparrow\uparrow\rangle$ or $|\downarrow\downarrow\rangle$ on the basis of a coin toss (see Fig. 6.3), because they manifest themselves in both the $\{\uparrow, \downarrow\}$ and $\{+, -\}$ directions.

Entanglement is a phenomenon that exists only in quantum mechanics, so we can say that quantum mechanics allows for stronger correlations between systems than classical physics. These stronger correlations have some very interesting consequences. One is quantum teleportation, in which the quantum state of a system can be transported over arbitrary distances without the need for a quantum system to

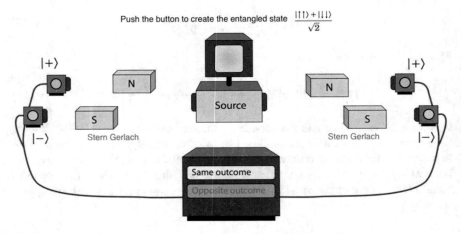

Fig. 6.3 Creating quantum correlations (see supplementary material 3)

carry it along. The other consequence is that we can build more powerful computers using entanglement. We consider these applications in the next two sections.

6.3 Quantum Teleportation

Probably the most extraordinary use of the quantum correlations present in entanglement is *quantum teleportation*.[1] It is the process of transferring the quantum state from one particle, held by Alice, to another particle, held by Bob. In cartoon form, it would look something like Fig. 6.4 (don't worry about what is going on inside the machine at this point). Furthermore, this transfer must succeed without either Alice or Bob gaining any information about the quantum state—say the electron spin. To make things extra hard, the three spins must not change places, so Alice cannot take spin 1 and bring it to Bob. Finally, Alice and Bob may be arbitrarily far apart, for example with Alice here on Earth and Bob somewhere in the Andromeda galaxy two million light years away. Surely, this is impossible.

Your first inclination might be to measure the spin of Alice's electron, and send the measurement result to Bob, who prepares the spin of his electron according to the measurement outcome. However, this will not work well. Alice needs to choose a direction to measure the spin of her electron. Once she has chosen a direction, she cannot measure any of the other two directions and expect to get a meaningful answer about the original spin direction (see Sect. 3.4). To see this, assume that the state of the electron is

[1] C. H. Bennett et al., *Teleporting an Unknown Quantum State via Dual Classical and Einstein-Podolsky-Rosen Channels*, Phys. Rev. Lett. **70** 1895, 1993.

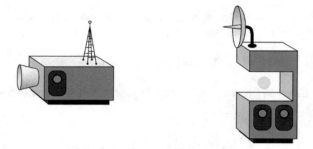

Fig. 6.4 Cartoon of a teleportation machine (see supplementary material 4)

$$|\psi\rangle = \cos\theta|\uparrow\rangle + i\sin\theta|\downarrow\rangle. \tag{6.18}$$

A measurement of the spin in the z-direction will give outcome $\uparrow$ with probability $p_\uparrow = \cos^2\theta$ and outcome $\downarrow$ with probability $p_\downarrow = \sin^2\theta$. After the measurement, the electron will be in the spin state $|\uparrow\rangle$ or $|\downarrow\rangle$, respectively. Measuring the spin next in the x- or y-direction will give one of two outcomes, but always with probability of $1/2$. This does not give us any more information about θ. Only the first measurement gives information about θ, but since the outcome is one bit of information (up or down), it can never contain enough to fully capture the value of θ. So a measure-and-send approach will not work.

However, we can achieve our goal of sending the state using quantum teleportation. Before we give a detailed description of this remarkable phenomenon we will need to construct a few more entangled states. The state space of two electrons is spanned by four orthogonal basis states, as given in (6.5). Alternatively, we can create a basis out of four entangled states:

$$|\Phi^+\rangle = \frac{1}{\sqrt{2}}|\uparrow\uparrow\rangle + \frac{1}{\sqrt{2}}|\downarrow\downarrow\rangle, \quad \text{and} \quad |\Psi^+\rangle = \frac{1}{\sqrt{2}}|\uparrow\downarrow\rangle + \frac{1}{\sqrt{2}}|\downarrow\uparrow\rangle,$$

$$|\Phi^-\rangle = \frac{1}{\sqrt{2}}|\uparrow\uparrow\rangle - \frac{1}{\sqrt{2}}|\downarrow\downarrow\rangle, \quad \text{and} \quad |\Psi^-\rangle = \frac{1}{\sqrt{2}}|\uparrow\downarrow\rangle - \frac{1}{\sqrt{2}}|\downarrow\uparrow\rangle. \tag{6.19}$$

This is called the Bell basis, after John Bell, and the states $|\Phi^\pm\rangle$ and $|\Psi^\pm\rangle$ are called the Bell states. They are maximally entangled states, in the sense that they describe the strongest possible correlations between two electron spins. We can also use these states as eigenvectors of an observable called the Bell operator. When we measure the Bell operator (called a Bell measurement), the outcomes tell us which of the Bell states the electrons are in. This is a joint measurement of the two spins, and does not reveal anything about the individual spins.[2]

---

[2] This is not as strange as it sounds: when you look at macroscopic object, say a pencil, you do not see the individual atoms in the pencil, but rather the collective state of the atoms that make up the pencil.

We are now ready to describe the process of quantum teleportation: Alice holds an electron, labelled 1, in the spin state

$$|\psi\rangle_1 = a|\uparrow\rangle_1 + b|\downarrow\rangle_1.$$ (6.20)

In addition, Alice and Bob each hold one of a pair of entangled electrons, labelled 2 and 3, in the state

$$|\Phi^+\rangle_{23} = \frac{1}{\sqrt{2}}|\uparrow_2, \uparrow_3\rangle + \frac{1}{\sqrt{2}}|\downarrow_2, \downarrow_3\rangle.$$ (6.21)

Let's say that Alice holds electron 2 and Bob holds electron 3. Next, we combine the state of the three electrons and expand it as

$$|\Upsilon\rangle = |\psi\rangle_1 |\Phi^+\rangle_{23} = (a|\uparrow\rangle_1 + b|\downarrow\rangle_1) \left(\frac{1}{\sqrt{2}}|\uparrow\uparrow\rangle + \frac{1}{\sqrt{2}}|\downarrow\downarrow\rangle \right)$$

$$= \frac{1}{\sqrt{2}} (a|\uparrow\uparrow\uparrow\rangle + a|\uparrow\downarrow\downarrow\rangle + b|\downarrow\uparrow\uparrow\rangle + b|\downarrow\downarrow\downarrow\rangle),$$ (6.22)

where we dropped the electron labels and pulled a common factor $1/\sqrt{2}$ out in front in the last line. This is the state of the system of three electrons (keep the order inside the kets the same!).

Alice now performs a Bell measurement on her two spins (1 and 2), which gives a measurement outcome that allows us to tell which one of the Bell states $|\Phi^\pm\rangle$ or $|\Psi^\pm\rangle$ the electrons are in. We write the spin states of the two electrons held by Alice $|\uparrow\uparrow\rangle$, $|\uparrow\downarrow\rangle$, $|\downarrow\uparrow\rangle$, and $|\downarrow\downarrow\rangle$ in the Bell basis:

$$|\uparrow\uparrow\rangle_{12} = \frac{|\Phi^+\rangle_{12} + |\Phi^-\rangle_{12}}{\sqrt{2}} \quad \text{and} \quad |\uparrow\downarrow\rangle_{12} = \frac{|\Psi^+\rangle_{12} + |\Psi^-\rangle_{12}}{\sqrt{2}},$$

$$|\downarrow\downarrow\rangle_{12} = \frac{|\Phi^+\rangle_{12} - |\Phi^-\rangle_{12}}{\sqrt{2}} \quad \text{and} \quad |\downarrow\uparrow\rangle_{12} = \frac{|\Psi^+\rangle_{12} - |\Psi^-\rangle_{12}}{\sqrt{2}}.$$ (6.23)

This is the "reverse" relation compared to (6.19). You should verify that this equation holds. We can use these substitutions for electrons 1 and 2 to write the state $|\Upsilon\rangle$ before the measurement as

$$|\Upsilon\rangle = \frac{1}{2} \big[|\Phi^+\rangle_{12}(a|\uparrow\rangle_3 + b|\downarrow\rangle_3) + |\Phi^-\rangle_{12}(a|\uparrow\rangle_3 - b|\downarrow\rangle_3)$$

$$+ |\Psi^+\rangle_{12}(b|\uparrow\rangle_3 + a|\downarrow\rangle_3) + |\Psi^-\rangle_{12}(b|\uparrow\rangle_3 - a|\downarrow\rangle_3) \big].$$ (6.24)

Alice finds one of four possible outcomes in her Bell measurement, and for each outcome we can read off the state of Bob's electron:

$$\begin{aligned}
\Phi^+ &: \quad |\psi\rangle_3 = a|\uparrow\rangle + b|\downarrow\rangle, \\
\Phi^- &: \quad |\psi\rangle_3 = a|\uparrow\rangle - b|\downarrow\rangle, \\
\Psi^+ &: \quad |\psi\rangle_3 = a|\downarrow\rangle + b|\uparrow\rangle, \\
\Psi^- &: \quad |\psi\rangle_3 = a|\downarrow\rangle - b|\uparrow\rangle.
\end{aligned} \tag{6.25}$$

From these outcomes, it is clear that the state held by Bob is different for the different measurement outcomes of Alice's Bell measurement. Let this sink in for a moment: After setting up the entangled state between Alice and Bob, who may be literally light years apart, Bob has done absolutely nothing to his spin, yet its state is different depending on Alice's measurement outcome! This should shock you: the information contained in a and b somehow made it from Alice to Bob without any communication, only a local measurement by Alice! Is there some violation of causality? No! It is a direct consequence of the stronger-than-classical correlations present in entangled states.

In order to turn the state of Bob's spin into the original state that entered Alice's device, Bob needs to apply a correction to his electron spin. Since he does not know the outcome of Alice's measurement, he must wait for a signal from Alice that tells him which correction to apply. This will take two classical bits, because there are four measurement outcomes. The correction operators that Bob needs to apply are:

$$\Phi^+ : \begin{pmatrix} 1 & 0 \\ 0 & 1 \end{pmatrix}, \quad \Phi^- : \begin{pmatrix} 1 & 0 \\ 0 & -1 \end{pmatrix}, \quad \Psi^+ : \begin{pmatrix} 0 & 1 \\ 1 & 0 \end{pmatrix}, \quad \Psi^- : \begin{pmatrix} 0 & 1 \\ -1 & 0 \end{pmatrix}. \tag{6.26}$$

Only after this correction does Bob have his electron in the original state

$$|\psi\rangle_3 = a|\uparrow\rangle_3 + b|\downarrow\rangle_3. \tag{6.27}$$

Notice how the label "1" in (6.20) has turned into a "3" in (6.27). The fact that Bob has to wait for Alice's measurement outcome means that causality is not violated. The process of teleportation is shown in Fig. 6.5.

To appreciate how remarkable this protocol really is, let's recall some of its most distinctive properties. First, no matter is transported, only the state of the system. In this sense teleportation is completely different from the transporter in Star Trek, to which it is often compared. In Star Trek, it seems that matter appears out of nowhere in the form of the away team at the surface of a planet. This cannot happen in physics, because matter is energy, and it would violate energy conservation. Second, neither Alice nor Bob learns anything about the input state $|\psi\rangle$. And finally, quantum teleportation cannot be used to signal faster than light. So there is no problem with quantum mechanics and Einstein's theory of relativity coexisting as far as teleportation is concerned. Still, quantum mechanics allows us to transfer a potentially large amount of information (contained in a and b) at the cost of sending only two bits of information.

Fig. 6.5 Inside the teleportation machine (see supplementary material 5)

Richard Feynman famously said that the double slit experiment captures everything that is mysterious about quantum mechanics: "it contains the *only* mystery". However, this is not quite true, because it does not contain entanglement. Instead, *quantum teleportation* is the process that contains the true mystery about quantum mechanics.[3]

6.4   *Mathematical Intermezzo*: **Qubits and Computation**

Ordinary (or "classical") computers operate using bits, which are units of information that are labelled 0 or 1. These bits of information are always carried by physical systems whose states are correspondingly labelled 0 and 1. For example, in a memory chip, the information carrier is a capacitor that either holds charge or not (bit value 0 or 1, respectively). The extension to quantum mechanics is immediate: instead of a classical physical system with states that can be either 0 or 1, we use quantum systems, whose states can be any superposition of 0 or 1. In our notation we write

$$|\psi\rangle = a|0\rangle + b|1\rangle. \tag{6.28}$$

We call this a quantum bit, or *qubit*. The actual physical system can be an electron spin with $|0\rangle = |\uparrow\rangle$ and $|1\rangle = |\downarrow\rangle$, a two-level atom with $|0\rangle = |g\rangle$ and $|1\rangle = |e\rangle$, a photon in two possible paths with $|0\rangle = |\text{path } 1\rangle$ and $|1\rangle = |\text{path } 2\rangle$, or any other system that has a two dimensional state space in quantum mechanics. Here, we will use generic qubits for the description of a quantum computation. Translating the abstract computation into a physical implementation is then a task for scientists and engineers who know how to manipulate photons, electrons, atoms, etc.

The idea behind computation is that an input bit string evolves according to some computational rules based on logical operators like AND, OR and NOT into an output bit string. For example

---

[3] R. Feynman, *Lectures on Physics*, Volume I, 37–2, 1963.

$$\text{AND}(1, 1) = 1 \quad \text{and} \quad \text{AND}(1, 0) = 0 \quad \text{etc.,} \tag{6.29}$$

or

$$\text{NOT}(0) = 1 \quad \text{and} \quad \text{NOT}(1) = 0. \tag{6.30}$$

These computational rules can encode any problem you want to solve. If you find these rule a little tricky, you may think of 0 and 1 as FALSE and TRUE. The expression $\text{AND}(1, 1) = 1$ then means that AND returns TRUE if both propositions in its argument are true. So the proposition "the Sun shines and I am happy" is true if "the Sun shines" is true and "I am happy" is true.

One more thing we need to learn is how to add bits. There is no value 2 in a bit, just 0 and 1. Nevertheless, we often want to add two bits. How do we do this? The answer is to add "modulo 2", where 2 turns into 0. The standard addition "+" is replaced by a "$\oplus$" to indicate that we used addition modulo 2. For example, we have

$$
\begin{aligned}
0 \oplus 0 &= 0 \\
0 \oplus 1 &= 1 \\
1 \oplus 0 &= 1 \\
1 \oplus 1 &= 0.
\end{aligned}
\tag{6.31}
$$

Adding a bit value of 1 to another bit is effectively the same as "flipping" the bit value from 0 to 1 and from 1 to 0 (incidentally, this is the NOT operator). You are already very familiar with addition modulo 12 because you use it implicitly when you read the time: four hours after 11 am is three pm. Doing the calculation explicitly, we add 4 to 11 modulo 12:

$$11 \oplus 4 = 15 \bmod 12 = 3. \tag{6.32}$$

Addition modulo two works along the same lines, but simpler because the numbers are smaller.

Equation (6.31) describes the "exclusive or" operator XOR between two bits. It differs from the OR operator in that it is 0 when both bits are 1:

$$
\begin{aligned}
\text{XOR}(0, 0) &= 0 \\
\text{XOR}(0, 1) &= 1 \\
\text{XOR}(1, 0) &= 1 \\
\text{XOR}(1, 1) &= 0.
\end{aligned}
\tag{6.33}
$$

We can also interpret the XOR as a controlled NOT, where the outcome of the XOR is a bit flip of the first bit if the second bit has value 1 (or vice versa; the

XOR is symmetric in the input). You can now see how this operation is important in computations, because it does something to one bit depending on the value of another bit.

6.5 ❧ Quantum Computers

Quantum mechanics has many applications in everyday life. For example, quantum mechanics is indispensable for the understanding of the semiconductor devices we use in all modern computers. But an even more profound use for quantum mechanics is quantum computing.

To show the basic idea behind quantum computing, we choose a very simple calculation: Suppose that we have a function f, which takes as input a single bit and produces another single bit as output. There are exactly four such functions:

f_1	f_2	f_3	f_4
$f(0) = 0$	$f(0) = 1$	$f(0) = 0$	$f(0) = 1$
$f(1) = 0$	$f(1) = 0$	$f(1) = 1$	$f(1) = 1$

These functions can be divided into two classes called "constant": $f(0) = f(1)$, and "balanced": $f(0) \neq f(1)$. The functions f_1 and f_4 above are constant, while the functions f_2 and f_3 are balanced. To check whether a function is balanced or constant, you normally have to evaluate the function twice: once with input 0, and once with input 1. We compare the outcomes, which then tells us whether f is constant or balanced. There is no way to do this by evaluating f fewer than two times, either on paper or with a classical computer. However, on a quantum computer we can tell whether f is constant or balanced by evaluating f only once.[4]

In the remainder of this section, we will show how a quantum computer can tell the difference between constant and balanced functions with only a single application of the function f. The algorithm was invented by David Deutsch[5] in 1989, and uses only two qubits that are initially prepared in the state $|01\rangle \equiv |0\rangle_1 |1\rangle_2$. First, we apply a so-called Hadamard gate to each qubit. This is just a unitary operator that changes the state of a qubit according to

$$|0\rangle \underset{H}{\rightarrow} \frac{1}{\sqrt{2}} |0\rangle + \frac{1}{\sqrt{2}} |1\rangle , \tag{6.34}$$

and

$$|1\rangle \underset{H}{\rightarrow} \frac{1}{\sqrt{2}} |0\rangle - \frac{1}{\sqrt{2}} |1\rangle . \tag{6.35}$$

---

[4] Perhaps this is not a very interesting calculation, but it serves as a simple example that demonstrates the power of quantum mechanics for computing.

[5] D. Deutsch, *Quantum Computational Networks*, Proc. Roy. Soc. London A **425** 73, 1989.

We can write this unitary operator in matrix form as

$$H = \frac{1}{\sqrt{2}} \begin{pmatrix} 1 & 1 \\ 1 & -1 \end{pmatrix}, \tag{6.36}$$

where H stands for "Hadamard" here, and not "Hamiltonian" as earlier. Those are two different operators, and you should not confuse them!

Notice how the Hadamard is mathematically identical to the beam splitter transformation in Chap. 2. If we label the Hadamard gate with subscripts according to the qubits they operate on, we can write the evolved state as

$$H_1 H_2 |0\rangle |1\rangle = \frac{1}{2} (|0\rangle + |1\rangle)(|0\rangle - |1\rangle). \tag{6.37}$$

Next, we apply the unitary operator U_f that implements the function f. We do not know how to write this as a unitary operator, because we do not know which of the four functions f_1, f_2, f_3, or f_4 we are implementing. We therefore need to be clever about it.

The function f affects the quantum state of the two qubits in a very specific way. Since we want to use unitary operators in the quantum computer, the computation must be reversible (remember that unitary transformations U always have an inverse U^{-1}, and can therefore always be reversed; that's what the inverse does). For a computation, we can make everything reversible by always keeping track of the input. This unitary operator U_f can be implemented for our function f in the following way:

$$|0\rangle |0\rangle \underset{U_f}{\rightarrow} |0\rangle |0 \oplus f(0)\rangle,$$

$$|0\rangle |1\rangle \underset{U_f}{\rightarrow} |0\rangle |1 \oplus f(0)\rangle,$$

$$|1\rangle |0\rangle \underset{U_f}{\rightarrow} |1\rangle |0 \oplus f(1)\rangle,$$

$$|1\rangle |1\rangle \underset{U_f}{\rightarrow} |1\rangle |1 \oplus f(1)\rangle. \tag{6.38}$$

We can write this more compactly as

$$|x\rangle |y\rangle \underset{U_f}{\rightarrow} |x\rangle |y \oplus f(x)\rangle, \tag{6.39}$$

where x and y can take the values 0 and 1.

We want to know what the function does to the input state in (6.37). To this end, we calculate the effect of f on $|0\rangle(|0\rangle - |1\rangle)$ and $|1\rangle(|0\rangle - |1\rangle)$ separately:

$$|0\rangle(|0\rangle - |1\rangle) \underset{U_f}{\to} |0\rangle(|0 \oplus f(0)\rangle - |1 \oplus f(0)\rangle),$$

$$|1\rangle(|0\rangle - |1\rangle) \underset{U_f}{\to} |1\rangle(|0 \oplus f(1)\rangle - |1 \oplus f(1)\rangle). \quad (6.40)$$

You may have noticed that we ignored the normalisation factor of 1/2. Since it is an overall factor, it does not affect the calculation, but strictly speaking we should carry it along if we want to calculate probabilities.

Since the outcome of $f(0)$ and $f(1)$ is a single bit, we can infer that

$$|0 \oplus f(0)\rangle - |1 \oplus f(0)\rangle = |0\rangle - |1\rangle \quad \text{if} \quad f(0) = 0, \quad (6.41)$$

or

$$|0 \oplus f(0)\rangle - |1 \oplus f(0)\rangle = -|0\rangle + |1\rangle \quad \text{if} \quad f(0) = 1. \quad (6.42)$$

We can write this compactly as

$$|0 \oplus f(0)\rangle - |1 \oplus f(0)\rangle = (-1)^{f(0)}(|0\rangle - |1\rangle), \quad (6.43)$$

where we use the trick that $(-1)^0 = 1$ and $(-1)^1 = -1$. Similarly, we find

$$|0 \oplus f(1)\rangle - |1 \oplus f(1)\rangle = (-1)^{f(1)}(|0\rangle - |1\rangle). \quad (6.44)$$

The state of the two qubits after applying the function f is therefore

$$\left[(-1)^{f(0)}|0\rangle + (-1)^{f(1)}|1\rangle\right](|0\rangle - |1\rangle). \quad (6.45)$$

Now if f is constant such that $f(0) = f(1)$, then up to an overall minus sign the first qubit will be in the state $|0\rangle + |1\rangle$. However, when f is balanced, $f(0) \neq f(1)$, and the state of qubit 1 becomes $|0\rangle - |1\rangle$. We can distinguish between these two states by applying another Hadamard to qubit 1, and measure whether the qubit is 0 or 1. If we find outcome 0 the function f is constant, and if we find 1, f is balanced. The steps of this calculation are shown in Fig. 6.6.

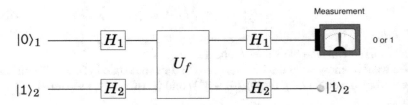

Fig. 6.6 Deutsch' algorithm in circuit form, with the Hadamard operators H_1 and H_2, and the single function call U_f (see supplementary material 6)

On a classical computer, we would have had to call the function twice, namely $f(0)$ and $f(1)$. However, on a quantum computer we call the function f only once, since we apply the unitary operator that gives us (6.39) only once.[6] So a quantum computer can solve problems with fewer calls to the function f. This becomes even more dramatic if we have a constant or balanced function with N input bits. The generalisation by Deutsch and Richard Jozsa[7] of the Deutsch algorithm, called the Deutsch–Jozsa algorithm, still allows you to find out whether the function is constant or balanced with a *single* call to the function f, but on a classical computer you may need as many as $N/2 + 1$ calls to the function f (if we take the worst case scenario). The quantum computer gives an enormous advantage when N is large, because it requires dramatically fewer steps in the computation (only one call to f).

You may wondering why is this interesting. The Deutsch–Jozsa algorithm is merely a toy problem, with little practical value. However, in 1994 Peter Shor[8] showed that a quantum computer can factorise large integers in fewer steps than a classical computer. He designed an algorithm for factoring an integer N that requires on the order of $(\log N)^3$ computational steps in a quantum computer to give the right result. By contrast, the best known classical algorithm to date requires on the order of N steps. So a quantum computer is exponentially faster than a classical computer in factoring integers. As an example, a classical computer can factor the number 15 in roughly 15 steps, while a quantum computer can factor it in $(\log_2 15)^3 = 60$ steps. This is not an improvement at all! But consider factoring a very large number of 2997 digits, say as shown in Fig. 6.7. The number of steps on a quantum computer is about 9×10^7. If the quantum computer has the same clock speed as a typical laptop, say 2 GHz, this calculation takes about 14 seconds. On the other hand, factoring the 2997 digit number on a classical computer with a clock speed of 2 GHz will take over 30 billion years, which is more than twice the age of the universe!

Shor's algorithm is important because the difficulty of factoring lies at the heart of modern cryptography. If we can build a quantum computer, we can crack pretty much all cryptographic codes that are used today. No wonder governments around the world are interested in building a quantum computer! Other applications of quantum computers include the design of new molecules, drugs, and other materials with predefined properties. Currently, we cannot calculate the properties of large atom configurations on classical computers because of their quantum mechanical character. However, these types of calculations would be easy for a quantum computer, since all it does is encode the rules of interaction of quantum systems, just like classical computers encode the rules that govern the behaviour of classical systems.

---

[6] You may wonder how we can implement U_f if we do not know f, but the point is that U_f is provided by some third party, or it is determined by some intermediate step in our computation.

[7] D. Deutsch and R. Jozsa, *Rapid solutions of problems by quantum computation*, Proc. Roy. Soc. London A **439** 553, 1992.

[8] P. W. Shor, *Polynomial-Time Algorithms for Prime Factorization and Discrete Logarithms on a Quantum Computer*, SIAM J. Comput. **26** 1484, 1997.

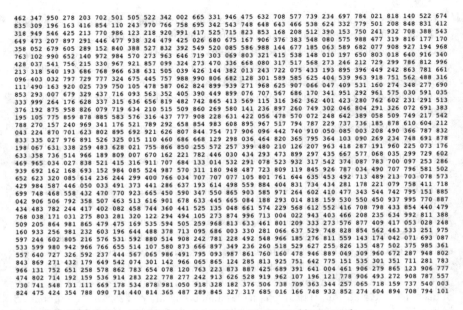

Fig. 6.7 A 2997 digit number that may be used in a cryptographic protocol

It is very difficult to build a large enough quantum computer to perform these tasks, and currently many people around the world are trying to figure out how to make it work. One of the main problems is the fragility of the qubits, which we study in the next chapter.

6.6 ✇ The No-Cloning Theorem and Quantum Cryptography

One of the most fundamental operations on a computer is the copying of bits. When a computer moves a bit from the working memory to the hard drive, it first copies the bit to the hard drive, and then erases the original in the memory. So for a brief moment in time there are two copies of the bit in the computer. In this section we will explore the process of copying of quantum bits, and in the next chapter we will discuss the erasure of quantum bits.

Mathematically, when we copy a bit we have two systems: the original bit and the destination bit, which may be in some initial state i. The copying process is then described as

$$(0, i) \rightarrow (0, 0) \quad \text{and} \quad (1, i) \rightarrow (1, 1).$$ (6.46)

We can set up a similar operation for quantum bits:

$$|0, i\rangle \to |0, 0\rangle \quad \text{and} \quad |1, i\rangle \to |1, 1\rangle . \tag{6.47}$$

This operation must be done with a unitary operator since it is an evolution of the qubit states, so we can write

$$U |0, i\rangle = |0, 0\rangle \quad \text{and} \quad U |1, i\rangle = |1, 1\rangle . \tag{6.48}$$

Next, we consider how to copy an arbitrary qubit state:

$$|\psi, i\rangle \to |\psi, \psi\rangle . \tag{6.49}$$

However, it is not possible to construct a unitary operation U that accomplishes this for any qubit state $|\psi\rangle$. To see this, we write $|\psi\rangle$ as a general superposition of $|0\rangle$ and $|1\rangle$:

$$|\psi\rangle = a |0\rangle + b |1\rangle , \tag{6.50}$$

with $|a|^2 + |b|^2 = 1$ the only restriction on a and b. Whichever way we implement the copying procedure U in (6.48), it *must* copy $|0\rangle$ and $|1\rangle$ correctly, so we can write

$$U |\psi, i\rangle = U(a |0\rangle + b |1\rangle) |i\rangle = U(a |0, i\rangle + b |1, i\rangle)$$
$$= a |0, 0\rangle + b |1, 1\rangle . \tag{6.51}$$

If U is a proper copying operation, then it must also produce

$$U |\psi, i\rangle = |\psi, \psi\rangle . \tag{6.52}$$

However,

$$|\psi, \psi\rangle = (a |0\rangle + b |1\rangle)(a |0\rangle + b |1\rangle)$$
$$= a^2 |0, 0\rangle + ab |0, 1\rangle + ab |1, 0\rangle + b^2 |1, 1\rangle . \tag{6.53}$$

This is not the same as $a |0, 0\rangle + b |1, 1\rangle$ for arbitrary a and b, so our U in (6.51) does not produce the correct quantum state, and is not a good copying operation for arbitrary qubits. This is the no-cloning theorem. It says that it is impossible to make a perfect copy of an unknown quantum state. Of course, when the state $|\psi\rangle$ is known, we can carefully design U so that it makes a copy of that particular state. But then it will not copy other states correctly.

There are two immediate consequences of the no-cloning theorem. First, the error correction that makes sure computers operate correctly requires that we make copies of fragile data. If we cannot copy the unknown quantum bits in a quantum computer, the necessary error correction is going to be a lot more complicated. Remarkably, quantum error correction is still possible. It is currently a very active area of research. Quantum computers already exist in various forms today, but the greatest challenge in scaling them up to powerful general purpose computing machines is the difficultly of implementing the quantum error correction.

Secondly, the fact that unknown qubits cannot be copied may be exploited in cryptography. If Alice and Bob wish to communicate in private, they can send quantum bits instead of classical bits. If an eavesdropper, Eve, tries to listen in, her attempts to copy the message will be noticed by Alice and Bob. Either she will get the information of the intercepted quantum bit, in which case Bob cannot have it due to no-cloning, or she does not get the information and may remain undetected. In the first case Bob raises the alarm and Eve is detected, and in the second case Alice and Bob shared a quantum bit without sharing any information with Eve. In the remainder of this section we will see exactly how Alice and Bob can set up an encrypted communication channel. This protocol was first proposed by Charlie Bennett and Giles Brassard in 1984, and is known as the BB84 protocol.

In order to understand BB84, we first need to know how cryptography works. Imagine I want to send a bit string with a message, for example "the password is 123456" in binary. By the way, don't use such an easily crackable password! In ASCII, this phrase is a series of zeros and ones. Imagine (part of) the message looking something like

$$m = 1001\ 1101\ 0010\ 1001\ ,$$

where m stands for "message". Next, we would like to scramble this message such that it makes sense only to those who know how to unscramble it. For this we need a key k that is another bit string of the same length as m:

$$k = 0100\ 1001\ 1010\ 1110\ ,$$

We scramble m by adding k modulo 2, leading to the encrypted message c:

$$
\begin{aligned}
m &= 1001\ 1101\ 0010\ 1001 \\
k &= \underline{0100\ 1001\ 1010\ 1110} \quad + \\
c = m \oplus k &= 1101\ 0100\ 1000\ 0111 \, .
\end{aligned}
\tag{6.54}
$$

If an eavesdropper intercepts c, then without knowing k it is not possible to reconstruct m. The message m can be anything, because k can be anything! Only Alice and Bob, who know k, can reconstruct m from c by adding k to c:

$$m = c \oplus k \, . \tag{6.55}$$

You should verify that adding the key k twice to the message m will produce m again. Equivalently, it means that $k \oplus k = \mathbf{0}$, the bit string of all zeros. To keep the communication secure, every time a message is sent, Alice and Bob should create a new key k, and never recycle old keys. Such a key is called a one-time pad. Recycling keys renders the cryptographic protocol insecure. There have been instances where using a key only twice allowed intelligence services to decrypt diplomatic communications. The key k also needs to be perfectly random, otherwise Eve may be able to guess parts of it. So the task is to come up with a method to share a random secret key k between Alice and Bob without anybody else gaining any knowledge about k. Remarkably, thanks to the no-cloning theorem this can be done in quantum mechanics. This procedure is called quantum key distribution, or QKD for short.

We have to set a few ground rules: first, we assume that Alice and Bob know that they are talking to each other, and not to imposters. That would be an authentication problem, which is different from the problem of secret key distribution we consider here. Second, we assume that a potential eavesdropper is limited only by the laws of physics, and that quantum physics is a correct description of reality. That way, if we can show that certain tasks are impossible in quantum mechanics, we deduce that they are impossible, full stop. The argument breaks if quantum mechanics turns out to be wrong in some very specific ways.

Alice and Bob proceed as follows: Alice sends Bob a series of qubits, for example polarised photons through optical fibres. Her photon source can send out the following four polarisation states:

$$|H\rangle, \quad |V\rangle, \quad |D\rangle = \frac{|H\rangle + |V\rangle}{\sqrt{2}}, \quad \text{and} \quad |A\rangle = \frac{|H\rangle - |V\rangle}{\sqrt{2}}, \tag{6.56}$$

where D and A denote the diagonal and anti-diagonal (linear) polarisations. Bob receives these photons, and measures either in the polarisation basis $\{|H\rangle, |V\rangle\}$ or $\{|D\rangle, |A\rangle\}$. The essential operation is now that for each individual photon Alice chooses *randomly* which of the four states she prepares and sends to Bob, recording exactly what she sent for future reference. Bob, for his part, chooses *randomly* in which polarisation basis he chooses to measure each incoming photon. Alice and Bob choose the preparation and measurements locally, so no knowledge of their procedure leaks out. This is critical for the security of the protocol (eavesdropping on the preparation or measurement is called a "side-chain" attack).

Alice will have to record two bits for each photon she sends, since there are four possibilities. One bit records whether the state is from the $\{|H\rangle, |V\rangle\}$ basis ("0") or from the $\{|D\rangle, |A\rangle\}$ basis ("1")). The second bit records whether the photon was $|H\rangle$ or $|D\rangle$ ("0"), or $|V\rangle$ or $|A\rangle$ ("1"). Therefore, the bit string 10 denotes the state $|D\rangle$, and the bit string 01 denotes the state $|V\rangle$, etc. Similarly, Bob will record which basis he used to measure each incoming photon, as well as the outcome for each measurement using the same binary convention. After sending N photons, both Alice and Bob have two bit strings, one indicating the basis and one indicating the bit value in that basis. They need to keep the bit values private, but they can share the basis bit string publicly after they have completed the key distribution process. They compare where they used the same basis, and in those cases where Alice and Bob prepared and measured in the same basis the values in their private bit strings should agree. Those bit values form the secret key k.

As an example, consider the string of twelve photons Alice sends to Bob shown in the table below. Alice records the basis as a bit string of zeros and ones, and separately records the bit value:

Alice's photon	H	V	D	V	A	H	H	D	A	V	D	A	
Basis	0	0	1	0	1	0	0	1	1	0	1	1	←
Bit value	0	1	0	1	1	0	0	0	1	1	0	1	
Bob's measurement	H	D	D	A	V	H	A	H	A	V	H	A	
Basis	0	1	1	1	0	0	1	0	1	0	0	1	←
Bit value	0	0	0	1	1	0	1	0	1	1	0	1	
Secret key	0	–	0	–	–	0	–	–	1	1	–	1	

Bob measures the polarisation of the photons and records the basis and the bit values as well. The bit values are kept secret, but Alice and Bob publicly compare the values of the basis. Where they are identical they pick the bit value as the next bit in the secret key. The bit values for those photons where Alice and Bob used different bases are discarded. So the final key is 000111.

To see why this is secure, we consider what happens when Eve tries to intercept the photons. From the no-cloning theorem we know that Eve cannot make a perfect copy of every photon that Alice sends to Bob. To gain information about the polarisation of each photon, Eve can measure the photon randomly in the $\{|H\rangle, |V\rangle\}$ basis or the $\{|D\rangle, |A\rangle\}$ basis, just like Bob. She then prepares a photon with the same polarisation indicated by her measurement and sends it to Bob. In principle, Eve can do this incredibly fast, so Bob doesn't realise she is on the line executing a so-called "man in the middle" attack.

However, Eve has to pick random polarisation bases, since she does not know which bases Alice or Bob choose for their measurements. So there will be situations where Alice and Bob happen to pick the same basis, but Eve picks the "wrong" basis. In that case, Alice and Bob think that they have a "good" bit value that will contribute to the secret key, but Eve's measurement has messed this up. For example, Alice sends a photon $|H\rangle$ and Bob measures in the $\{|H\rangle, |V\rangle\}$ basis. He should be measuring $|H\rangle$. Eve, having measured in the $\{|D\rangle, |A\rangle\}$ basis, found the polarisation $|D\rangle$ or $|A\rangle$, so she sends $|D\rangle$ or $|A\rangle$ to Bob. Unfortunately for Eve, the probability that Bob measures the required $|H\rangle$ is only one half:

$$\Pr(H) = |\langle H|D\rangle|^2 = \frac{1}{2} \quad \text{and} \quad \Pr(H) = |\langle H|A\rangle|^2 = \frac{1}{2}. \tag{6.57}$$

Therefore, Eve's man in the middle attack leads to errors in the secret key. Bob would sometimes measure $|V\rangle$ where he should measure $|H\rangle$. Eve is undetectable in this attack when she picks the same basis as Alice or Bob.

Alice and Bob have to compare a part of their secret key to make sure there are no discrepancies. They have to announce part of the key publicly, since they do not yet have a private communication channel (that's why they try to generate the secret key!). If Eve was on the line, on average a quarter of the bits in the secret key are wrong. When Alice and Bob find that many errors, Eve is detected and they cannot use the key for private communication. They will need to use a different channel.

In any practical implementation of the BB84 protocol there will always some discrepancies due to experimental imperfections. As long as the error rate is small enough (no larger than about 10%), Alice and Bob can recover a perfectly good secret key by a further process called privacy amplification. Importantly, these errors may be due to Eve trying to eavesdrop on part of the key, and the privacy amplification ensures that she *still* will not learn anything about the key. Many countries around the world are developing quantum key distribution technology to protect private communication against eavesdroppers of all kinds. Once quantum computers become powerful enough the traditional key distribution that is broken by prime factoring is no longer secure, and when this happens global trade and defence will need an alternative way to establish secure communication.

6.7 ✎ Entanglement Generation Over Large Distances

In the future, when we have quantum computers we will almost certainly want to connect them over large distances, creating the Quantum Internet. Also, we want to establish secret keys between parties that are hundreds or even thousands of kilometres apart. The obvious way to send quantum bits over long distances is to encode them in photons. However, while photons do like to travel fast, they have a tendency of being absorbed or scattered in the optical fibres they travel through. To get an idea of the severity of this effect, consider the attenuation length L_0 of an optical fibre, which is the length of fibre at which the probability of a photon passes unabsorbed is $1/e \simeq 37\%$. The probability that a photon passes through a fibre of length L without being absorbed is then

$$\Pr(L) = e^{-L/L_0}. \tag{6.58}$$

Very good fibres can have an attenuation length of tens of kilometres, so the probability of a photon making it through a fibre 1 000 km long with an attenuation length of, for example, 20 km is

$$\Pr(1\,000\ \text{km}) = e^{-1\,000/20} = e^{-50} \approx 10^{-22}. \tag{6.59}$$

This is vanishingly small. Each bit would have to be sent around 10^{22} times in order for one photon to make it through. A typical 1 mW laser pointer operating at 650 nm has an output of around 10^{15} photons per second. To pass on average one photon per second through this fibre, the power needs to be increased by a factor of ten million, leading to a laser power of ten megawatts. A laser this bright will *evaporate* the fibre.

So we need a different method to get photons across large distances. Classical fibre-optic communication uses so-called repeaters. These devices operate by amplifying the signal at set intervals along the fibre. However, amplification is a type of copying, and this is impossible for quantum bits as we have seen in the previous section. The solution is an altogether different approach, called a *quantum repeater*.

One way to get a qubit across a very large distance is to set up an entangled pair between the end points, Alice and Bob, and use quantum teleportation to teleport the qubit from Alice to Bob. At first, this does not seem to solve the problem, because how do we set up entanglement between Alice and Bob in the first place? However, since we know exactly what kind of quantum state we want to share between Alice and Bob (as opposed to sending an unknown qubit), we can circumvent the no-cloning theorem.

First, we consider a variation of quantum teleportation called entanglement swapping. Imagine three parties, Alice, Bob and Charlie. Charlie is located halfway between Alice and Bob, and we are going to assume that the distance d between Charlie and Alice, and between Charlie and Bob is small enough that photons in fibres will not get lost too often. Charlie shares an entangled qubit pair with Alice, and also with Bob. Therefore, Charlie holds two qubits while Alice and Bob each hold one qubit. Moreover, both entangled pairs are in the state $(|00\rangle + |11\rangle)/\sqrt{2}$. The total quantum state held by Alice, Bob, and Charlie is then

$$|\Psi\rangle = \frac{|00\rangle_{AC_1} + |11\rangle_{AC_1}}{\sqrt{2}} \frac{|00\rangle_{C_2B} + |11\rangle_{C_2B}}{\sqrt{2}}, \qquad (6.60)$$

where the labels A, B, C_1 and C_2 denote the qubits held by Alice, Bob, and Charlie. Next, Charlie will teleport his qubit C_1 to Bob by making a Bell measurement on the qubits C_1 and C_2. We rewrite the state $|\Psi\rangle$ as

$$|\Psi\rangle = \frac{1}{2} \left(|00\rangle_{AC_1} |00\rangle_{C_2B} + |00\rangle_{AC_1} |11\rangle_{C_2B} + |11\rangle_{AC_1} |00\rangle_{C_2B} + |11\rangle_{AC_1} |11\rangle_{C_2B} \right)$$

$$= \frac{1}{2} \left(|00\rangle_C |00\rangle_{AB} + |01\rangle_C |01\rangle_{AB} + |10\rangle_C |10\rangle_{AB} + |11\rangle_C |11\rangle_{AB} \right), \qquad (6.61)$$

where we abbreviated $C = C_1C_2$. Next, we can write the computational basis again in terms of the Bell states, as we did in (6.23):

$$|00\rangle_C = \frac{|\Phi^+\rangle_C + |\Phi^-\rangle_C}{\sqrt{2}} \quad \text{and} \quad |01\rangle_C = \frac{|\Psi^+\rangle_C + |\Psi^-\rangle_C}{\sqrt{2}},$$

$$|11\rangle_C = \frac{|\Phi^+\rangle_C - |\Phi^-\rangle_C}{\sqrt{2}} \quad \text{and} \quad |10\rangle_C = \frac{|\Psi^+\rangle_C - |\Psi^-\rangle_C}{\sqrt{2}}. \qquad (6.62)$$

The four-qubit state $|\Psi\rangle$ then becomes

$$|\Psi\rangle = \frac{1}{2} |\Phi^+\rangle_C \left(\frac{|00\rangle_{AB} + |11\rangle_{AB}}{\sqrt{2}} \right) + \frac{1}{2} |\Phi^-\rangle_C \left(\frac{|00\rangle_{AB} - |11\rangle_{AB}}{\sqrt{2}} \right)$$

$$+ \frac{1}{2} |\Psi^+\rangle_C \left(\frac{|01\rangle_{AB} + |10\rangle_{AB}}{\sqrt{2}} \right) + \frac{1}{2} |\Psi^-\rangle_C \left(\frac{|01\rangle_{AB} - |10\rangle_{AB}}{\sqrt{2}} \right). \qquad (6.63)$$

After the Bell measurement, the remaining two-qubit states held by Alice and Bob are

$$|\Phi^+\rangle_C : \quad \frac{|00\rangle_{AB} + |11\rangle_{AB}}{\sqrt{2}}$$

$$|\Phi^-\rangle_C : \quad \frac{|00\rangle_{AB} - |11\rangle_{AB}}{\sqrt{2}}$$

$$|\Psi^+\rangle_C : \quad \frac{|01\rangle_{AB} + |10\rangle_{AB}}{\sqrt{2}}$$

$$|\Psi^-\rangle_C : \quad \frac{|01\rangle_{AB} - |10\rangle_{AB}}{\sqrt{2}}. \qquad (6.64)$$

In other words, Alice and Bob now share an entangled qubit pair, even though their qubits have never interacted. More importantly, we have established entanglement over a distance $2d$. Alice can use this shared entanglement with Bob to teleport a qubit of her choice to him.

We can repeat this procedure many times, with not just Charlie in between Alice and Bob, but also Dave, Erin, Freddie, etc. This way, we can extend the reach of quantum communication arbitrarily far, using N legs with $N - 1$ entanglement swapping protocols. One important caveat is that we must first verify that no photons have been lost at each leg of the protocol. However, if we try to establish just one entangled photon pair for each leg, then the probability that *all* legs have a photon pair is p^N, with p the probability that a photon makes across the distance d. This gets us right back where we started, since p^N approaches zero exponentially fast when $p < 1$. Instead, for each leg we must attempt to share several entangled photon pairs so that we can be confident that with near certainty at least one pair made it through (see Fig. 6.8). This means that the receiving party, e.g., Charlie, must check that the photon successfully made it through the optical fibre *before* deciding which photons are to be used in the Bell measurement. This creates a complicated chain of classical communication between the repeater stations, which must be carefully taken into account. In particular, the photons must be stored in quantum memories while Charlie and the other repeater stations work out on which photon pairs they must perform their Bell measurements.

Once Alice and Bob established shared entanglement over a distance Nd through the use of $N - 1$ repeaters, they will have to make sure that no errors have crept into their entangled state. In practice, with so many repeater stations this is virtually unavoidable. They must therefore take the noisy entangled pairs they share, and perform so-called entanglement distillation to obtain near-perfect entangled pairs.

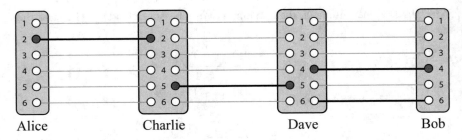

Fig. 6.8 Alice and Bob share entanglement by using two repeater stations, Charlie and Dave. Alice sends Charlie enough photons such that one photon is likely to make it through. In this case they share the entangled photon pair 2. Charlie and Dave do the same, and they establish that a photon from pair 5 made it across unabsorbed. Finally, Dave and Bob were luck and managed to share two entangled photon pairs, of which they pick pair number 4 (via some pre-established decision process). Charlie will now perform a Bell measurement on photons 2 and 5, and Dave will perform a Bell measurement on photons 5 and 4. After they share the measurement outcomes with everybody else, Alice and Bob share an entangled state over a much larger distance than the distance between repeater stations

Exercises

1. Consider two electrons with spin 1/2. Indicate whether the following states are separable or entangled:

   (a) $|\Psi\rangle = (|\uparrow\downarrow\rangle + |\downarrow\uparrow\rangle + \sqrt{2}|\downarrow\downarrow\rangle)/2$
   (b) $|\Psi\rangle = \frac{1}{2}|\uparrow\uparrow\rangle + \frac{1}{2}|\uparrow\downarrow\rangle + \frac{1}{2}|\downarrow\uparrow\rangle - \frac{1}{2}|\downarrow\downarrow\rangle$
   (c) $|\Psi\rangle = \frac{1}{\sqrt{15}}|\uparrow\uparrow\rangle + \frac{2}{\sqrt{15}}|\uparrow\downarrow\rangle + \frac{\sqrt{2}}{\sqrt{15}}|\downarrow\uparrow\rangle + \frac{2\sqrt{2}}{\sqrt{15}}|\downarrow\downarrow\rangle$

   If the state is separable, give the states of the individual systems.
2. Calculate the eigenvalues and the eigenstates of the bit flip operator X, and show that the eigenstates form an orthonormal basis. Calculate the expectation value of X for $|\psi\rangle = 1/\sqrt{3}|0\rangle + i\sqrt{2/3}|1\rangle$.
3. Consider a qubit in an arbitrary pure state $|\psi\rangle_1 = a|0\rangle_1 + b|1\rangle_1$, and a second qubit in the state $|+\rangle_2 = (|0\rangle_2 + |1\rangle_2)/\sqrt{2}$.

   (a) Calculate the two-qubit state after applying a CZ gate, defined by

   $$|00\rangle \rightarrow |00\rangle, \qquad |10\rangle \rightarrow |10\rangle,$$
   $$|01\rangle \rightarrow |01\rangle, \qquad |11\rangle \rightarrow -|11\rangle. \qquad (6.65)$$

   (b) Apply a Hadamard gate to both qubits and then perform a measurement in the {0, 1} basis. What is are the resulting states for qubit 2? This is sometimes called "local teleportation".
   (c) The Hadamard followed by a measurement of 0, 1 is the same as a measurement of +, −. Show that a measurement in the equatorial plane of the

Bloch sphere induces a rotation around the axis defined by $\{0, 1\}$ on the second qubit.

(d) Show that by daisy-chaining three local teleportation events we can induce any single qubit gate on the input qubit $|\psi\rangle$.

4. Show that the following states form an orthonormal basis for two qubits:

$$|\phi_1\rangle = \frac{1}{\sqrt{2}}|0, 1\rangle + \frac{1}{\sqrt{2}}|1, 0\rangle , \tag{6.66}$$

$$|\phi_2\rangle = \frac{1}{\sqrt{2}}|0, -\rangle + \frac{1}{\sqrt{2}}|1, +\rangle , \tag{6.67}$$

$$|\phi_3\rangle = \frac{1}{\sqrt{2}}|+, 1\rangle + \frac{1}{\sqrt{2}}|-, 0\rangle , \tag{6.68}$$

$$|\phi_4\rangle = \frac{1}{\sqrt{2}}|+, -\rangle + \frac{1}{\sqrt{2}}|-, +\rangle . \tag{6.69}$$

5. Show that we can get any Bell state in (6.19) from any other using only a *local* unitary operation on one of the qubits.

6. Consider the two-qubit state $|\Phi^+\rangle = (|00\rangle + |11\rangle)/\sqrt{2}$. The evolution $U = \exp(i\omega t |1\rangle\langle 1|)$ is applied to both qubits. Show that the time it takes for $|\Phi^+\rangle$ to evolve back to itself is twice as short as when U is applied to only one qubit.

7. Construct a matrix that relates the Bell basis in (6.19) to the spin basis in (6.23), and show that this matrix is unitary.

8. Consider an eavesdropping attack on the BB84 protocol where Eve tries to copy the qubit sent by Alice to Bob, only to make her measurement on the qubit *after* Alice and Bob announced their respective preparation and measurement bases. She uses the copying operation from (6.48).

(a) Show that Eve can gain information about the key without being detected when Alice sends photons with polarisation $|H\rangle$ or $|V\rangle$.

(b) Show that this same procedure will reveal her presence 50% of the time when Alice sends photons with polarisation $|D\rangle$ or $|A\rangle$.

9. For a repeater with N legs, m entangled photon pairs per leg, and a probability p that a photon makes it across the fibre in one leg, derive an expression for the failure probability of the repeater.

10. For the example of fibre attenuation length of 20 km, a distance of 1 000 km between Alice and Bob, and a maximum failure rate to establish a photon of 1%, calculate how many photons on average are needed if there are 99 repeaters between Alice and Bob. Compare this to the number of photons in the 10 MW laser in the text.

Chapter 7
Decoherence

In this chapter, we ask what happens when a quantum system is in contact with its environment. To describe this, we need a more general concept of the quantum state, namely the density matrix. We study the density matrix, and how it describes both classical and quantum uncertainty. This also naturally leads to a description of quantum systems at a non-zero temperature. We conclude this chapter with a short discussion on how entropy arises in quantum mechanics.

7.1 Classical and Quantum Uncertainty

In Chap. 1 we considered three simple experiments that went right to the heart of quantum mechanics. In the first two experiments we studied how a very weak laser triggered photodetectors, and we found that light comes in indivisible packets. This led us to hypothesise that light consists of particles, called photons. In the third experiment we set up a Mach-Zehnder interferometer that ensured the photon ended up in detector D_2, and never in detector D_1. When we use high-intensity laser light, this behaviour is completely explained by the classical wave theory of light, and is called interference. We then arrived at the uncomfortable conclusion that a single particle (the photon) can exhibit interference. To explore this further, we modified the third experiment to add quantum non demolition detectors in the paths of the interferometer. This served to tell us which path the photons took inside the interferometer. Consequently, the interference pattern was destroyed: the photons no longer end up only in detector D_2, but randomly trigger D_1 or D_2. We concluded

Supplementary Information The online version contains supplementary material available at https://doi.org/10.1007/978-3-031-16165-0_7.

143
P. Kok, *A First Introduction to Quantum Physics*, Undergraduate Lecture Notes in Physics, https://doi.org/10.1007/978-3-031-16165-0_7

that photons are neither particles, nor waves, and make up a new type of physical objects called quanta with characteristics of both.

In this section, we will study this last experiment a little further to establish what kind of knowledge we can have about quanta. We will arrive at the notion that there are two kinds of uncertainty: classical and quantum uncertainty. To see this, consider the experiment where we measure the path taken by the photon. In this case we lose the interference and find photons hitting both detector D_1 and detector D_2. We can imagine that this experiment is some setup on a bench, with lasers, beam splitters, mirrors, and detectors that are merrily clicking away. Suppose that the outcomes of the photodetectors D_1 and D_2 as well as the which-path QND detectors are recorded in two data files on a computer, one for the photodetectors, and one for the QND detectors. After a while of taking data we are getting a bit thirsty. Since our presence is not required for the correct running of our automated experiment, we decide to leave the lab briefly for a cup of coffee.

When we come back, we discover that disaster has struck! One of the data files on the hard drive of our computer is corrupted. The file with recorded data for D_1 and D_2 is fine, and we see that the photons indeed kept going randomly to D_1 and D_2. However, the data file for the QND detectors has been corrupted and can no longer be read out. This means that we no longer have the information which path the photons took. Nevertheless, as the data file of detectors D_1 and D_2 showed, we did not recover the interference (detector D_1 kept firing, as well as detector D_2). This is completely in agreement with common sense, because if we deliberately destroy the which-path information ourselves we do not expect to change events that we already observed. That would wreak havoc with causality!

However, this thought experiment has interesting implications. In one experiment we do not measure the path of the photon and obtain interference in the detector outcomes. We have no knowledge about the path of the photon. In the other experiment we lose the information about the path of the photon, and consequently have no knowledge about the path either. However, we have lost the interference in the detector outcomes. Since the two experiments have different implications for the interference in detectors D_1 and D_2, these must be two different kinds of uncertainty. The case where we lose the information about the path we call classical uncertainty, because we know that the photon has taken a well-defined path, and we are just uncertain as to which it was. On the other hand, when we do not make a path measurement in the first place, we cannot say that the photon took either one of the two possible paths. The path of the photon is indeterminate. This is quantum uncertainty.

We can ask the question whether it is important that a measurement was made. What if the data file was not corrupted, but instead the outcomes weren't send to the computer in the first place, perhaps due to a faulty cable? Maybe the electronics in the measurement apparatus was not working properly. It is easy to see that we can push the fault in the measurement apparatus closer and closer to the key aspect of what makes the which-path detector a measurement, namely the interaction of the apparatus with the photon. If there is an interaction of the photon with another system, and the occurrence of that interaction is stored *anywhere*, from a classical data file to the quantum state of the environment, then we lose the destructive interference in D_1. The loss of interference in the detectors indicates that something has changed in the state of the photon as a result of the interaction with the measurement apparatus. This change is called *decoherence*.

7.2 The Density Matrix

In the previous section we have seen that the interaction of a photon with a measurement apparatus or with the environment will cause the disappearance of interference, and the uncertainty about the path of the photon in the Mach–Zehnder interferometer is no longer quantum but classical. In this section we will see how we can describe classical uncertainties mathematically in terms of an object called the density matrix, and later in this chapter we will show how to describe the situation of the previous section.

We introduce classical uncertainty in our description of the state of a photon in the following way: Every time we send a single photon into a Mach–Zehnder interferometer, we toss a coin on whether or not we remove the first beam splitter BS1. When the beam splitter is present, detector D_2 will always click, as before. When the beam splitter is not present, the photon will always be in the upper arm of the interferometer (technically it is no longer a Mach–Zehnder interferometer), and the second beam splitter BS2 will make the photon end up in detectors D_1 and D_2, each with probability $\frac{1}{2}$. This is shown in Fig. 7.1.

The uncertainty in the measurement outcomes in the absence of BS1 is a quantum uncertainty, since prior to the detection the photon is in the superposition of moving towards D_1 and moving towards D_2. Mathematically, this state is

$$|\neg\text{BS1}\rangle = \frac{1}{\sqrt{2}}|D_1\rangle + \frac{1}{\sqrt{2}}|D_2\rangle = \frac{1}{\sqrt{2}}\begin{pmatrix} 1 \\ 1 \end{pmatrix}, \tag{7.1}$$

where the symbol $\neg$ means "not". On the other hand, when BS1 is present, there is no uncertainty about which detector the photon will trigger:

$$|\text{BS1}\rangle = |D_2\rangle = \begin{pmatrix} 1 \\ 0 \end{pmatrix}. \tag{7.2}$$

The probability that detector D_2 will be triggered by the photon is different in both cases, and we can calculate this as

$$\Pr(D_2|\text{BS1}) = |\langle D_2|\text{BS1}\rangle|^2 = 1$$

$$\Pr(D_2|\neg\text{BS1}) = |\langle D_2|\neg\text{BS1}\rangle|^2 = \frac{1}{2}. \tag{7.3}$$

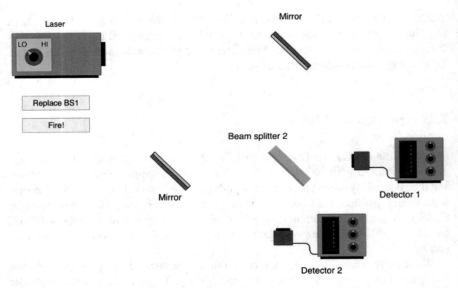

Fig. 7.1 A Mach–Zehnder interferometer with a removable beam splitter (see supplementary material 1)

The vertical bar in Pr $(\cdot|\cdot)$ should be read as "given that", so Pr $(a|b)$ is the probability that a happens, given that b occurred. This is a so-called conditional probability.

So far, we have looked only at the detection probability in D_2 for each separate arrangement of the interferometer. However, what if we do not know whether BS1 was removed or not, but we only know that a (fair) coin was used to determine if BS1 is present? In that case, we have a classical uncertainty about the setup. The classical rules of probability then dictate that we add the probabilities Pr $(D_2|BS1)$ and Pr $(D_2|\neg BS1)$, weighed by the probabilities of BS1 being present or absent:

$$\Pr(D_2) = \frac{1}{2}\Pr(D_2|BS1) + \frac{1}{2}\Pr(D_2|\neg BS1),\qquad(7.4)$$

where we used that

$$\Pr(BS1) = \Pr(\neg BS1) = \frac{1}{2}.\qquad(7.5)$$

Note that these probabilities are not conditioned; they are the straight-up probabilities of whether the first beam splitter is there or not.

We can rewrite (7.4) by substituting the expressions of the modulo-squared scalar products in (7.3), and we arrive at

$$\text{Pr}(D_2) = \frac{1}{2}|\langle D_2|\text{BS1}\rangle|^2 + \frac{1}{2}|\langle D_2|\neg\text{BS1}\rangle|^2$$

$$= \frac{1}{2}\langle D_2|\text{BS1}\rangle\langle\text{BS1}|D_2\rangle + \frac{1}{2}\langle D_2|\neg\text{BS1}\rangle\langle\neg\text{BS1}|D_2\rangle$$

$$= \langle D_2|\left(\frac{1}{2}|\text{BS1}\rangle\langle\text{BS1}| + \frac{1}{2}|\neg\text{BS1}\rangle\langle\neg\text{BS1}|\right)|D_2\rangle$$

$$= \langle D_2|\rho|D_2\rangle . \tag{7.6}$$

where

$$\rho = \frac{1}{2}|\text{BS1}\rangle\langle\text{BS1}| + \frac{1}{2}|\neg\text{BS1}\rangle\langle\neg\text{BS1}| \tag{7.7}$$

is called the density matrix, or density operator. You see that it plays the role of the quantum state of the system in the calculation of the probability of the measurement outcome D_2, and that it includes the classical uncertainty in the form of the factors $\frac{1}{2}$. We can write ρ in matrix notation by using

$$|\text{BS1}\rangle\langle\text{BS1}| = \begin{pmatrix} 1 \\ 0 \end{pmatrix}\begin{pmatrix} 1 & 0 \end{pmatrix} = \begin{pmatrix} 1 & 0 \\ 0 & 0 \end{pmatrix}, \tag{7.8}$$

and

$$|\neg\text{BS1}\rangle\langle\neg\text{BS1}| = \frac{1}{2}\begin{pmatrix} 1 \\ 1 \end{pmatrix}\begin{pmatrix} 1 & 1 \end{pmatrix} = \frac{1}{2}\begin{pmatrix} 1 & 1 \\ 1 & 1 \end{pmatrix}. \tag{7.9}$$

The density matrix is then

$$\rho = \frac{1}{2}|\text{BS1}\rangle\langle\text{BS1}| + \frac{1}{2}|\neg\text{BS1}\rangle\langle\neg\text{BS1}|$$

$$= \frac{1}{2}\begin{pmatrix} 1 & 0 \\ 0 & 0 \end{pmatrix} + \frac{1}{2}\times\frac{1}{2}\begin{pmatrix} 1 & 1 \\ 1 & 1 \end{pmatrix} = \frac{1}{4}\begin{pmatrix} 3 & 1 \\ 1 & 1 \end{pmatrix}. \tag{7.10}$$

The probability of finding the photon in detector D_2 when we are uncertain about the presence or absence of BS1 is then

$$\text{Pr}(D_2) = \langle D_2|\rho|D_2\rangle = \frac{1}{4}\begin{pmatrix} 1 & 0 \end{pmatrix}\begin{pmatrix} 3 & 1 \\ 1 & 1 \end{pmatrix}\begin{pmatrix} 1 \\ 0 \end{pmatrix} = \frac{1}{4}\begin{pmatrix} 1 & 0 \end{pmatrix}\begin{pmatrix} 3 \\ 1 \end{pmatrix} = \frac{3}{4}. \tag{7.11}$$

Similarly, we can calculate the probability of finding the photon in detector D_1 using the same density matrix:

$$\text{Pr}(D_1) = \langle D_1|\rho|D_1\rangle = \frac{1}{4}\begin{pmatrix} 0 & 1 \end{pmatrix}\begin{pmatrix} 3 & 1 \\ 1 & 1 \end{pmatrix}\begin{pmatrix} 0 \\ 1 \end{pmatrix} = \frac{1}{4}\begin{pmatrix} 0 & 1 \end{pmatrix}\begin{pmatrix} 1 \\ 1 \end{pmatrix} = \frac{1}{4}. \tag{7.12}$$

A quick check reveals that the probability of finding the photon in either D_1 or D_2 is

$$\frac{1}{4} + \frac{3}{4} = 1,\tag{7.13}$$

as it should be, since there are no other possibilities for the photon to behave (and we assumed lossless detectors). Note that while the density matrix describes the state of a quantum system, it is no longer a vector. This is because vectors do not have the room to mathematically encode both the quantum and the classical uncertainty.

Perhaps the coin we used to decide whether to include BS1 was not fair. We can assign probabilities that are different from 1/2, for example p_1 and p_2, with $p_1 + p_2 = 1$. If for brevity we write the two different photon states as $|\psi_1\rangle$ and $|\psi_2\rangle$, the density matrix becomes

$$\rho = p_1|\psi_1\rangle\langle\psi_1| + p_2|\psi_2\rangle\langle\psi_2|,\tag{7.14}$$

where $|\psi_1\rangle$ and $|\psi_2\rangle$ do not have to be orthogonal (for example, $|BS1\rangle$ and $|\neg BS1\rangle$ are not orthogonal; you should verify this!).

In the extreme case where $p_1 = 1$ and $p_2 = 0$, there is no classical uncertainty, and all probabilistic measurement outcomes are due to quantum uncertainty. There is a quick way to test whether a density matrix has classical uncertainty. Calculate the square of ρ :

$$\rho^2 = (p_1|\psi_1\rangle\langle\psi_1| + p_2|\psi_2\rangle\langle\psi_2|)^2\tag{7.15}$$
$$= p_1^2|\psi_1\rangle\langle\psi_1| + p_2^2|\psi_2\rangle\langle\psi_2| + p_1 p_2 \left(\langle\psi_1|\psi_2\rangle|\psi_1\rangle\langle\psi_2| + \langle\psi_2|\psi_1\rangle|\psi_2\rangle\langle\psi_1|\right).$$

In the special case where $p_1 = 1$ and $p_2 = 0$ we see that

$$\rho^2 = \rho.\tag{7.16}$$

In Chap. 5 we learned that this is the defining characteristic of a projection operator (along with being a Hermitian operator, which ρ also is). So if there is no classical uncertainty, the quantum state of the system can be described by a projection operator. The converse is also true: if a quantum system is described by a projection operator, then there is no classical uncertainty in the system, and all probabilistic measurement outcomes are due to quantum uncertainty. When $\rho^2 = \rho$, we call the state of the system pure, and when $\rho^2 \neq \rho$ we call the state mixed:

$$\rho^2 = \rho \;\rightarrow\; \rho \text{ is a pure state,}$$
$$\rho^2 \neq \rho \;\rightarrow\; \rho \text{ is a mixed state.}\tag{7.17}$$

The terminology "mixed" always refers to classical uncertainty, while the term "superposition" refers to quantum uncertainty. You should never call a superposition a "mixture" of possibilities, because that will lead to confusion.

Some quantum states are more mixed than others, in the sense that they contain a greater classical uncertainty. We can quantify this by calculating the trace of the square of the density matrix, which is the sum over the diagonals. For the density matrix in (7.10) we calculate

$$\rho^2 = \frac{1}{16} \begin{pmatrix} 3 & 1 \\ 1 & 1 \end{pmatrix} \begin{pmatrix} 3 & 1 \\ 1 & 1 \end{pmatrix} = \frac{1}{16} \begin{pmatrix} 10 & 4 \\ 4 & 2 \end{pmatrix}. \tag{7.18}$$

The trace of ρ^2 is then given by

$$\text{Tr}\left(\rho^2\right) = \frac{10}{16} + \frac{2}{16} = \frac{12}{16} = \frac{3}{4}. \tag{7.19}$$

This is called the *purity* of the state ρ. By contrast, the pure state for the photon coming out of the interferometer without BS1 can be written as the density matrix

$$\rho_{\neg BS1} = \frac{1}{2} \begin{pmatrix} 1 & 1 \\ 1 & 1 \end{pmatrix}. \tag{7.20}$$

Moreover, since there is no classical uncertainty about the absence of BS1 the quantum state is pure and $\rho_{\neg BS1} = \rho^2_{\neg BS1}$. The trace of the diagonal elements of $\rho^2_{\neg BS1}$ is therefore

$$\frac{1}{2} + \frac{1}{2} = 1. \tag{7.21}$$

pure states have a purity of 1.

There are also maximally mixed states. These are quantum states with maximum classical uncertainty, where we know nothing about the path of the photon. We can construct such a state by sending photons at detectors D_1 and D_2 based on the random outcome of a fair coin. The density matrix for this state is

$$\rho_{max} = \frac{1}{2} \begin{pmatrix} 1 & 0 \\ 0 & 1 \end{pmatrix}. \tag{7.22}$$

The purity of this state is $\frac{1}{2}$. It cannot be lower for systems with two orthogonal states.

Note that the trace of ρ is always equal to 1. This is because the eigenvalues of ρ are the probabilities of finding the measurement outcomes associated with the eigenvectors of ρ. In addition, since probabilities are always real numbers between zero and one, density matrices have real eigenvalues. That makes them Hermitian matrices, for which $\rho^\dagger = \rho$. Also, the expectation value of ρ with respect to any quantum state must be positive, since it is the probability of finding that state in a measurement of the quantum system. This property of the density matrix is called positivity. Every density matrix in quantum mechanics has the following properties:

1. $\mathrm{Tr}\,(\rho) = 1$ (normalisation),
2. $\rho^\dagger = \rho$ (Hermiticity),
3. $\langle\psi|\rho|\psi\rangle \geq 0$ for any $|\psi\rangle$ (positivity).

Any matrix that satisfies these requirements is a valid density matrix.

So far we have shown how we can calculate the probabilities of measurement outcomes given a density matrix ρ. However, we also need to know how we can calculate the expectation value for an observable if our system is in the state ρ. Let our density matrix be given by a generalisation for (7.14):

$$\rho = \sum_j p_j |\psi_j\rangle\langle\psi_j|, \tag{7.23}$$

where j runs over all possible states $|\psi_j\rangle$ that are prepared with probability p_j. The expectation value of an operator A with respect to one of the states $|\psi_j\rangle$ is given by

$$\langle A\rangle_j = \langle\psi_j|A|\psi_j\rangle. \tag{7.24}$$

When the preparation procedure is a probability distribution over all $|\psi_j\rangle$, the expectation value of A must be the weighted average over all $|\psi_j\rangle$. We therefore obtain

$$\langle A\rangle = \sum_j p_j \langle A\rangle_j = \sum_j p_j \langle\psi_j|A|\psi_j\rangle. \tag{7.25}$$

We can write this in a more suitable way for our purposes by using the expression in (5.70) in terms of the projector $P_j = |\psi_j\rangle\langle\psi_j|$:

$$\langle\psi_j|A|\psi_j\rangle = \mathrm{Tr}AP_j = \mathrm{Tr}A|\psi_j\rangle\langle\psi_j|. \tag{7.26}$$

We can then express the expectation value of A with respect to the mixed state ρ as

$$\langle A\rangle = \sum_j p_j \mathrm{Tr}A|\psi_j\rangle\langle\psi_j| = \mathrm{Tr}A\left(\sum_j p_j|\psi_j\rangle\langle\psi_j|\right) = \mathrm{Tr}A\rho. \tag{7.27}$$

This is a very compact expression that is easy to calculate: we need only multiply two matrices and calculate its trace, given by the sum over the diagonal elements. Other expressions, such as the variance of A, follow directly from this form.

7.3 Interactions with the Environment

Now that we know how to describe classical uncertainty in a quantum state, let's explore how this arises in the case where we place a quantum non-demolition detector in the arms of a Mach–Zehnder interferometer. As we have seen, inside the interferometer the photon is in the state

$$|\text{photon}\rangle = \frac{|\text{path 1}\rangle + |\text{path 2}\rangle}{\sqrt{2}}. \tag{7.28}$$

As the photon interacts with the QND detector, it has to change the state of the detector, otherwise we cannot use the device as a detector in any sense of a measurement (since we are monitoring the state of the detector). We can define such states without knowing anything about the details of the detectors, because all we need is that the states are distinguishable, that is, orthonormal. Furthermore, we can consider the two QND detectors in each path as a single physical device with three relevant orthonormal states:

$$
\begin{aligned}
|Q_0\rangle &: \quad \text{the "ready" state,} \\
|Q_1\rangle &: \quad \text{the "photon is in path 1" state,} \\
|Q_2\rangle &: \quad \text{the "photon is in path 2" state.}
\end{aligned} \tag{7.29}
$$

The composite system of the photon and the QND detector is then initially in the state $|\text{photon}\rangle|Q_0\rangle$.

If the interaction of the photon with the detector is to produce a useful readout, the evolution of the composite system must obey the following rules:

$$
\begin{aligned}
|\text{path 1}\rangle|Q_0\rangle &\quad\rightarrow\quad |\text{path 1}\rangle|Q_1\rangle \\
|\text{path 2}\rangle|Q_0\rangle &\quad\rightarrow\quad |\text{path 2}\rangle|Q_2\rangle,
\end{aligned} \tag{7.30}
$$

that is, the photon in path 1 changes the state of the detector from the "ready" state to the state that indicates that the photon is in path 1, and similarly for path 2, as required. Also, the detector does not change the path of the photon. This is all we need from a device that we may reasonably call a detector.

Now let's see how this affects the state of the photon in the Mach–Zehnder interferometer:

$$
\begin{aligned}
|\text{photon}\rangle|Q_0\rangle &= \left(\frac{|\text{path 1}\rangle + |\text{path 2}\rangle}{\sqrt{2}} \right) |Q_0\rangle \\
&= \frac{1}{\sqrt{2}} \left(|\text{path 1}\rangle|Q_0\rangle + |\text{path 2}\rangle|Q_0\rangle \right) \\
&\rightarrow \frac{|\text{path 1}\rangle|Q_1\rangle + |\text{path 2}\rangle|Q_2\rangle}{\sqrt{2}}.
\end{aligned} \tag{7.31}
$$

You see that the photon and the QND detector are now entangled. We can no longer write the separate states for the photon and for the detector. A moment's thought will convince you that entanglement must occur every time the photon is not initially in the state $|\text{path 1}\rangle$ or $|\text{path 2}\rangle$.

If we now apply the second beam splitter in the Mach–Zehnder interferometer to the photon, we obtain the following transformation:

$$|\text{path } 1\rangle \rightarrow \frac{|D_1\rangle + |D_2\rangle}{\sqrt{2}} \,, \tag{7.32}$$

and

$$|\text{path } 2\rangle \rightarrow \frac{-|D_1\rangle + |D_2\rangle}{\sqrt{2}} \,. \tag{7.33}$$

This transforms (7.23) into

$$\frac{1}{2} \left(|D_1\rangle|Q_1\rangle + |D_2\rangle|Q_1\rangle - |D_1\rangle|Q_2\rangle + D_2\rangle|Q_2\rangle \right) \,. \tag{7.34}$$

The terms including $|D_1\rangle$ no longer cancel, due to the presence of $|Q_1\rangle$ and $|Q_2\rangle$. This means that detector D_1 now also fires. There is no longer destructive interference that keeps D_1 dark. In the case where $|Q_1\rangle = |Q_2\rangle$ the QND detector does not extract any information about the path of the photon (i.e., it is not a good path detector), and we recover the cancellation of the terms including $|D_1\rangle$.

To illustrate this further, we consider another example. A quantum particle is in a superposition of two orthogonal states $|\phi_a\rangle$ and $|\phi_b\rangle$ with amplitudes a and b:

$$|\psi\rangle = a|\phi_a\rangle + b|\phi_b\rangle \,. \tag{7.35}$$

In terms of the density matrix, this can be written as

$$\rho_{\text{particle}} = |\psi\rangle\langle\psi| = \begin{pmatrix} |a|^2 & ab^* \\ a^*b & |b|^2 \end{pmatrix} \,. \tag{7.36}$$

The particle interacts with the environment in such a way that it leaves a trace of its presence in the form of a changed state of the environment. The undisturbed state of the environment is $|E_0\rangle$, while the disturbed environmental states due to the states of the particle $|\phi_a\rangle$ and $|\phi_b\rangle$ are $|E_a\rangle$ and $|E_b\rangle$. The state of the particle and the environment after the interaction is then

$$|\psi\rangle|E_0\rangle \rightarrow a|\phi_a\rangle|E_a\rangle + b|\phi_b\rangle|E_b\rangle \,. \tag{7.37}$$

When $|E_a\rangle$ and $|E_b\rangle$ are different, this is again an entangled state.

So far, we have considered only the state of the composite system of the photon and the QND detector, or the particle and its environment. However, often we are really interested in the state of the photon or the particle by itself. This means that we must somehow lose the detector or the environment from our description.

We know that in general the state must be given by a density matrix, rather than a pure state, since classical uncertainty is involved. So we start by writing the state of the particle and the environment in (7.37) in the form of a density matrix

$$\rho = (a|\phi_a\rangle|E_a\rangle + b|\phi_b\rangle|E_b\rangle)(a^*\langle\phi_a|\langle E_a| + b^*\langle\phi_b|\langle E_b|) \tag{7.38}$$

that we can write out in individual terms as

$$\begin{aligned}\rho =&|a|^2|\phi_a\rangle\langle\phi_a| \otimes |E_a\rangle\langle E_a| + ab^*|\phi_a\rangle\langle\phi_b| \otimes |E_a\rangle\langle E_b|\\ &+ a^*b|\phi_b\rangle\langle\phi_a| \otimes |E_b\rangle\langle E_a| + |b|^2|\phi_b\rangle\langle\phi_b| \otimes |E_b\rangle\langle E_b| ,\end{aligned} \tag{7.39}$$

where we used the symbol $\otimes$ to make a clear distinction between operators $|\phi_a\rangle\langle\phi_b|$ on the particle and operators $|E_a\rangle\langle E_b|$ on the environment.

One way to "forget" about the environment is to measure it and discard the measurement outcome. In our examples there are two possible measurement outcomes after the interaction, namely E_a and E_b. The probabilities of these outcomes are given by

$$p_a = \langle E_a|\rho_E|E_a\rangle \quad \text{and} \quad p_b = \langle E_b|\rho_E|E_b\rangle , \tag{7.40}$$

where ρ_E is the density operator for the environment. However, we do not have ρ_E, we have the density matrix ρ of both the particle and the environment. But we can still take the expectation value of ρ with respect to $|E_a\rangle$ and $|E_b\rangle$. The result is not a probability, since we retain the kets and bras of the particle:

$$\begin{aligned}\langle E_a|\rho|E_a\rangle =&|a|^2|\phi_a\rangle\langle\phi_a|\langle E_a|E_a\rangle\langle E_a|E_a\rangle + ab^*|\phi_a\rangle\langle\phi_b|\langle E_a|E_a\rangle\langle E_b|E_a\rangle\\ &+ a^*b|\phi_b\rangle\langle\phi_a|\langle E_a|E_b\rangle\langle E_a|E_a\rangle + |b|^2|\phi_b\rangle\langle\phi_b|\langle E_a|E_b\rangle\langle E_b|E_a\rangle\end{aligned} \tag{7.41}$$

Since $\langle E_a|E_a\rangle = 1$ and $\langle E_a|E_b\rangle = \langle E_b|E_a\rangle = 0$, we find that

$$\langle E_a|\rho|E_a\rangle = |a|^2|\phi_a\rangle\langle\phi_a| . \tag{7.42}$$

Similarly, we find that $\langle E_0|\rho|E_0\rangle = 0$, and

$$\langle E_b|\rho|E_b\rangle = |b|^2|\phi_b\rangle\langle\phi_b| . \tag{7.43}$$

Therefore, the density operator that describes the particle is given by

$$\rho_{\text{particle}} = |a|^2|\phi_a\rangle\langle\phi_a| + |b|^2|\phi_b\rangle\langle\phi_b| = \begin{pmatrix} |a|^2 & 0 \\ 0 & |b|^2 \end{pmatrix} , \tag{7.44}$$

where the trace of ρ_{particle} is given by $|a|^2 + |b|^2 = 1$, as required. The procedure we used to reduce the state of the composite system to the state of a subsystem is called the partial trace, or the "trace over the environment". Since the off-diagonal terms in (7.44) are zero, the quantum uncertainty between $|\psi_a\rangle$ and $|\psi_b\rangle$ present in (7.36) has vanished. Only classical uncertainty about the states $|\psi_a\rangle$ and $|\psi_b\rangle$

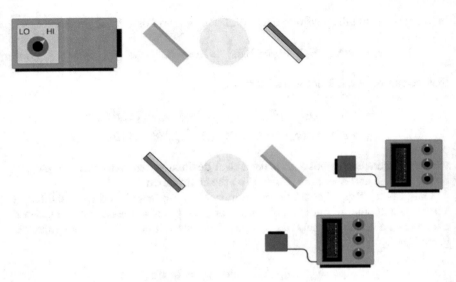

Fig. 7.2 A Mach–Zehnder interferometer where the photon interacts with the environment (see supplementary material 2)

remains. The interaction of a photon in a Mach–Zehnder interferometer interacting with the environment is shown in Fig. 7.2.

When the interaction with the environment is weak, we retain some quantum uncertainty in the state of the particle. Suppose that the interaction is such that the environment does not evolve to orthogonal states, but that the state of the environment is perturbed only slightly:

$$
\begin{aligned}
|\phi_a\rangle|E_0\rangle &\rightarrow |\phi_a\rangle \left(\sqrt{1 - |\epsilon|^2}\,|E_0\rangle + \epsilon|E_a\rangle\right) \\
|\phi_b\rangle|E_0\rangle &\rightarrow |\phi_b\rangle \left(\sqrt{1 - |\epsilon|^2}\,|E_0\rangle + \epsilon|E_b\rangle\right),
\end{aligned} \tag{7.45}
$$

where ϵ is some (potentially complex) parameter that depends on the actual interaction strength between the particle and the environment. This means that the environment does not gain complete information about the path of the particle. The state $|\psi\rangle = a|\phi_a\rangle + b|\phi_b\rangle$ after the interaction is then given by

$$
|\psi\rangle|E_0\rangle \rightarrow \sqrt{1 - |\epsilon|^2}\,|\psi\rangle|E_0\rangle + \epsilon\,(a|\phi_a\rangle|E_a\rangle + b|\phi_b\rangle|E_b\rangle). \tag{7.46}
$$

When $\epsilon = 0$ the particle does not interact with the environment at all, and the state of the particle and the environment remains separable. For any other value of ϵ the composite system exhibits entanglement. This example also allows us to model phenomenologically a quantum system that decoheres over time, by including a time dependence t in ϵ. Since $0 \leq |\epsilon| \leq 1$, we can choose that for late times $\epsilon \rightarrow 1$. One such function is

$$\epsilon(t) = \sqrt{1 - e^{-\gamma t}}. \tag{7.47}$$

This describes a particle that slowly loses its coherence. The parameter γ has units of inverse time, and it is the decoherence rate of the particle.

We can apply the partial trace over the environment to the composite state in (7.46), which gives for the density operator of the particle

$$\rho_{\text{particle}} = \begin{pmatrix} |a|^2 & (1 - |\epsilon|^2)\, ab^* \\ (1 - |\epsilon|^2)\, a^*b & |b|^2 \end{pmatrix}. \tag{7.48}$$

When we substitute the form of ϵ given by the decoherence model in (7.47) we obtain

$$\rho_{\text{particle}} = \begin{pmatrix} |a|^2 & ab^*\, e^{-\gamma t} \\ a^*b\, e^{-\gamma t} & |b|^2 \end{pmatrix}. \tag{7.49}$$

This demonstrates the exponential suppression of the off-diagonal terms of the density matrix that is associated with decoherence. The form of ϵ must be derived from a detailed model of the interaction Hamiltonian between the particle and the environment. This requires a lot more knowledge about modelling the interaction with the environment than we have developed here, and it is currently an area of active research.

Suppose that the state $|\phi_a\rangle = |g\rangle$ is the ground state on an atom and the state $|\phi_b\rangle = |e\rangle$ is the excited state. The constant (weak) interaction of the atom with the environment is increasingly likely to record the energy state of the atom, and this results in increasingly suppressed off-diagonal terms in the density matrix as quantum uncertainty gives way to classical uncertainty. This is a problem that designers of quantum computers must overcome to build a working machine, and typically the qubits (which may be atoms) can be kept sufficiently free of decoherence for only a short amount of time.

7.4 Quantum Systems at Finite Temperature

Another important physical application for density operators is when a physical system is held at a finite temperature instead of zero temperature. In such situations, the system is kept in thermal contact with a bath at temperature T, which means that there is some kind of interaction between the system and the bath. Typically, the thermal bath is a very large system, and will therefore be difficult—if not impossible—to characterise completely. Fortunately, when lots of particles in a bath interact with a quantum system, their interactions will average, and this can be described by referring only to the density matrix of the system itself. In effect, we are looking at the state of the system after the thermal bath (and its interaction with our system) has been traced out.

A thermal bath will inject energy into our quantum system that is proportional to T. This has the effect of populating the higher energy levels. At the same time, as we have seen in the previous section, ignoring the quantum state of the system that interacts with our quantum system, in this case the thermal bath, destroys the coherences of our system. We therefore expect that when we let our system come to thermal equilibrium with a bath, the resulting state should look something like

$$\rho(T) = \sum_n p_n(T)|E_n\rangle\langle E_n|, \qquad (7.50)$$

where $|E_n\rangle$ are the energy eigenstates, and $p_n(T)$ is the probability of finding the system in energy state $|E_n\rangle$ given the temperature T. Our task is now to find these probabilities.

When our system is held at temperature T, there will be some chance that upon a measurement it is found in energy state $|E_n\rangle$. However, the larger the energy E_n, the smaller the probability that our system is in the state $|E_n\rangle$. We relate temperature and energy by Boltzmann's constant $k_B = 1.38 \times 10^{-23}$ J K^{-1}, such that $k_B T$ has units of energy. The probability that the system is in state $|E_n\rangle$ when the energy of the bath $k_B T$ is smaller than E_n is exponentially suppressed, because the random fluctuations of the bath must all coincide to push the system in the higher energy state. So we can write

$$p_n(T) \propto \exp\left(-\frac{E_n}{k_B T}\right), \qquad (7.51)$$

where the $\propto$ symbol denotes "proportional to". This is called the Boltzmann factor. We will find it convenient to define $\beta = (k_B T)^{-1}$ in what follows. When we substitute this expression into the thermal density matrix, we obtain

$$\tilde{\rho}(T) = \sum_n e^{-\beta E_n}|E_n\rangle\langle E_n|, \qquad (7.52)$$

which is not a normalised density matrix (hence the tilde on ρ). Since the energy eigenstates are orthonormal states, the normalised thermal density matrix becomes simply

$$\rho(T) = \frac{\sum_n e^{-\beta E_n}|E_n\rangle\langle E_n|}{\sum_n e^{-\beta E_n}}. \qquad (7.53)$$

This is the quantum state of a system in thermal equilibrium with a bath at temperature T. Note that the nature of the interaction between the system and the bath is not part of the description of $\rho(T)$, which is very convenient!

We can write (7.53) in a much more elegant way using the fact that the energy eigenstates and eigenvalues together form the spectral decomposition of the Hamiltonian:

$$H = \sum_n E_n |E_n\rangle\langle E_n| \,. \tag{7.54}$$

We can express the exponential function $e^{-\beta E_n}$ in a power series

$$e^{-\beta E_n} = \sum_{k=0}^{\infty} \frac{(-\beta)^k E_n^k}{k!} \,. \tag{7.55}$$

Substituting this into (7.52), we obtain

$$
\begin{aligned}
\sum_n e^{-\beta E_n} |E_n\rangle\langle E_n| &= \sum_n \sum_{k=0}^{\infty} \frac{(-\beta)^k E_n^k}{k!} |E_n\rangle\langle E_n| \\
&= \sum_{k=0}^{\infty} \frac{(-\beta)^k}{k!} \left(\sum_n E_n^k |E_n\rangle\langle E_n| \right) \\
&= \sum_{k=0}^{\infty} \frac{(-\beta)^k}{k!} \left(\sum_n E_n |E_n\rangle\langle E_n| \right)^k \\
&= \sum_{k=0}^{\infty} \frac{(-\beta)^k}{k!} H^k = \exp(-\beta H) \,.
\end{aligned}
\tag{7.56}
$$

The step from the second to the third line is allowed because the energy eigenstates are orthonormal: $\langle E_n|E_m\rangle = \delta_{nm}$, with $\delta_{nm} = 0$ if $n \neq m$ and $\delta_{nm} = 1$ if $n = m$. We can apply a very similar argument to prove that

$$\sum_n e^{-\beta E_n} = \mathrm{Tr}\, e^{-\beta H} \,. \tag{7.57}$$

The normalised thermal density matrix then becomes

$$\rho(\beta) = \frac{e^{-\beta H}}{\mathrm{Tr}\, e^{-\beta H}} = \frac{1}{Z} e^{-\beta H} \,, \tag{7.58}$$

where we defined the so-called *partition function* $Z = \mathrm{Tr}\, e^{-\beta H} = \sum_n e^{-\beta E_n}$. This is an important quantity in statistical physics, where it is used to calculate a variety of thermodynamical quantities.

As an example, we can calculate the average energy in our thermal system. According to (7.27), the expectation value of the Hamiltonian is given by

$$\langle H \rangle = \text{Tr} H \rho(\beta),\tag{7.59}$$

which can now be written as

$$\langle H \rangle = \text{Tr} H \rho(\beta) = \text{Tr}\frac{H e^{-\beta H}}{Z} = \frac{1}{Z}\text{Tr} H e^{-\beta H}$$

$$= \frac{1}{Z}\text{Tr}-\frac{d}{d\beta} e^{-\beta H} = -\frac{1}{Z}\frac{d}{d\beta}\text{Tr}\, e^{-\beta H}$$

$$= -\frac{1}{Z}\frac{dZ}{d\beta} = -\frac{d \ln Z}{d\beta}.\tag{7.60}$$

So we can find the average energy straight from the partition function Z by a simple derivative with respect to β. Note that Z ultimately depends only on the energy eigenvalues E_n and the temperature T.

In statistical physics, we typically are interested in a macroscopically large number of particles. In principle we can describe this using the density matrix for a large number of systems. The Hamiltonian must then be replaced with the many-particle Hamiltonian. This involves numerous technical steps that are beyond the scope of this book, e.g., what to do when we have identical particles. The partition function then generally depends on the energy eigenvalues of the system, the temperature, the number of particles, and other thermodynamic quantities. The resulting quantum theory predicts behaviour such as superconductivity, superfluidity, and Bose-Einstein condensation, and is an active area of current research.

7.5 ✒ Entropy and Landauer's Principle

An important concept in physics is the *entropy* of a system. It is one of the fundamental concepts in thermodynamics, together with energy, temperature, heat and work. Loosely speaking, it is a measure of the disorder of a system. The second law of thermodynamics says that the entropy in an isolated system can never decrease, and this has been related to the arrow of time. In this section we show how the entropy of a quantum system can be obtained from the density matrix.

Ludwig Boltzmann was the first person to think really deeply about the meaning of entropy at the microscopic (or atomic) scale. Boltzmann realised that the entropy is given by the number of ways in which you can arrange the atoms in a gas (the micro-states) without changing the macroscopic quantities, such as the temperature and the pressure (the macro-states). Let us denote by W the total number of micro-states that give rise to the same macro-state. A typical macroscopic amount of gas has an enormous number of molecules (approximately 6×10^{23} molecules per mole of gas), and the total number of ways these molecules can be distributed is therefore extremely large. To make the numbers manageable, we take the logarithm of W. The entropy S of a gas is therefore defined as

$$S = k_B \log W ,\tag{7.61}$$

where we typically take the natural log, defined by

$$\log e^a = a .\tag{7.62}$$

Later we will use the logarithm with base 2, defined as $\log_2 2^a = a$, which means that the entropy is measured in bits. We will also work in units where $k_B = 1$, which amounts to a redefinition of the temperature scale.

Notice that we can think of the entropy as a lack of information, or uncertainty, about the microscopic state. For an ideal gas we can assume that all micro-states are equally likely, and the probability p_i for the system to be in the state i is then given by

$$p_i = \frac{1}{W} \quad \text{and} \quad \sum_{i=1}^{W} p_i = 1\tag{7.63}$$

We can manipulate Boltzmann's equation to obtain

$$S = \log W = \sum_i p_i \log W = \sum_i p_i \log \frac{1}{p_i}$$
$$= -\sum_i p_i \log p_i ,\tag{7.64}$$

where $k_B = 1$, and we used $\log a^b = b \log a$ in the last line. This is the formula for the entropy that we will use here. It is important to note that this formula is true also when the p_i are different, which is what we will be using shortly. This expression for the entropy was derived by Claude Shannon in the context of information theory. It was later pointed out to him that this is in fact the same as the Boltzmann entropy $S = k_B \log W$ for an ideal gas when all the probabilities are the same.

The uncertainty about the microscopic state is a classical uncertainty, and we can interpret the probabilities p_i as the eigenvalues of the density matrix that describes the system:

$$\rho = \begin{pmatrix} p_1 & 0 & 0 \\ 0 & \ddots & 0 \\ 0 & 0 & p_W \end{pmatrix} .\tag{7.65}$$

Since W can be huge, this may be a very large matrix. When ρ is diagonal, we can write

$$\log \rho = \begin{pmatrix} \log p_1 & 0 & 0 \\ 0 & \ddots & 0 \\ 0 & 0 & \log p_W \end{pmatrix},$$ (7.66)

and multiplying this by ρ using matrix multiplication yields

$$\rho \log \rho = \begin{pmatrix} p_1 \log p_1 & 0 & 0 \\ 0 & \ddots & 0 \\ 0 & 0 & p_W \log p_W \end{pmatrix}.$$ (7.67)

The entropy is then the sum of the diagonal elements of this matrix, with an overall minus sign. This is just the trace of the matrix, so we can write

$$S = -\text{Tr}\,(\rho \log \rho),$$ (7.68)

which is the general relationship between the entropy of a system and its density matrix. It also holds when ρ is not a diagonal matrix, but then we must be more careful in taking the logarithm. The expression for S remains true also when W is small.

A curious consequence of this relationship between the entropy and the density matrix of a system is that erasing information will increase the entropy of a system. To see this, consider again the photon in a Mach–Zehnder interferometer with QND detectors that tell us which path inside the interferometer the photon took. After the measurement, we know that the photon is in, say, the upper path:

$$|\text{photon}\rangle = |\text{upper}\rangle = \begin{pmatrix} 1 \\ 0 \end{pmatrix}.$$ (7.69)

The density matrix of this pure state is given by

$$\rho = |\text{upper}\rangle \langle \text{upper}| = \begin{pmatrix} 1 & 0 \\ 0 & 0 \end{pmatrix}.$$ (7.70)

The eigenvalues of this density matrix are 1 and 0, which means that the entropy of this density matrix is given by

$$S = -0 \log 0 - 1 \log 1 = 0 + 0 = 0,$$ (7.71)

since the logarithm of 1 is zero, and the logarithm of 0 is multiplied by zero (in other words, we define $0^0 = 1$). Therefore, the pure state $|\text{upper}\rangle$ has zero entropy. This argument can be used to show that *all pure states have zero entropy*.

When we erase the information about the photon path, we become completely uncertain (classically) about the path, and the density matrix must be the maximum uncertainty state:

Fig. 7.3 Locating a particle in a box (see supplementary material 3)

$$\rho = \begin{pmatrix} \frac{1}{2} & 0 \\ 0 & \frac{1}{2} \end{pmatrix} \qquad (7.72)$$

The eigenvalues of this matrix are both $\frac{1}{2}$, and the entropy becomes

$$S = -\frac{1}{2}\log_2\frac{1}{2} - \frac{1}{2}\log_2\frac{1}{2} = 1. \qquad (7.73)$$

In other words, erasing the path information (one bit: "upper" or "lower") of the photon increases the entropy by 1 bit.

We can use this calculation to solve an interesting paradox: Suppose that we construct a container with a single molecule or atom. One of the walls of the container is a piston that moves in and out of the container without friction (see Fig. 7.3). We can then create an engine as follows: At first, we do not know where the molecule in the container is, so we make a measurement. If we find the molecule in the half closest to the piston we wait a little bit longer, before we measure again. If we find the molecule in the half furthest away from the piston, we quickly move the piston into the container (to the halfway point). We can do this without doing any work, because the molecule is not there to oppose the motion of the piston. We then let the molecule push the piston outwards, which amounts to work done by the molecule on the piston. The energy for this work is supplied by the heat of the molecule (its kinetic energy), and we therefore convert heat into work. The piston returns to its initial position, so we can repeat this experiment until all the heat from the molecule is converted into work. This is called Szilárd's engine.[1]

However, this means that the heat of the molecule goes to zero after many rounds of the experiment, and its entropy goes to zero (the molecule ends up in a pure state). Meanwhile, the work done by the molecule can be supplied to a task that does not increase entropy (such as pushing an stationary object up a hill). So the overall entropy seems to decrease, and this violates the second law of thermodynamics!

How do we solve this paradox? We have glossed over one crucial aspect of the thought experiment: after we measure the position of the molecule, we make the decision whether to pull in the piston or not. To make this decision we must record the measurement outcome. This can be done on a computer or in our head, but

---

[1] L. Szilard, *Über die Entropieverminderung in einem thermodynamischen System bei Eingriffen intelligenter Wesen*, Zeitschrift für Physik **53** 840, 1929.

Fig. 7.4 Erasing information will increase the entropy (see supplementary material 4)

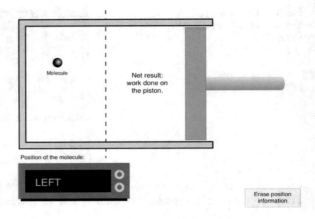

somewhere in the physical world the information about the molecule's position must be stored if we are to base a decision on it. Therefore, when we repeat this experiment a large number of times, we will build up a long string of measurement outcomes: "left", "right", "left", "left", "right", and so on. On the other hand, for the container and piston to operate in a proper cycle, we must bring the whole engine back to it's initial state. This means that we need to erase the memory of the measurement outcomes.

However, we have seen that the erasure of the path information of a photon leads to an entropy increase of $S = 1$ bit. Similarly, the erasure of the measurement outcome "left" or "right" must be stored in a physical bit whose quantum state can be denoted by the pure state $|\text{left}\rangle$ or $|\text{right}\rangle$. This state has zero entropy. To erase this bit we need to bring the memory to an initial state $|i\rangle$ from both $|\text{left}\rangle$ and $|\text{right}\rangle$. Since these two orthogonal states map onto the *same* state $|i\rangle$, the transformation cannot be unitary, and one bit of information must be lost in the process. This corresponds to an entropy of $S = 1$. This increase in entropy turns out to be exactly enough to save the second law of thermodynamics. In general, we can say that the erasure of one bit of information must fundamentally lead to an increase in entropy:

$$S = 1. \tag{7.74}$$

This is known as Landauer's principle.[2] It is a surprising connection between information theory and heat (see Fig. 7.4).

> Szilárd's engine is closely related to Maxwell's demon, which is a hypothetical being that sits at the gate in a separation between two gasses. By letting through fast molecules in one direction, and slow molecules in the other, the demon can create a temperature difference between the two gasses without doing work or injecting energy. This is also an apparent

[2] R. Landauer, *Irreversibility and heat generation in the computing process*, IBM Journal of Research and Development **5** 183, 1961.

violation of the second law of thermodynamics, and can similarly be resolved by considering the memory of the demon.

7.6 ॰ Entanglement Measures

Entanglement plays an important role in quantum information tasks such as teleportation and quantum computing, and it is also a key concept when we wish to describe the interaction of a system with its environment, or indeed with any other quantum system. This leads naturally to the question of how much entanglement there is in a given quantum state. This has turned out to be very difficult to answer in general, because the structure of quantum states is so varied and complex once you move beyond a system of two qubits.

A typical strategy for creating a quantitative entanglement measure is to choose a task that requires entanglement to succeed—such as quantum teleportation—and see how well it performs when instead of the required entangled state we use another quantum state. For the example of teleportation, we aim to transfer the quantum state $|\psi\rangle = a|0\rangle + b|1\rangle$ from an incoming particle to an outgoing particle (see Sect. 6.3) using a maximally entangled state $|\Phi^+\rangle = (|00\rangle + |11\rangle)/\sqrt{2}$. If instead of $|\Phi^+\rangle$ we use some other state (which may not even be entangled!), we can see how much the teleported state ρ differs from $|\psi\rangle$. A standard way to compare two quantum states is via the *fidelity*:

$$ F = \langle\psi|\rho|\psi\rangle, \tag{7.75} $$

where ρ may be pure or mixed. If you compare the fidelity with the Born rule, you see that F is the probability of mistaking ρ for ψ in a measurement. When $\rho = |\psi\rangle\langle\psi|$ the fidelity is $F = 1$, and when ρ is maximally mixed (which is an example of a separable state) the fidelity is $F = \frac{1}{2}$. So for qubits the fidelity lies somewhere in the interval $[\frac{1}{2}, 1]$.

The optimal quantum teleportation fidelity is well-defined as an entanglement measure, but it may not be the most natural number for a given application. Another widely-used entanglement measure is the *entanglement of formation*, which corresponds to the average number of Bell states required to create the state in question using only local operations and classical communication. There is no reason to believe that for the same state the entanglement of formation takes the same numerical value as the teleportation fidelity, and indeed they are typically different.

The entanglement of formation E_f is defined for bipartite states, and while we can calculate the entanglement of formation for any bipartite quantum state, mixed or pure, the calculation for mixed states is considerably more involved than for a pure state, so here we will consider the case of two qubits in a pure state $|\Psi\rangle$. Our starting point is that the entropy for any pure quantum state is zero. From the previous section we saw that zero entropy means that there is no information missing in the description of our quantum system. However, when we trace out one system and consider the state of the remaining system, its entropy is generally no longer zero. Only when the state is fully separable is the entropy zero. To see this, write the joint separable state as

$$|\Psi\rangle = |\psi\rangle|\phi\rangle \,, \tag{7.76}$$

and construct the density matrix

$$\rho = |\psi, \phi\rangle\langle\psi, \phi| \,. \tag{7.77}$$

Taking the partial trace over system 2 yields the reduced density matrix of system 1:

$$\rho_1 = \text{Tr}[2]|\psi, \phi\rangle\langle\psi, \phi| = \text{Tr}|\phi\rangle\langle\phi||\psi\rangle\langle\psi| = |\psi\rangle\langle\psi| \,, \tag{7.78}$$

since the trace over the pure state $|\phi\rangle\langle\phi|$ is 1. The reduced density matrix ρ_1 is the pure state $|\psi\rangle\langle\psi|$, so the entropy of this state is zero, and consequently the entanglement of formation is $E_f = 0$.

Another important example is a maximally entangled two-qubit state

$$|\Psi\rangle = \frac{1}{\sqrt{2}}|00\rangle + \frac{1}{\sqrt{2}}|11\rangle \,. \tag{7.79}$$

The density matrix for this state can be written as

$$\rho = \frac{1}{2}\left(|00\rangle\langle00| + |00\rangle\langle11| + |11\rangle\langle00| + |11\rangle\langle11|\right) \,. \tag{7.80}$$

Taking the partial trace over the second qubit yields

$$
\begin{aligned}
\rho_1 = \text{Tr}[2]\rho &= \sum_{j=0}^{1} {}_2\langle j|\rho|j\rangle_2 \\
&= \frac{1}{2}\left(\langle0|00\rangle\langle00|0\rangle + \langle0|00\rangle\langle11|0\rangle + \langle0|11\rangle\langle00|0\rangle + \langle0|11\rangle\langle11|0\rangle\right) + \\
&\quad + \frac{1}{2}\left(\langle1|00\rangle\langle00|1\rangle + \langle1|00\rangle\langle11|1\rangle + \langle1|11\rangle\langle00|1\rangle + \langle1|11\rangle\langle11|1\rangle\right) \\
&= \frac{1}{2}|0\rangle\langle0| + \frac{1}{2}|1\rangle\langle1| \,.
\end{aligned}
\tag{7.81}
$$

This is a diagonal matrix with eigenvalues $p_1 = \frac{1}{2}$ and $p_2 = \frac{1}{2}$, and the entropy of this reduced density matrix is

$$S(\rho_1) = -p_1 \log_2 p_1 - p_2 \log_2 p_2 = -\frac{1}{2}\log_2\frac{1}{2} - \frac{1}{2}\log_2\frac{1}{2} = -\log_2 2^{-1} = 1 \,. \tag{7.82}$$

Therefore, the entropy of formation of two qubits in a maximally entangled Bell state is $E_f = 1$. You can verify that the entropy of formation for the other Bell states is also 1.

As a more general example, consider the two-qubit pure state

$$|\Psi\rangle = a_{00}|00\rangle + a_{01}|01\rangle + a_{10}|10\rangle + a_{11}|11\rangle = \sum_{j,k=0}^{1} a_{jk}|jk\rangle . \tag{7.83}$$

We can write this in density matrix form as

$$|\Psi\rangle\langle\Psi| = \sum_{j,k=0}^{1}\sum_{l,m=0}^{1} a_{jk}a_{lm}^*|jk\rangle\langle lm| . \tag{7.84}$$

Taking the partial trace over the second qubit yields the density matrix ρ_1

$$\rho_1 = \text{Tr}[2]|\Psi\rangle\langle\Psi| = \sum_{n=0}^{1} {}_2\langle n| \left(\sum_{j,k=0}^{1}\sum_{l,m=0}^{1} a_{jk}a_{lm}^*|jk\rangle\langle lm|n| \right) |n\rangle_2$$

$$= \sum_{n,j,k,l,m=0}^{1} \langle n|k\rangle_2 |j\rangle_1 a_{jk}a_{lm}^* \langle l|_1\langle m|n\rangle_2$$

$$= \sum_{n,j,k,l,m=0}^{1} \delta_{nk}\delta_{mn}\, a_{jk}a_{lm}^*|j\rangle_1\langle l|_1$$

$$= \sum_{j,l,n=0}^{1} a_{jn}a_{ln}^*|j\rangle_1\langle l|_1 . \tag{7.85}$$

Therefore the matrix element $[\rho_1]_{jl}$ is given by

$$[\rho_1]_{jl} = \sum_{n=0}^{1} a_{jn}a_{ln}^* = a_{j0}a_{l0}^* + a_{j1}a_{l1}^* . \tag{7.86}$$

In matrix form this becomes

$$\rho_1 = \begin{pmatrix} a_{00}a_{00}^* + a_{01}a_{01}^* & a_{00}a_{10}^* + a_{01}a_{11}^* \\ a_{10}a_{00}^* + a_{11}a_{01}^* & a_{10}a_{10}^* + a_{11}a_{11}^* \end{pmatrix} . \tag{7.87}$$

To find the entropy of this matrix we must find its eigenvalues p_1 and p_2 and substitute them into (7.64). Since this leads to a lengthy general expression we consider a numerical example

$$|\Psi\rangle = \frac{2}{\sqrt{58}}|00\rangle + \frac{5}{\sqrt{58}}|01\rangle + \frac{5}{\sqrt{58}}|10\rangle - \frac{2i}{\sqrt{58}}|11\rangle. \tag{7.88}$$

Verify that this is a normalised quantum state. The reduced density matrix ρ_1 then becomes

$$\rho_1 = \begin{pmatrix} \frac{2}{\sqrt{58}}\frac{2}{\sqrt{58}} + \frac{5}{\sqrt{58}}\frac{5}{\sqrt{58}} & \frac{2}{\sqrt{58}}\frac{5}{\sqrt{58}} + \frac{5}{\sqrt{58}}\frac{2i}{\sqrt{58}} \\ \frac{5}{\sqrt{58}}\frac{2}{\sqrt{58}} - \frac{2i}{\sqrt{58}}\frac{5}{\sqrt{58}} & \frac{5}{\sqrt{58}}\frac{5}{\sqrt{58}} - \frac{2i}{\sqrt{58}}\frac{2i}{\sqrt{58}} \end{pmatrix}$$

$$= \frac{1}{58}\begin{pmatrix} 29 & 10+10i \\ 10-10i & 29 \end{pmatrix}. \tag{7.89}$$

This is a Hermitian matrix, as it should be, and therefore has real eigenvalues. When we calculate the eigenvalues, we find that

$$p_1 = \frac{1}{2} + \frac{5\sqrt{2}}{29} \quad \text{and} \quad p_2 = \frac{1}{2} - \frac{5\sqrt{2}}{29}. \tag{7.90}$$

Another quick check confirms that these eigenvalues sum to 1, as required since these can be interpreted as probabilities. The numerical value of the entropy is

$$S(\rho_1) = -p_1 \log_2 p_1 - p_2 \log_2 p_2 = 0.82 = E_f. \tag{7.91}$$

This demonstrates that there is indeed entanglement in the state $|\Psi\rangle$, since $S = E_f > 0$. However, it is not a maximally entangled state, since $E_f = 0.82 < 1$.

Finally, we consider the *concurrence* as entanglement measure. We will construct the concurrence for a general two-qubit pure state $|\Psi\rangle$. First, we take the complex conjugate of $|\Psi\rangle$ and act on both qubits with the σ_y Pauli operator to form the new state

$$|\tilde{\Psi}\rangle = \sigma_{y,1} \otimes \sigma_{y,2}|\Psi^*\rangle, \tag{7.92}$$

where $\sigma_{y,j}$ denotes σ_y acting on qubit $j = 1, 2$, and $|\Psi^*\rangle$ is the complex conjugate of $|\Psi\rangle$. The concurrence C is then given by the overlap of this state with $|\Psi\rangle$:

$$C(\Psi) = \left|\langle\Psi|\tilde{\Psi}\rangle\right|. \tag{7.93}$$

As an example, consider again the state in (7.88). Using the relations

$$\sigma_y|0\rangle = i|1\rangle \quad \text{and} \quad \sigma_y|1\rangle = -i|0\rangle, \tag{7.94}$$

we find that $|\tilde{\Psi}\rangle$ is

$$|\tilde{\Psi}\rangle = -\frac{2}{\sqrt{58}}|11\rangle + \frac{5}{\sqrt{58}}|10\rangle + \frac{5}{\sqrt{58}}|01\rangle - \frac{2i}{\sqrt{58}}|00\rangle \,. \qquad (7.95)$$

Next, we calculate the scalar product between $|\Psi\rangle$ and $|\tilde{\Psi}\rangle$:

$$\langle\Psi|\tilde{\Psi}\rangle = -\frac{4i}{58} + \frac{25}{58} + \frac{25}{58} - \frac{4i}{58} = \frac{1}{29}(25 - 4i) \,. \qquad (7.96)$$

This yields the concurrence

$$C(\Psi) = \left|\langle\Psi|\tilde{\Psi}\rangle\right| = \sqrt{\left(\frac{25}{29}\right)^2 + \left(\frac{4}{29}\right)^2} = 0.87 \,. \qquad (7.97)$$

Indeed, this is a different value than the entanglement of formation $E_f = 0.82$ for the *same* state $|\Psi\rangle$ in (7.88). This is not a problem, because the concurrence and the entanglement of formation measure slightly different things about $|\Psi\rangle$. There is, however, a relationship between E_f and C, given by

$$E_f = S_2\left(\frac{1 + \sqrt{1 - C^2}}{2}\right), \qquad (7.98)$$

where we used the entropy $S_2(p)$ for the probability distribution of two probabilities p and $1 - p$:

$$S_2(p) = -p \log_2 p - (1 - p) \log_2(1 - p) \,. \qquad (7.99)$$

This function is often called $H(p)$, but we have already overloaded H as the Hamiltonian and the Hadamard operation, so we use S_2 instead. When we calculate E_f using $C = 0.87$, we find from $\frac{1}{2}(1 + \sqrt{1 - C^2}) = 0.74$ that indeed

$$E_f = S_2(0.74) = 0.82 \,, \qquad (7.100)$$

as required.

It is not so clear what the operational meaning of the concurrence is. It was constructed to provide a much easier calculation of the entanglement of formation for mixed states,[3] but is now often used in its own right. There are many other entanglement measures, and understanding the nature of quantum entanglement is still an area of active research.

---

[3] W. K. Wootters, Phys. Rev. Lett. **80**, 2245 (1998).

Exercises

1. Calculate the purity of the state

$$\rho = \frac{1}{3}\begin{pmatrix} 1 & 1 \\ 1 & 2 \end{pmatrix}.$$

   Is it a pure state or a mixed state?

2. Two qubits, 1 and 2, are prepared in the initial states

$$|\psi\rangle_1 = \frac{1}{\sqrt{2}}(|0\rangle_1 + |1\rangle_1) \qquad \text{and} \qquad |\phi\rangle_2 = |0\rangle_2 ,$$

   and the interaction between the two qubits is described by the Hamiltonian

$$H = \hbar g |1\rangle_1\langle 1| \otimes (|1\rangle_2\langle 0| + |0\rangle_2\langle 1|) .$$

   (a) Calculate the state of the joint two-qubit system after an interaction time T.
   (b) What are the quantum states of the *individual* qubits after they have interacted for a time T?
   (c) Calculate the entropy $S(\rho) = -\text{Tr}\rho \log_2 \rho$ of the individual qubit states ρ.
   (d) For what value of T are the two qubits maximally entangled?

3. (a) A spin-$\frac{1}{2}$ particle is in the state $\rho = \frac{1}{3}|\uparrow\rangle\langle\uparrow| + \frac{2}{3}|\downarrow\rangle\langle\downarrow|$. Calculate the purity of the spin, given by $\text{Tr}\rho^2$.
   (b) The spin evolves in time according to the Hamiltonian

$$H = \hbar\omega X \qquad \text{with} \qquad X = |\uparrow\rangle\langle\downarrow| + |\downarrow\rangle\langle\uparrow| .$$

   Calculate the state $\rho(t)$ at time t.
   (c) Sketch the probability of finding the measurement outcome "$\uparrow$" in a measurement of the spin as a function of time.
   (d) How will the purity of the spin change when all spin coherence is lost?

4. Quantum teleportation with three entangled qubits.

   (a) A so-called GHZ state for three qubits can be written as

$$|\text{GHZ}\rangle = \frac{|000\rangle + |111\rangle}{\sqrt{2}} .$$

   After losing the third qubit, what is the state of the remaining two qubits? Calculate the entropy $S(\rho) = -\text{Tr}\rho \log_2 \rho$ of this state.
   (b) A so-called W state for three qubits can be written as

$$|\text{W}\rangle = \frac{|001\rangle + |010\rangle + |100\rangle}{\sqrt{3}} .$$

After losing the third qubit, what is the state of the remaining two qubits? Calculate the entropy of this state.

(c) The first qubit of the GHZ state is used as part of the entanglement channel for teleporting a fourth qubit in the state $\alpha|0\rangle + \beta|1\rangle$. Calculate the remaining two-qubit state after applying the teleportation protocol. Assuming perfect equipment, can the remaining two qubits retrieve the original qubit with certainty?

(d) Instead of the GHZ state, use the W state for the teleportation protocol in part (c). Can the remaining two qubits retrieve the original qubit with certainty?

5. The density matrix.

(a) Show that $\frac{1}{2}|0\rangle + \frac{1}{2}|+\rangle$ is not a properly normalized state.

(b) Show that $\text{Tr}\rho = 1$, and then prove that any density operator has unit trace and is Hermitian.

(c) Show that density operators are *convex*, i.e., that $\rho = w_1\rho_1 + w_2\rho_2$ with $w_1 + w_2 = 1$ ($w_1, w_2 \geq 0$), and ρ_1, ρ_2 again density operators.

6. Two qubits, labeled A and B, are prepared in the entangled state $|\Psi\rangle = \frac{3}{5}|00\rangle + \frac{4}{5}|11\rangle$.

(a) Show that the correlations in the $\{|0\rangle, |1\rangle\}$ basis are perfect, but the correlations in the $\{|+\rangle, |-\rangle\}$ basis are not (think first carefully how you would define a correlation). The states $|+\rangle$ and $|-\rangle$ are defined by

$$|\pm\rangle = \frac{|0\rangle \pm |1\rangle}{\sqrt{2}} .$$

(b) Calculate the entanglement entropy $\mathscr{E}$ of $|\Psi\rangle$. This is a measure of the amount of entanglement in $|\Psi\rangle$, and is given by

$$\mathscr{E} = S(\text{Tr}[B]|\Psi\rangle\langle\Psi|) \quad \text{with} \quad S(\rho) = -\text{Tr}\rho \log_2 \rho .$$

(c) We teleport a state $\alpha|0\rangle + \beta|1\rangle$ using the entangled state $|\Psi\rangle$. What is the best and worst average teleportation fidelity $\langle F\rangle$, using otherwise ideal components? The fidelity F is defined by $F = \left|\langle\psi_{\text{ideal}}|\psi_{\text{teleported}}\rangle\right|^2$.

7. The *partial* transpose of a two-qubit state is calculated by writing the density matrix ρ as

$$\rho = \sum_{jklm} \rho_{jk,lm}|j, k\rangle\langle l, m| \quad \text{with} \quad j, k, l, m = 0, 1$$

and swapping k and m in the kets and bras (the full transpose would be swapping both k and m *and* j and l). If the eigenvalues of the resulting matrix include negative values, then the original state was entangled. Calculate the partial transpose of the states given in Exercise 1 of Chap. 6 (making the identification $|0\rangle \leftrightarrow |\uparrow\rangle$ and $|1\rangle \leftrightarrow |\downarrow\rangle$) and determine their eigenvalues. Which states are entangled? Does it agree with your previous answer?

8. An electron with spin 1/2 is prepared in a state

$$\rho = \frac{1}{2} \begin{pmatrix} 1 & e^{-i\omega t - \gamma t} \\ e^{i\omega t - \gamma t} & 1 \end{pmatrix},$$

where ω and γ are real positive numbers, and t denotes time. Sketch the time evolution of the state in the Bloch sphere. What is the physical interpretation of ω and γ?

9. Any single-qubit mixed state can be written in terms of the Pauli matrices as

$$\rho = \frac{\mathbb{I} + \mathbf{r} \cdot \boldsymbol{\sigma}}{2},$$

where $\boldsymbol{\sigma} = (\sigma_x, \sigma_y, \sigma_z)$ and $\mathbf{r} = (r_x, r_y, r_z)$, $|\mathbf{r}| \leq 1$. Moreover, any single qubit observable can be expressed as

$$A = a_0 \mathbb{I} + \mathbf{a} \cdot \boldsymbol{\sigma},$$

with a_0 and $\mathbf{a} = (a_x, a_y, a_z)$ real numbers. Show that the expectation value of A with respect to ρ can be written as

$$\langle A \rangle = \frac{1}{2} a_0 + \frac{1}{2} \mathbf{a} \cdot \mathbf{r}.$$

10. The Boltzmann entropy of a system is given by the partition function and the average energy $\langle E \rangle$ as

$$S = k_B \left(\ln Z + \beta \langle E \rangle \right).$$

Show that the Shannon entropy in the second line of (7.64) for a thermal state corresponds to the Boltzmann entropy when we choose $k_B = 1$.

11. A two-level atom has energy separation $\hbar\omega$ between the ground state and the excited state. The atom is brought into thermal equilibrium with a bath at temperature T.

    (a) Find the thermal state of the atom.
    (b) Calculate the average energy and the entropy of the atom.
    (c) Let's assume that we prepare the atom in the excited state, and we do not let it thermalise with the bath. What would be the effective temperature of the atom? You can use the expression for the Boltzmann entropy in Exercise 10.

12. Does a thermal state evolve over time?
13. Prove (7.57).
14. Express the variance of an observable A in terms of the trace and the density operator, similar to (7.27).

Chapter 8
Motion of Particles

In this chapter, we will explore how to describe the position and momentum of a particle in quantum mechanics, and how we set up its Schrödinger equation. To this end, we need to work out how to describe the momentum and energy for a particle. We will conclude with a description of a bizarre phenomenon called quantum tunnelling, which is now widely used in advanced microscopes, and with a qualitative overview of the quantum mechanical foundations of chemistry.

8.1 A Particle in a Box

So far, we have looked at physics problems in quantum mechanics for which each possibility can be simply labelled: spin up or down, energy E_n, a photon coming from the left or from the top, etc. Our approach works extremely well for these cases, but what about the situation where a particle is moving continuously through space? This is a very common problem in physics, and we want to treat it quantum mechanically as well.

To solve this problem, we are going to extend the quantum theory we have developed so far to cover continuous properties such as position and momentum, where the value does not take on discrete values but can take any real value in an interval. Consider a particle sealed in a one-dimensional box—a long narrow tube—of length L. The position along the tube is given by x, with $0 \leq x \leq L$. We want to develop the mathematical theory for the state of a particle at a position x in the tube, but we will have to do this step by step.

Supplementary Information The online version contains supplementary material available at https://doi.org/10.1007/978-3-031-16165-0_8.

First, we consider the very crude question whether the particle is in the left or the right side of the tube. Using the formalism of quantum mechanics we have developed so far, we can construct quantum states for this situation:

$$|\text{left}\rangle = \begin{pmatrix} 1 \\ 0 \end{pmatrix} \quad \text{and} \quad |\text{right}\rangle = \begin{pmatrix} 0 \\ 1 \end{pmatrix} . \tag{8.1}$$

Just like the states for the spin of an electron or the energy level of an atom, we can construct superpositions of these basis states:

$$|\psi\rangle = a|\text{left}\rangle + b|\text{right}\rangle = \begin{pmatrix} a \\ b \end{pmatrix} , \tag{8.2}$$

with $|a|^2 + |b|^2 = 1$. The probabilities of finding the particle in the left and right compartments of the tube are then given by

$$\text{Pr}(\text{left}) = |\langle\text{left}|\psi\rangle|^2 = |a|^2 \quad \text{and} \quad \text{Pr}(\text{right}) = |\langle\text{right}|\psi\rangle|^2 = |b|^2 , \tag{8.3}$$

exactly according to the rules of quantum mechanics we developed earlier. We can imagine that the tube can be taken apart into different compartments to see in which compartment the particle resides, for example by shaking and listening to the particle rattling inside. This would not give us any extra information about the exact position of the particle inside the compartment.

Instead of two compartments of length $L/2$, we can divide the tube into three smaller compartments with length $L/3$. The corresponding quantum states are then given by

$$|\text{left}\rangle = \begin{pmatrix} 1 \\ 0 \\ 0 \end{pmatrix} , \quad |\text{middle}\rangle = \begin{pmatrix} 0 \\ 1 \\ 0 \end{pmatrix} , \quad |\text{right}\rangle = \begin{pmatrix} 0 \\ 0 \\ 1 \end{pmatrix} . \tag{8.4}$$

Note that this is now a three-dimensional vector, because there are three perfectly distinguishable measurement outcomes. The particle in the tube can again be in any superposition of these states, such as

$$|\psi\rangle = a|\text{left}\rangle + b|\text{middle}\rangle + c|\text{right}\rangle = \begin{pmatrix} a \\ b \\ c \end{pmatrix} , \tag{8.5}$$

with $|a|^2 + |b|^2 + |c|^2 = 1$. The probabilities of finding the particle in the different compartments are then

$$\text{Pr (left)} = |\langle \text{left}|\psi\rangle|^2 = |a|^2,$$
$$\text{Pr (middle)} = |\langle \text{middle}|\psi\rangle|^2 = |b|^2,$$
$$\text{Pr (right)} = |\langle \text{right}|\psi\rangle|^2 = |c|^2, \tag{8.6}$$

The generalisation to N compartments should now be obvious. We can divide the tube into N compartments, each of length L/N. The corresponding quantum states $|n\rangle$ are labelled by a number n that indicates the compartment (running from 1 to N). The particle can be in any superposition of these states, so each $|n\rangle$ has an amplitude c_n (replacing the a, b, c, etc.). The state of a particle in a box of N compartments is therefore given by

$$|\psi\rangle = \sum_{n=1}^{N} c_n |n\rangle \quad \text{with} \quad \sum_{n=1}^{N} |c_n|^2 = 1. \tag{8.7}$$

This is the generalisation of (8.5). Since N is very large, it is no longer convenient to write the states $|n\rangle$ in terms of vectors, so from now on we will have to rely on kets and bras.

The probability of finding the particle in the mth compartment is then given by

$$\text{Pr} (m) = |\langle m|\psi\rangle|^2 = \left| \sum_{n=1}^{N} c_n \langle m|n\rangle \right|^2 = |c_m|^2, \tag{8.8}$$

where we used that $\langle m|n\rangle = 0$ if $m \neq n$, and $\langle m|n\rangle = 1$ if $m = n$. The states $|n\rangle$ can be considered position states with a precision ℓ that is given by the size of the compartment $\ell = L/N$.

You may expect that we can make N arbitrarily large to get position states with arbitrary high precision. This is correct, but in order to get a true continuum of position states, we have to be very careful with the mathematics. We will go through it at a snail's pace, so we don't make any mistakes by relying on false intuitions. To derive the continuum of position states, we start with the normalisation condition that all probabilities defined by (8.8) must sum to one:

$$\sum_m \text{Pr} (m) = \sum_m |c_m|^2 = 1. \tag{8.9}$$

We can graphically place these probabilities next to each other in a histogram, where each bar has width (see Fig. 8.1). The normalisation condition means that the areas of all rectangles in the histogram must sum to one.

Next, we can identify the total horizontal length of the histogram with the length L of the box, which means that each bar in the histogram has a width ℓ. However, the area of each bar is the vertical height times the horizontal length, or $\text{Pr} (n) \times \ell$ for the nth bar. The total area A of the bars in the histogram is then given by the sum over all the bars

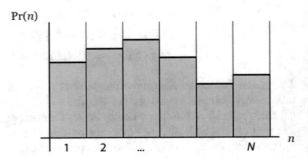

Fig. 8.1 Discrete position probabilities

$$A = \sum_{n=1}^{N} \text{Pr}(n) \times \ell = \ell \left(\sum_{n=1}^{N} \text{Pr}(n) \right) = \ell. \tag{8.10}$$

So $A = \ell$, instead of 1, and it has units of length! This means that we have broken the normalisation condition. We can fix this by dividing all probabilities by ℓ:

$$p(m) = \frac{\text{Pr}(m)}{\ell}. \tag{8.11}$$

The $p(m)$ are no longer probabilities, because they are not dimensionless and do not take values between 0 and 1 (ℓ can be very small, pushing $p(m)$ beyond 1).

The $p(m)$ form a so-called *probability density function*. The name is explained by the fact that $p(m)$ is probability $\text{Pr}(m)$ per length ℓ, just like a regular mass density is mass M per volume V. The normalisation condition can then be written in two ways:

$$\sum_{n=1}^{N} \text{Pr}(n) = 1 \quad \text{and} \quad \sum_{n=1}^{N} p(n)\ell = 1. \tag{8.12}$$

If we double the number of compartments while keeping the length of the box L constant, the width of each compartment halves: $\ell \to \ell/2$. Increasing N further will make ℓ increasingly smaller, and leads to a histogram shown in Fig. 8.2.

When we let $N \to \infty$, the width of each bar narrows to an infinitesimal length $\ell \to dx$, and the probability density $p(n)$ tends to a distribution $p(x)$ that depends on a continuous variable $x \in \mathbb{R}$ instead of an integer variable $n \in \mathbb{N}$. This is shown in Fig. 8.2. Because n and x belong to different types of numbers ($\mathbb{R}$ versus $\mathbb{N}$) we change the notation $n \to x$. In addition, while n runs from 1 to N, the position variable x runs from 0 to L.

The sum over the bars in a histogram gives the area under the curve $p(x)$ when we let $N \to \infty$. You may recall that this is exactly how integration is defined! So we can rewrite the right-hand side of (8.12) as

Fig. 8.2 Towards a continuum of probability densities (see supplementary material 2)

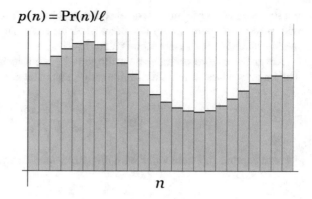

$p(n) = \Pr(n)/\ell$

n

$$\int_0^L p(x)\,dx = 1. \tag{8.13}$$

The probability that the particle is found in the region between $x = a$ and $x = b$ is then given by the area under the curve between a and b:

$$\Pr([a, b]) = \int_a^b p(x)\,dx. \tag{8.14}$$

Now that we have constructed a continuous probability density function $p(x)$ from the discrete probabilities $\Pr(n)$, we need to extend the description of the discrete set of states $|n\rangle$ to a continuum set of states $|x\rangle$. We will see that this leads to a peculiarity in the normalisation of such states.

Since $\Pr(n) = |\langle n|\psi\rangle|^2$, we can write

$$p(n) = \frac{\Pr(n)}{\ell} = \left|\frac{\langle n|\psi\rangle}{\sqrt{\ell}}\right|^2. \tag{8.15}$$

In other words, rather than considering the normalised states $|n\rangle$, we define a new (unnormalised) state

$$|x_n\rangle = \frac{|n\rangle}{\sqrt{\ell}}, \tag{8.16}$$

such that $p(n) = |\langle x_n|\psi\rangle|^2$. In the continuum limit $N \to \infty$, the nth compartment state $|n\rangle$ becomes the position eigenstate $|x\rangle$, where x becomes a real number. However, in this limit we have $\ell \to 0$, which means that the normalisation factor of $|x\rangle$ becomes $1/\sqrt{\ell}$, which goes to infinity as $\ell \to 0$. In other words, the state $|x\rangle$ cannot be normalised! This is something we have to live with, and luckily most of the time it will not cause any trouble. However, it does hint at difficulties when we are dealing with physics at increasingly smaller scales, for example when we want to combine quantum mechanics with general relativity.

Nevertheless, we still have an orthogonality relationship for position eigenstates. The inner product of two different position eigenstates $|x\rangle$ and $|x'\rangle$ must be zero.

$$\langle x|x'\rangle = \begin{cases} 0 & x \neq x', \\ \infty & x = x'. \end{cases} \tag{8.17}$$

We can capture this behaviour with the Dirac delta function δ, such that

$$\langle x|x'\rangle = \delta(x - x'), \tag{8.18}$$

and δ is defined as $\delta(u) = 0$ if $u \neq 0$, and $\delta(u) = \infty$ if $u = 0$. Delta "function" is a bit of a misnomer, because technically it is not a function but a distribution. We cannot properly define it the way we just did, and instead we must define it via a limiting procedure. We will discuss this in more detail in the next mathematical intermezzo.

How do we use delta functions in a calculation? Nothing could be easier! Suppose you encounter the integral

$$\int f(x)\,\delta(x - a)\,dx. \tag{8.19}$$

When you see something like this, you don't have to evaluate the integral at all, just replace the argument x of the function f in the integrand with the value $x = a$ that makes the argument of the delta function zero. This works for any function where a lies in the domain of integration

$$\int f(x)\,\delta(x - a)\,dx = f(a). \tag{8.20}$$

If a is outside the domain of integration, the integral is zero. This means that delta functions are your friend: they may look complicated, but they make integrals disappear without difficult or impossible calculations!

Let's see how this works for the state of a quantum particle. We can write the quantum state $|\psi\rangle$ as a superposition over position states:

$$|\psi\rangle = \int dx\, c(x)|x\rangle, \tag{8.21}$$

with $\int dx\, |c(x)|^2 = 1$ so that $|\psi\rangle$ is properly normalised. Notice how $|\psi\rangle$ can still be normalised, even if $|x\rangle$ cannot. Now, we calculate the inner product of $|\psi\rangle$ with some position eigenstate $|x'\rangle$, which gives

$$\langle x'|\psi\rangle = \langle x'| \int dx\, c(x)|x\rangle$$
$$= \int dx\, c(x)\langle x'|x\rangle = \int dx\, c(x)\,\delta(x - x')\,. \tag{8.22}$$

This is an integral with a Dirac delta function, so we can evaluate it by choosing $x = x'$ for the integrand:

$$\langle x'|\psi\rangle = c(x')\,. \tag{8.23}$$

Even though $|x'\rangle$ (and therefore $\langle x'|$) is not normalisable, the scalar product $\langle x'|\psi\rangle$ is a perfectly well-defined complex number. Traditionally, we choose the symbol ψ for the amplitudes $c(x)$ of a particle's position, so the label in the ket $|\psi\rangle$ matches the label of the amplitudes:

$$\langle x|\psi\rangle = \psi(x) \quad \text{and} \quad |\psi\rangle = \int dx\, \psi(x)|x\rangle\,. \tag{8.24}$$

We can also combine the two equations in (8.24) to obtain an expression for the identity operator in terms of the position eigenstates

$$|\psi\rangle = \int dx\, |x\rangle\, \psi(x) = \int dx\, |x\rangle\langle x|\psi\rangle = \left(\int dx\, |x\rangle\langle x| \right) |\psi\rangle\,, \tag{8.25}$$

and therefore

$$\int dx\, |x\rangle\langle x| = \mathbb{I}\,. \tag{8.26}$$

This is the completeness relation for the position of a particle. The symbol $\psi(x)$ is called the *wave function* of the particle, and it plays a central role in the quantum mechanical treatment of a particle, because it can be used to calculate everything we need. Notice that it is technically not the same as the state; it is the complex amplitude at position x.

How do we use the wave function in quantum mechanical calculations? As the simplest example, we can calculate the average position $\langle x\rangle$ of a particle. For any average, you multiply the value a_i with the probability p_i of that value, and then sum over all values:

$$\langle a\rangle = \sum_i a_i\, p_i\,. \tag{8.27}$$

For the position of a quantum mechanical particle, we multiply each possible infinitesimal position interval $[x, x + dx)$ by the probability $p(x)dx = |\psi(x)|^2 dx$ of finding the particle in that interval, and integrate ("sum") over all positions. This becomes

$$\langle x \rangle = \int_0^L x |\psi(x)|^2 dx .$$ (8.28)

Averages of other functions of position $f(x)$ can also be calculated according to the same rule:

$$\langle f(x) \rangle = \int_0^L f(x) |\psi(x)|^2 dx .$$ (8.29)

For example, we can calculate the square of the position of a particle by evaluating

$$\langle x^2 \rangle = \int_0^L x^2 |\psi(x)|^2 dx .$$ (8.30)

Next, we note that the position of a particle is an observable. It is a physical quantity that can be measured and yields real numbers as measurement outcomes. According to Chap. 5, this means there should be a Hermitian operator $\hat{x}$ associated with the position (technically, we should call this a self-adjoint operator). Indeed, we can derive the form of the operator by requiring the right measurement statistics:

$$\langle \hat{x} \rangle = \int_{-\infty}^{+\infty} x \, p(x) \, dx = \int_{-\infty}^{+\infty} x |\langle x|\psi \rangle|^2 \, dx$$

$$= \int_{-\infty}^{+\infty} dx \, x \langle \psi|x \rangle \langle x|\psi \rangle = \langle \psi| \left(\int_{-\infty}^{+\infty} dx \, x |x \rangle \langle x| \right) |\psi \rangle .$$ (8.31)

Since this must be true for any state $|\psi \rangle$, we find that

$$\hat{x} = \int_{-\infty}^{+\infty} dx \, x |x \rangle \langle x| .$$ (8.32)

This is the position operator. You can check yourself that it is self-adjoint: $\hat{x}^\dagger = \hat{x}$. By inspection you see that the states $|x \rangle$ are the (non-normalisable) eigenstates of $\hat{x}$ with eigenvalues x.

Let us consider now a specific example of a particle in a box of length L. Suppose that the wave function is given by

$$\psi(x) = \sqrt{\frac{2}{L}} \sin \left(\frac{2\pi x}{L} \right) .$$ (8.33)

You can check that the wave function is properly normalised by calculating the integral

$$\int_a^b |\psi(x)|^2 \, dx = \frac{2}{L} \int_a^b \sin^2\left(\frac{2\pi x}{L}\right) dx = \frac{1}{\pi} \int_{u=\frac{2\pi a}{L}}^{v=\frac{2\pi b}{L}} \sin^2 w \, dw$$

$$= \frac{1}{2\pi} \int_u^v [1 - \cos(2w)] \, dw = \frac{1}{2\pi} \left[w - \frac{1}{2}\sin(2w) \right]_{w=u}^{w=v}$$

$$= \frac{b}{L} - \frac{a}{L} - \frac{1}{4\pi}\sin\left(\frac{2\pi b}{L}\right) - \frac{1}{4\pi}\sin\left(\frac{2\pi a}{L}\right). \tag{8.34}$$

When $a = 0$ and $b = L$, we find

$$\int_{a=0}^{b=L} |\psi(x)|^2 \, dx = 1, \tag{8.35}$$

which means that the state is properly normalised. We can plot the probability density $p(x) = |\psi(x)|^2$ as a function of position, which is shown in Fig. 8.3.

The probability of finding the particle in the left-hand side of the box is given by the area under the curve between $x = 0$ and $x = L/2$. By reflection symmetry around the point $x = L/2$, you can see immediately that this should be one half. Indeed, when we calculate the probability, we find

$$\int_0^{+\frac{1}{2}L} |\psi(x)|^2 dx = \frac{2}{L} \int_0^{+\frac{1}{2}L} \sin^2\left(\frac{2\pi x}{L}\right) dx = \frac{1}{2}. \tag{8.36}$$

Fig. 8.3 The probability density $|\langle x|\psi\rangle|^2$ for the position of a particle in a box in the state $\psi(x)$ of (8.33)

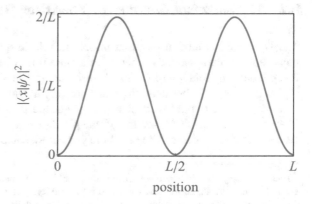

You also see from the plot that the probability that we find the particle at the position $x = 0$ is zero. This raises the question: How does the particle get from the left-hand side of the box to the right-hand side if it has to go through a forbidden point? To answer this, remember that in Chap. 1 we could not say that the photon really went through one arm of the interferometer or the other, since extracting that information via a measurement destroyed the interference in the detectors. When a particle is in a quantum superposition of two states, you cannot say that it really is in one or the other of those two states.

Similarly, here the particle is in a quantum superposition of position states, so we cannot say that the quantum particle really has a position that we just don't know about. The particle does not have a position prior to a measurement! You can therefore not simply talk about the trajectory of a quantum particle in a meaningful way (without adding extra ingredients such as decoherence to make the particle behave more like a classical object). The answer to the question is thus that the question itself is ill-posed, because it mistakenly assumes that quantum particles have trajectories.

The general rule for calculating expectation values with a wave function can now be determined as follows: Suppose we have a state $|\psi\rangle$ and an operator A. Operating on the state produces a new state, which we can denote by $|A\psi\rangle = A|\psi\rangle$. Using the completeness relation, the expectation value of the operator A then becomes

$$\langle A \rangle = \langle \psi | A\psi \rangle = \int_{-\infty}^{+\infty} dx \, \langle \psi | x \rangle \langle x | A\psi \rangle = \int_{-\infty}^{+\infty} dx \, \psi^*(x) A \, \psi(x). \qquad (8.37)$$

The operator A on the right hand side may be a function of x, or even a more complicated operator like a derivative. We will see how this comes about in the next section.

8.2 *Mathematical Intermezzo*: The Dirac Delta Function

The Dirac delta function plays an important role in quantum mechanics—and in physics in general—so let's spend a little more time with it. You have already seen that the defining relation (8.20) can be used to effortlessly evaluate integrals. There is also a variety of other definitions that are very useful to know.

First, we need some kind of limiting procedure to give a precise meaning to infinity[1] (∞) in (8.17). Notice that if you plot $\delta(x - a)$ against the x-axis, you have zero everywhere, except at $x = a$ where you have a spike of infinite height. One way

---

[1] There are many ways in which we can encounter infinity, and many different *types* of infinity. The symbol ∞ is a shorthand for how we ended up at infinity; you cannot use this symbol in regular arithmetic without paying attention to the context of how the infinity came about. This is why we have to introduce limiting procedures that provide this context.

to approximate this is with a very tall, narrow rectangular function of height h and with w at position a, and zero everywhere else. Let's use this approximation of the Dirac delta function with the simplest possible function $f(x)$ in (8.20), namely the constant function $f(x) = c$. We then have

$$\int f(x)\,\delta(x-a)\,dx \approx \int_{a}^{a+w} ch\,dx = ch\,[x]_{x=a}^{x=a+w} = chw = c\,, \qquad (8.38)$$

and we require $hw = 1$. This is true for all functions $f(x)$, so the area under the rectangle is 1. We can now define the Dirac delta function as the limit of a rectangle with width w and height $h = 1/w$ at $x = a$:

$$\delta(x-a) = \lim_{w\to 0} \begin{cases} 0 & x < a \ \ \text{or} \ \ x > a+w\,, \\ \frac{1}{w} & a \le x \le a+w\,. \end{cases} \qquad (8.39)$$

Another useful definition is

$$\delta(a) = \frac{1}{2\pi} \int_{-\infty}^{\infty} e^{iax}\,dx\,. \qquad (8.40)$$

To get a feel for this definition, notice that for each value of e^{iax} there is another value $e^{iax+i\pi} = -e^{iax}$. So all factors cancel each other, except when $a = 0$. In that case all $e^{iax} = 1$, and add to infinity. Various constructions for the delta function are shown in Fig. 8.4.

Fig. 8.4 The Dirac delta function (see supplementary material 4)

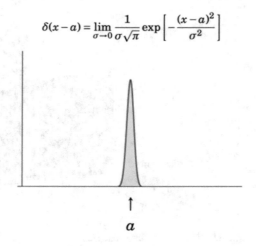

$$\delta(x-a) = \lim_{\sigma\to 0} \frac{1}{\sigma\sqrt{\pi}} \exp\left[-\frac{(x-a)^2}{\sigma^2}\right]$$

a

8.3   The Momentum of a Particle

We have spent quite some time exploring how to describe the position of a particle, and the obvious next question is: how do we describe its velocity v? Normally, we take the time derivative of the position of a particle to get the velocity:

$$v = \frac{dx}{dt}. \tag{8.41}$$

However, the position operator in (8.32) has no time dependence, and we therefore cannot use this traditional definition. Instead we must find another operator that captures the observable of velocity. We will find it more convenient to consider the momentum p instead, and use the relation $p = mv$ (where m is the mass of the particle) if we want to find the velocity. The reason for this is that the momentum is a more fundamental measure of the motion of a particle because it takes into account the inertial mass.

Before we consider the momentum of quantum particles, we briefly consider the famous double-slit experiment by Thomas Young, shown in Fig. 8.5. When we shine light of a specific wavelength onto a screen with two parallel narrow slits, the light behind the screen will give an interference pattern:

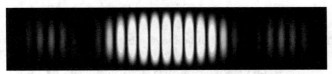

The width and fringe separation of the interference pattern depends on the width of the slit a and the distance d between the slits. Crucially, the interference pattern $I(x)$ also depends on the wavelength λ of the light:

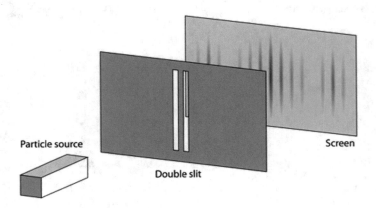

Particle source

Double slit

Screen

Fig. 8.5   The double slit experiment

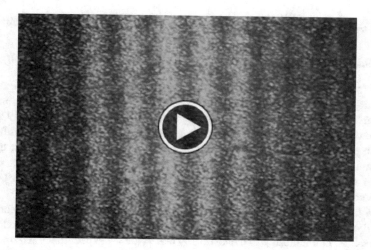

Fig. 8.6 Building up the diffraction pattern with individual electrons. Courtesy of the Central Research Laboratory, Hitachi, Ltd. Japan (see supplementary material 6)

$$I(x) = I_0 \cos^2\left(\frac{\pi d x}{\lambda R}\right) \text{sinc}^2\left(\frac{\pi a x}{\lambda R}\right) , \tag{8.42}$$

where x is the horizontal position on the screen, and R is the distance between the slits and the screen. The constant I_0 is related to the intensity of the light source.

What happens when we use electrons instead of photons? To find out, we prepare electrons with momentum p and send them through the slits and record the positions of the electrons on a screen behind the slits. We find exactly the same type of interference pattern that we obtained using light, shown in Fig. 8.6 (performed in 1989 by A. Tonomura and co-workers at the Central Research Laboratory, Hitachi, Ltd. Japan).

So electrons with momentum p behave in some way as waves with a wavelength λ that can be inferred from (8.42). When we compare the momentum p of the electrons with the wavelength λ inferred from the interference pattern, we experimentally find the relation

$$p = \frac{h}{\lambda} , \tag{8.43}$$

where h is Planck's constant. Does relation (8.43) hold also for photons? We know that light has a wavelength λ, but does it have momentum? Light can push objects just by reflecting off their surface. This radiation pressure on the object causes a change in its momentum, and the conservation of the total momentum requires that the momentum change must be balanced by something. This implies that light has momentum too, even though photons are massless particles. Moreover, if we calculate the radiation pressure, the momentum of a photon with wavelength λ is exactly given by (8.43). We therefore extrapolate that (8.43) captures a fundamental truth about nature:

Particles with momentum p can behave as waves with wavelength $\lambda = h/p$.

This is a very profound result, which became known as the *wave-particle duality*. Each particle has characteristics of a wave, and vice versa. This correspondence between waves and particles was first proposed by Louis de Broglie[2] in 1924, and has so far held up for every type of particle and wave out there with this identification between particle momentum and wavelength. It is a direct manifestation of the fact that quantum mechanics describes objects as quanta. Notice also that this is a relation between wavelength and momentum, not velocity, which is why we prefer to work with momentum.[3]

We will now explore the properties of the momentum states, and give an expression for the momentum operator. The state of an electron with momentum p is denoted as $|p\rangle$. Since the momentum p can take on any real value we expect that the momentum states behave similarly to the position eigenstates. We therefore take

$$\langle p|p'\rangle = \delta(p - p')\,. \tag{8.44}$$

We now wish to find the relationship between the position eigenstates $|x\rangle$ and the momentum states $|p\rangle$.

Since every state, including $|p\rangle$, can be written as a superposition of position states, we can write

$$|p\rangle = \int_{-\infty}^{+\infty} dx\, \psi_p(x)|x\rangle\,, \tag{8.45}$$

where the complex numbers $\psi_p(x) = \langle x|p\rangle$ encode the phases and amplitudes of the wave with momentum p at position x. Normally, the amplitudes and phases are written as $A\cos\phi$ or $A\sin\phi$, but because we work with complex phases we write $A\,e^{i\phi}$. The phase ϕ in the interval $[0, 2\pi)$ indicates where we are along the wave, and can be written as

$$\phi = 2\pi\frac{x}{\lambda}\,. \tag{8.46}$$

We can then write

$$\psi_p(x) = A\,e^{i\phi} = A\,e^{2\pi ix/\lambda} = A\,e^{ipx/\hbar} = \langle x|p\rangle\,, \tag{8.47}$$

---

[2] L. de Broglie, *Recherches sur la théorie des quanta*, PhD thesis, Paris 1924; Annales de Physique **3** 22, 1925.

[3] Also, photons in free space can have varying momentum, but their velocity is always c, which doesn't tell us very much.

where the last equality follows directly from the definition in equation (8.45). We also used $\hbar = h/2\pi$. You can calculate the value for A by evaluating

$$\int dx \, \langle p'|x\rangle\langle x|p\rangle \tag{8.48}$$

using both (8.44) and (8.47), and equating the results. This leads to the fundamental relationship

$$\langle x|p\rangle = \frac{1}{\sqrt{2\pi\hbar}} \, e^{ipx/\hbar} \,. \tag{8.49}$$

Notice that this inner product is never zero, and has a constant modulus for all possible values of x and p. Only the phase ϕ depends on x and p.

Next, we want an expression for the momentum operator $\hat{p}$. We require that the momentum operator satisfies the eigenvalue equation

$$\hat{p}|p\rangle = p|p\rangle \,. \tag{8.50}$$

When we apply the completeness relation of (8.26) for position to both sides of this equation we obtain

$$\hat{p} \int dx \, |x\rangle\langle x|p\rangle = p \int dx \, |x\rangle\langle x|p\rangle \tag{8.51}$$

or

$$\hat{p} \int dx \, \frac{e^{ipx/\hbar}}{\sqrt{2\pi\hbar}}|x\rangle = p \int dx \, \frac{e^{ipx/\hbar}}{\sqrt{2\pi\hbar}}|x\rangle \,. \tag{8.52}$$

In other words, the operator $\hat{p}$ on the left brings the value p down from the exponential in the integral. The simplest mathematical operator that accomplishes this is the derivative:

$$\frac{d}{dx} e^{ipx/\hbar} = \frac{i}{\hbar} \, p \, e^{ipx/\hbar} \,. \tag{8.53}$$

We will have to correct for the factor $i/\hbar$ with a factor $-i\hbar$, which gives for the momentum operator

$$\hat{p} = -i\hbar\frac{d}{dx} \,. \tag{8.54}$$

A quick check of the units shows that the operator $\hat{p}$ has indeed dimensions of momentum.

In order to calculate the average momentum $\langle \hat{p} \rangle$ for a particle in the quantum state $\psi(x)$, we have to sandwich the operator $\hat{p}$ between $\psi^*(x)$ and $\psi(x)$, and integrate over x. The operator acts to the right on the wave function $\psi(x)$. We can do this for the wave function in (8.33) of a particle in a box:

$$
\begin{aligned}
\langle \hat{p} \rangle &= -i\hbar \int_0^L dx\, \psi^*(x) \frac{d}{dx} \psi(x) \\
&= -\frac{2i\hbar}{L} \int_0^L dx\, \sin\left(\frac{2\pi x}{L}\right) \frac{d}{dx} \sin\left(\frac{2\pi x}{L}\right) \\
&= -\frac{4\pi i\hbar}{L^2} \int_0^L dx\, \sin\left(\frac{2\pi x}{L}\right) \cos\left(\frac{2\pi x}{L}\right) \\
&= -\frac{2\pi i\hbar}{L^2} \int_0^L dx\, \sin\left(\frac{4\pi x}{L}\right) \\
&= -\frac{2\pi i\hbar}{L^2} \left[\frac{L}{4\pi} \cos\left(\frac{4\pi x}{L}\right) \right]_0^L = 0 .
\end{aligned}
\tag{8.55}
$$

This makes sense, since a nonzero momentum would imply that the particle moves outside of the boundaries of the box after a long enough time.

You may have noticed that we now have two ways of calculating probabilities, namely via states (using kets and bras), and via wave functions (using integration over functions). Historically, these two approaches were developed side-by-side by Heisenberg[4] and Schrödinger,[5] respectively (even though the ket and bra notation was due to Dirac). The relation between these two approaches is collected in Table 8.1.

8.4   *Mathematical Intermezzo*: **Fourier Transforms**

Sometimes it is easier to solve a problem in quantum mechanics using the position operator, while at other times it is easier to solve them using the momentum operator. In general, when a system behaves in a wave-like manner, it is going to be easier to use momentum states, while systems that behave like particles are better described using position eigenstates. There is an important relationship between the states $|x\rangle$ in

---

[4] W. Heisenberg, *Über quantentheoretische Umdeutung kinematischer und mechanischer Beziehungen*, Zeitschrift für Physik **33** 879, 1925.

[5] E. Schrödinger, *An Undulatory Theory of the Mechanics of Atoms and Molecules*, Phys. Rev. **28** 1049, 1926.

Table 8.1 Comparison between matrix mechanics and wave mechanics

Matrix mechanics	Wave mechanics
Quantum state $|\psi\rangle$	Wave function $\psi(x)$
Position state $|x_0\rangle$	$\psi(x) = \delta(x - x_0)$
Momentum state $|p_0\rangle$	$\psi(x) = \frac{1}{\sqrt{2\pi\hbar}}\, e^{ip_0 x/\hbar}$
$\hat{x} = \int x\,|x\rangle\langle x|\,dx$	$\hat{x} = x$
$\hat{p} = \int p\,|p\rangle\langle p|\,dp$	$\hat{p} = -i\hbar\frac{d}{dx}$
$\langle A\rangle = \langle\psi|A|\psi\rangle$	$\langle A\rangle = \int \psi^*(x)\,A\,\psi(x)\,dx$

position space (which is infinite-dimensional, *not* three-dimensional!) and the states $|p\rangle$ in momentum space. This is given by the *Fourier transform*.

Suppose we have a quantum state $|\psi\rangle$, which can be expressed in terms of positions x via (8.24). But what if we want to express this state in terms of the momenta p? We can manipulate (8.24) as follows:

$$
\begin{aligned}
|\psi\rangle &= \int dx\,\psi(x)|x\rangle = \mathbb{I}\int dx\,\psi(x)|x\rangle \\
&= \left(\int dp\,|p\rangle\langle p|\right)\int dx\,\psi(x)|x\rangle = \int dp\int dx\,\psi(x)|p\rangle\langle p|x\rangle \\
&= \frac{1}{\sqrt{2\pi\hbar}}\int dp\int dx\,\psi(x)\,e^{-ipx/\hbar}|p\rangle \\
&= \int dp\,\Psi(p)|p\rangle\,,
\end{aligned}
\tag{8.56}
$$

where we defined

$$
\Psi(p) = \frac{1}{\sqrt{2\pi\hbar}}\int dx\,\psi(x)\,e^{-ipx/\hbar}\,.
\tag{8.57}
$$

The complex function $\Psi(p)$ of the momentum variable p is the *Fourier transform* of the complex function $\psi(x)$ of the position variable x. We can also reverse this relation to obtain the inverse Fourier transform from momentum to position space

$$
\psi(x) = \frac{1}{\sqrt{2\pi\hbar}}\int dp\,\Psi(p)\,e^{ipx/\hbar}\,.
\tag{8.58}
$$

You can prove this yourself using the expression in (8.40) for the Dirac delta function. The Fourier transform allows us to quickly relate the quantum state in the position space to the quantum state in momentum space by evaluating the integrals in (8.57) and (8.58). The Fourier transform is used in all of physics, but it is especially important in quantum mechanics.

8.5   The Energy of a Particle

Now that we know how to describe the position and momentum of a particle, we can finally derive what is arguably the most famous and important equation in quantum mechanics: the Schrödinger equation for a particle in a potential. As we have seen in Chap. 4, this is an equation that relates the time evolution of the quantum state to the energy. We therefore have to first construct the energy operator, or Hamiltonian.

You know already that the energy of a particle has two parts: kinetic energy and potential energy. The kinetic energy is the part of the energy that you get from movement, and can be written as

$$E_{\text{kinetic}} = \frac{1}{2}mv^2 = \frac{p^2}{2m}, \tag{8.59}$$

where m is the mass of the particle. In order to find the quantum mechanical equivalent, we just replace[6] the momentum p with the momentum operator $\hat{p}$. It is as simple as that:

$$p \rightarrow \hat{p} = -i\hbar\frac{d}{dx}. \tag{8.60}$$

The kinetic energy therefore becomes an operator as well:

$$E_{\text{kinetic}} \rightarrow H_0 = \frac{\hat{p}^2}{2m} = -\frac{\hbar^2}{2m}\frac{d^2}{dx^2}. \tag{8.61}$$

To find the quantum mechanical description of the potential energy, we first recall that the quantum mechanical average of any function f of position x is just given by the integral

$$\langle f(x) \rangle = \int dx \, \psi^*(x) \, f(x) \, \psi(x). \tag{8.62}$$

Therefore, the average potential energy is similarly given by

$$\langle V \rangle = \int dx \, \psi^*(x) \, V(x) \, \psi(x), \tag{8.63}$$

and the quantum mechanical potential energy is just the function $V(x)$. The potential energy V is typically a function of position x, but not of momentum p. The potential does in general depend on the time t, for example when potentials are switched on

---

[6] The procedure of replacing classical quantities with operators is called *quantisation*. Once you know which operators the classical quantities correspond to you can quantise nearly every problem in physics.

for a finite period of time. We therefore write $V(x, t)$. The total energy operator H is then given by

$$H = -\frac{\hbar^2}{2m}\frac{d^2}{dx^2} + V(x, t).\tag{8.64}$$

This is the Hamiltonian for a particle moving in a one-dimensional potential.

Finally, we use this Hamiltonian to get the full Schrödinger equation in one dimension:

$$i\hbar\frac{d}{dt}\psi(x, t) = \left[-\frac{\hbar^2}{2m}\frac{d^2}{dx^2} + V(x, t)\right]\psi(x, t).\tag{8.65}$$

This is a differential equation for the wave function $\psi(x, t)$. For arbitrary potentials $V(x, t)$ such equations are typically difficult to solve, and must often be tackled numerically on a computer. If you want to go any further in your study of quantum mechanics, you must learn the mathematics of differential equations.

We will do one particularly simple example, where the potential is constant: $V(x, t) = V_0$. The Schrödinger equation in (8.65) can then be written as

$$i\hbar\frac{d\psi(x, t)}{dt} = -\frac{\hbar^2}{2m}\frac{d^2\psi(x, t)}{dx^2} + V_0\psi(x, t).\tag{8.66}$$

We can substitute a trial wave function

$$\psi(x, t) = A\exp\left(ikx - \frac{iEt}{\hbar}\right),\tag{8.67}$$

where k is a constant we can choose appropriately, E is the energy of the particle, and A the normalisation of the wave function. Taking the time and space derivatives of $\psi(x, t)$ gives

$$\frac{d\psi(x, t)}{dt} = -\frac{iE}{\hbar}\psi(x, t),\tag{8.68}$$

and

$$\frac{d^2\psi(x, t)}{dx^2} = -k^2\psi(x, t).\tag{8.69}$$

Putting this all back into (8.66) produces the equation

$$\left(E - \frac{\hbar^2 k^2}{2m} - V_0\right)\psi(x, t) = 0.\tag{8.70}$$

Since $\psi(x, t)$ is not zero in general, the term in brackets must be zero. This allows us to solve for k:

$$k = \pm \frac{\sqrt{2m(E - V_0)}}{\hbar} . \tag{8.71}$$

Notice that $\hbar k$ has units of momentum. We can therefore write $p = \pm\sqrt{2m(E - V_0)}$ and the wave function becomes[7]

$$\psi(x, t) = A \exp\left(\frac{ipx}{\hbar} - \frac{iEt}{\hbar}\right) . \tag{8.72}$$

This is the solution for a particle in a constant potential V_0 with energy E and momentum p. The constant A depends on the size of the space, and we sometimes refer to it as the amplitude of the wave function in that part of space. The momentum p can be both positive and negative, corresponding to movement in the positive and the negative x-direction, respectively. When $E < V_0$, the momentum is imaginary! This is not a mistake in the theory. In the next section we will explore what this means.

8.6 ✎ The Scanning Tunnelling Microscope

We have seen already in Chap. 1 that quantum mechanical particles can behave rather strangely, and sometimes defy our classical way of thinking. Another counterintuitive and highly nonclassical consequence of quantum mechanics is quantum tunnelling.

Consider the following simple physical process: You throw a tennis ball against a wall. The ball will bounce back towards you. Now repeat this with a quantum mechanical "ball", for example an atom. Most of the time the atom will bounce right back to you, just like the tennis ball. Every once in a while, however, the atom ends up behind the wall. Obviously, no classical ball can do this; the effect is purely quantum mechanical. To understand this, remember that the position of the atom is described by a wave function. The wall is described by a potential barrier. The potential is high at the location of the wall, and zero everywhere else. The atom bounces off the wall (most of the time) because its energy is lower than the potential barrier that is the wall.

The potential barrier determines the function $V(x)$ in the Hamiltonian in (8.65). Since the wall is there for the duration of the bouncing process, we can assume that V does not change in time. We therefore have for a wall of thickness a at position $x = 0$:

---

[7] If you have already taken a course on special relativity, you may recognise $px - Et$ as the product between the position and momentum four-vectors, even though our discussion has been completely non-relativistic. The quantity $px - Et$ is the accumulated phase of a wave, which is a Lorentz invariant scalar.

Fig. 8.7 Quantum
tunnelling

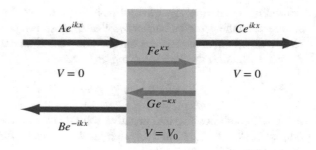

$$V(x) = \begin{cases} 0 & \text{if } x < 0 \text{ or } x > a, \\ V_0 & \text{if } 0 \leq x \leq a. \end{cases} \tag{8.73}$$

We calculated the the wave function of a particle with energy E and mass m in a constant potential V in the previous section, and when we set $t = 0$ in (8.72), we find

$$\psi(x) = A \exp\left(\frac{i}{\hbar}\sqrt{2m(E - V_0)}\, x\right), \tag{8.74}$$

where, for brevity, we will write $k = \sqrt{2m(E - V_0)}/\hbar$. The situation is shown in Fig. 8.7.

On the left-hand side of the potential barrier we have $V = 0$, so the wave function is just a complex wave with amplitude A:

$$\psi(x < 0) = A \exp(ikx). \tag{8.75}$$

However, since there may also be a reflected wave, we have to superpose another wave function (with a different amplitude B): $\psi(x < 0) = B \exp(-ikx)$, and the total wave function ψ_L on the left of the barrier is

$$\psi_L(x < 0) = A\,e^{ikx} + B\,e^{-ikx}. \tag{8.76}$$

The minus sign indicates that the reflected particle is moving in the opposite direction. On the right-hand side of the wall, we have again a solution ψ_R of a very similar form, but possibly with a different amplitude:

$$\psi_R(x > a) = C\,e^{ikx}. \tag{8.77}$$

There is no wave coming from the right in this example, so there is no contribution with $-k$ in the wave function in the region $x > a$.

Next, we wonder what the wave function looks like inside the barrier. Classically, this region is forbidden, so you might expect that $\psi(0 \leq x \leq a) = 0$. However, the

Schrödinger equation says otherwise! Inside the wall the energy of the particle E is smaller than the potential barrier V_0. That makes the square root in (8.74) imaginary, and the argument of the exponential becomes real. If we define $\kappa = \sqrt{2m(V_0 - E)}/\hbar$, the wave function ψ_M inside the region becomes

$$\psi_M(0 \leq x \leq a) = F\, e^{\kappa x} + G\, e^{-\kappa x}, \tag{8.78}$$

again with both positive and negative solutions for κ. We now have to find how the amplitudes A, B, C, F, and G are related. The wave function must be continuous everywhere, which means that

$$\psi_L(0) = \psi_M(0) \quad \text{and} \quad \psi_M(a) = \psi_R(a). \tag{8.79}$$

Similarly, the slopes of the wave functions in the different region must match at the boundaries, so that

$$\left.\frac{d\psi_L}{dx}\right|_{x=0} = \left.\frac{d\psi_M}{dx}\right|_{x=0} \quad \text{and} \quad \left.\frac{d\psi_M}{dx}\right|_{x=a} = \left.\frac{d\psi_R}{dx}\right|_{x=a}, \tag{8.80}$$

and we can solve this to get for the wave function at $x = 0$

$$A + B = F + G \quad \text{and} \quad ik(A_B) = \kappa(F - G), \tag{8.81}$$

and for the wave function at $x = a$

$$C\, e^{ika} = F\, e^{\kappa a} + G\, e^{-\kappa a}$$
$$ikC\, e^{ika} = \kappa(F\, e^{\kappa a} - G\, e^{-\kappa a}). \tag{8.82}$$

We can now solve for the variables A, B, C, F, and G using these equations and the normalisation to find the total wave function. However, it is more instructive to look at the ratios $R = |B|^2/|A|^2$ and $T = |C|^2/|A|^2$, because it tells us how much of the wave is reflected off the barrier, and how much is transmitted through the barrier. When we do this, after a fair amount of algebra we find that

$$R = \left[1 + \frac{4E(V_0 - E)}{V_0^2 \sinh^2(\kappa a)}\right]^{-1}, \tag{8.83}$$

and

$$T = \left[1 + \frac{V_0^2 \sinh^2(\kappa a)}{4E(V_0 - E)}\right]^{-1}. \tag{8.84}$$

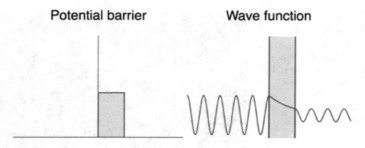

Fig. 8.8 The effect of the barrier on quantum tunnelling (see supplementary material 8)

This means that even though the energy of the particle is lower than the potential of the barrier, it still has a small chance of being found behind the barrier. This is quantum tunnelling, shown in Fig. 8.8.

> A similar phenomenon exists for classical electromagnetic waves that impinge on a conducting surface. When the conductivity is finite, the electromagnetic waves penetrate the conductor, and the amplitude will fall off exponentially with a characteristic length we call the skin depth. If the conducting material is thin enough, some of the wave will pass through. The remarkable thing in quantum mechanics is that this classical wave phenomenon also applies to particles due to the wave-particle duality.

There is a very beautiful application of quantum tunnelling, namely scanning tunnelling microscopy.[8] In this device, a very sharp needle (the tip of which is only a few atoms wide) is brought close to the surface of a material and given a positive electric potential. If the tip were to touch the surface, the electric potential would cause electrons to flow into the needle, creating a current. However, by keeping the needle slightly away from the surface, there is a small gap that acts as a potential barrier for the electrons on the surface. If the gap is big enough, the electrons will not have sufficient energy to jump the gap, and the only way to make it into the needle is via quantum tunnelling. The current measured in the needle is then directly proportional to the amount of tunnelling, given by T in (8.84). Since this depends on the separation a between the tip and the surface, we can scan the tip over the surface and obtain a current intensity profile. This is directly related to the profile of the surface, and can be used to "see" the surface of a material, for example silicon (shown in Fig. 8.9). This way, we can visualise features that are much too small to see with an optical microscope.

---

[8] G. Binnig and H. Rohrer, *Scanning tunneling microscopy*, IBM Journal of Research and Development **30** 355, 1986.

Fig. 8.9 An image of the surface of a silicon crystal from a scanning quantum tunnelling microscope. You can see the hexagonal crystal structure of silicon, as well as a lot of defects. Image taken by Dr. Ashley Cadby at the University of Sheffield

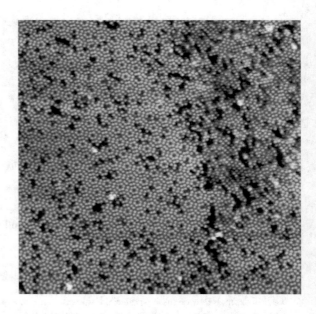

8.7 ✆ A Brief Glance at Chemistry

One of the great early triumphs of quantum mechanics is that it explains how chemistry works. While this requires a good working knowledge of how to solve partial differential equations, and is well beyond the scope of this book, we can still get a feel for how quantum mechanics explains the structure of atoms and how they can interact with each other. To see this, we need two ingredients: the quantum behaviour of electrons bound to the nucleus by the Coulomb force, and the Pauli Exclusion Principle.[9] We have not encountered this principle before, since it does not follow directly from the theory we have developed so far. Instead, it is a very deep result that follows from a relativistic description of quantum mechanics, which goes far beyond our current level. Luckily for us, we can state the principle quite easily for our purposes:

Two electrons can not be in the same quantum state at the same time.

This means that if we want to put two electrons in a box, they cannot both have the same energy, spin, position and momentum. At least one of these must be different. While the Pauli Exclusion Principle is not an actual force, in some sense it has the effect that two electrons repel each other. This repulsion is different from the Coulomb force, which is true for all types of particles with like charge. The Pauli Exclusion Principle specifically holds for indistinguishable particles with half-integer

---

[9] W. Pauli, *Über den Zusammenhang des Abschlusses der Elektronengruppen im Atom mit der Komplexstruktur der Spektren*, Zeitschrift für Physik **31** 765, 1925.

spin ($S = \hbar/2$, $S = 3\hbar/2$, $S = 5\hbar/2$, etc.), regardless of any other forces on the particles.

We now discuss in broad strokes how quantum mechanics describes the structure of atoms. The simplest atom—the hydrogen atom—consists of a negatively charged electron bound to a positively charged proton. Since the proton is almost two thousand times heavier than the electron it will not move much under the Coulomb force of the electron, and we can assume that the proton sits at the centre of our coordinate system, and the electron orbits the proton. The Coulomb force F acting on the electron due to the proton can be written as

$$F = -\frac{e^2}{4\pi \epsilon_0 r^2},$$
(8.85)

where e is the charge of the electron and the proton, r is the distance between the electron and the proton, and ϵ_0 is the permittivity of free space.

However, in quantum mechanics we do not really deal with forces, so unfortunately we cannot use (8.85) directly. Instead, we need the kinetic and potential energy of the electron, so we can construct the Hamiltonian for the Schrödinger equation. We need to translate the Coulomb force to the electrostatic potential $\Phi(r)$ created by the proton, as felt by the electron:

$$\Phi(r) = \frac{e}{4\pi \epsilon_0 r}.$$
(8.86)

The potential energy V of the electron in the electric field of the proton is the electron charge times the electrostatic potential of the proton: $V(r) = -e\,\Phi(r)$. The total Hamiltonian, including the kinetic energy of the electron, then becomes

$$H = \frac{\hat{p}^2}{2m} - \frac{e^2}{4\pi \epsilon_0 r},$$
(8.87)

with m the electron mass, and

$$\frac{\hat{p}^2}{2m} = -\frac{\hbar^2}{2m}\left(\frac{\partial^2}{\partial x^2} + \frac{\partial^2}{\partial y^2} + \frac{\partial^2}{\partial z^2}\right).$$
(8.88)

The symbol $\partial/\partial x$ indicates (partial) differentiation in the x direction, and so on. The Schrödinger equation for the hydrogen atom then becomes

$$i\hbar \frac{d}{dt}\psi(\mathbf{r}, t) = \left(\frac{\hat{p}^2}{2m} - \frac{e^2}{4\pi \epsilon_0 r}\right)\psi(\mathbf{r}, t).$$
(8.89)

Solving this equation is beyond the scope of this book, since it requires advanced knowledge of partial differential equations, and we have not developed this here.

Nevertheless, the Schrödinger equation can be solved, and in particular we can find the eigenstates ψ_n of the Hamiltonian from the eigenvalue equation

$$H \, \psi_n = E_n \, \psi_n \, , \tag{8.90}$$

where E_n are the energy eigenvalues for the electron.

When we solve for ψ_n in spherical coordinates r, θ, and ϕ, we find that it depends on two special functions, namely a radial function $R_{nl}(r)$ and the spherical harmonics $Y_{lm}(\theta, \phi)$ for the angular coordinates:

$$\psi_{nlm}(\mathbf{r}) = R_{nl}(r) \, Y_{lm}(\theta, \phi). \tag{8.91}$$

We will not study these special functions here, but they play a central role in many branches of physics. The wave function depends not only on the energy label n, but also on two other numbers l and m. Together with the spin $s = \pm 1/2$, these are the quantum numbers of the electron in the hydrogen atom. The energy quantum number n is also called the *principal* quantum number.

The second quantum number l can take integer values from 0 to ∞, and denotes the angular momentum of the electron in the orbit around the proton. Angular momentum a quantity analogous to ordinary momentum $p = mv$, but for rotational motion instead of linear motion. It is conserved for the electron in the hydrogen atom.

Finally, the third quantum number m is the magnetic quantum number. This becomes important when we place the atom in a magnetic field. The value of the magnetic quantum number is bounded by l, such that $-l \leq m \leq l$. The three quantum numbers n, l, and m completely determine the spatial state of the electron (not taking into account its spin), and we can write it in shorthand as $|n, l, m\rangle$. They are orthogonal, in the sense that

$$\langle n', l', m'|n, l, m\rangle = \delta_{nn'} \, \delta_{ll'} \, \delta_{mm'} \, , \tag{8.92}$$

where δ_{nm} is zero when $n \neq m$, and one when $n = m$. This is the Kronecker delta, and it is the discrete version of the Dirac delta function. The spatial part of the wave function in spherical coordinates can then be found from $\psi_{nlm}(\mathbf{r}) = \langle r, \theta, \phi|n, l, m\rangle$, analogous to (8.24).

The energy eigenvalues of the electron in the hydrogen atom also follow from solving the eigenvalue (8.90), and are given by

$$E_{nls} \approx -\frac{mc^2\alpha^2}{2} \frac{1}{n^2}. \tag{8.93}$$

The number $\alpha \simeq 1/137$ is the fine structure constant. This is still not completely in agreement with experimental values, because there is another effect in play—the so-called *spin-orbit* coupling—which modifies the eigenvalues slightly. However, quantum mechanics has no problem including this effect.

Fig. 8.10 Probability densities for the electron in hydrogen (see supplementary material 10)

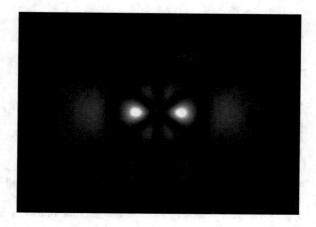

Next, we want to know what the orbitals look like. To find the actual shapes of the electronic orbitals in hydrogen, we have to calculate the square of the wave function, since it is the probability (density) for finding the electron at position **r**. Note that this is the best we can do in quantum mechanics: The electron does not orbit the nucleus in the same way a planet orbits the Sun. It does not have a path around the nucleus. We often speak of the electron "cloud", which is denser where it is more likely to find the electron. The probability densities for an electron in a hydrogen atom with the first five principal quantum numbers are plotted in Fig. 8.10. We restrict the plots to the electrons in the xz-plane for practical reasons.

So far, we have considered only a single electron in a bound state around a single proton. What about other atoms such as helium, carbon, and oxygen? If we treat the nucleus as a point at the centre of the atom (which is a very good approximation), then an atom with Z protons in the nucleus produces an electrostatic potential that is Z time stronger than the nucleus of hydrogen, which is a single proton. Any neutrons that may be present in the nucleus do not contribute to the electrostatic potential, since they are electrically neutral.

An atomic nucleus with Z protons can attract Z electrons before the atom becomes electrically neutral. Indeed, this is what tends to happen in ordinary matter: an atom captures Z electrons and will appear electrically neutral from a reasonably large distance (which in this case is only a few nanometers). However, all electrons have the same electrical charge $-e$, and like charges repel each other. As a crude approximation, we can assume that the electrons are so strongly attracted to the nucleus that they do not feel the repulsive Coulomb force of the other electrons. Of course, this is not true in reality, and the term in the Hamiltonian due to the electron-electron interaction will cause a shift in the energy levels of the atom. This can be observed in absorption and emission spectra (see Chap. 4 on how interactions change the energy eigenvalues of the Hamiltonian matrix). Here, we ignore this effect, because we merely seek a qualitative understanding of atoms and chemistry at this point. Nevertheless, quantum mechanics can deal with this complication too. So the electrons in an atom mostly ignore each other, and can be modelled approximately as individual electrons in a central potential of the form

$$V = -\frac{Ze}{4\pi \epsilon_0 r}. \tag{8.94}$$

Therefore, the solutions for the electrons in orbit around an atomic nucleus will look similar to the solutions we have found for the hydrogen atom. However, one crucial effect for electrons in atoms is far from negligible, and that is Pauli's Exclusion Principle.

If we want to put more than one electron in orbit around the nucleus, no two electrons can be in states that have equal n, l, m, and spin due to the Exclusion Principle. This means that the electrons with the same spin and principal quantum number n must have either different angular momentum l or a different magnetic quantum number m (or both). As we add electrons to orbits around the nucleus, they fill the accessible quantum states $|n, l, m\rangle$ and form a complex electronic structure.

The accessible quantum states are organised in "shells", corresponding to the different energy levels n. The angular momentum states are traditionally called "s" for $l = 0$, "p" for $l = 1$, "d" for $l = 2$, and "f" for $l = 3$. Larger values of l continue in alphabetical order ("g", "h", etc.). You may think that the electrons fill the shells one by one, but this is not the case. The energy eigenvalue in (8.93) depends on l as well as on the spin of the electron s, and this means that the filling order of the electron shells takes place as shown in Fig. 8.11. As a result, we first fill the lowest energy state $n = 1$, which can have only $l = 0$ and $m = 0$. We still have the spin value of the electron, $s = \pm 1/2$, so by choosing two electrons with opposite spin we can add two electrons to the lowest shell. We denote these electronic configurations by

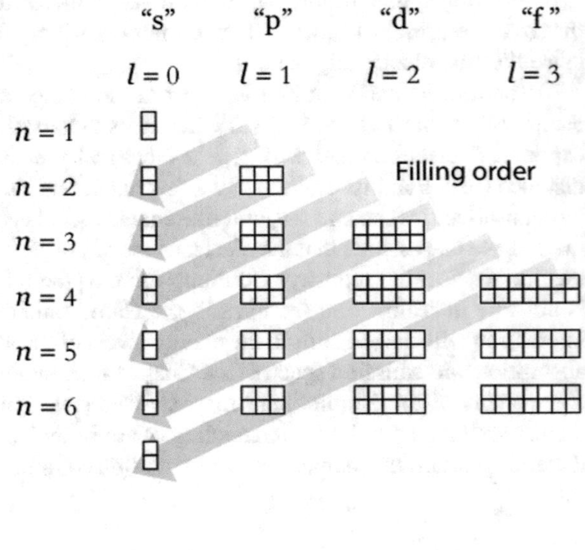

Fig. 8.11 Atom energy shell filling order

$$1s^1 \quad \text{and} \quad 1s^2 , \tag{8.95}$$

where the superscript indicates how many electrons populate the level 1s. These are the electronic configurations of hydrogen (H) and helium (He).

Next, we put three electrons in orbit around a nucleus with three protons. The first shell is filled, which means that we must move to the next shell. According to the filling order shown in Fig. 8.11 this is the $n = 2, l = 0$ state, which in our new notation can be written as

$$\text{Li} : 1s^2 \, 2s^1 , \tag{8.96}$$

Similarly, we can write for beryllium

$$\text{Be} : 1s^2 \, 2s^2 , \tag{8.97}$$

and for boron

$$\text{B} : 1s^2 \, 2s^2 \, 2p^1 . \tag{8.98}$$

Since for the heavier elements we end up with long strings of filled shells that do not change for subsequent elements, we often drop the specification of the lower filled shells. For example, we can denote the electronic structure of gold by

$$\text{Au} : [\text{Xe}] \, 6s^2 \, 4f^{14} \, 5d^9 , \tag{8.99}$$

where [Xe] is the electronic structure of the noble gas xenon:

$$\text{Xe} : 1s^2 \, 2s^2 \, 2p^6 \, 3s^2 \, 3p^6 \, 4s^2 \, 3d^{10} \, 4p^6 \, 5s^2 \, 4d^{10} \, 5p^6 . \tag{8.100}$$

Noble gases are elements that have a completely filled outer electronic shell (the p shell, except for helium), and are shown in the right-most column of the periodic table.

When atoms interact with other atoms, it is the outer electrons that determine their behaviour. These are the valence electrons. Atoms bond to each other when the valence electrons from one atom interact with the valence electrons from another atom. When the outer shell of an atom is full, there are no valence electrons, and the atom does not interact with other atoms much. Such an element is called inert, and these are the noble gases (Fig. 8.12).

The shape of the electronic clouds, shown in Fig. 8.10, determines how the atoms form into molecules. They explain why the angle of two hydrogen atoms with respect to an oxygen atom is 104.45° in a water molecule, and why diamonds have a tetrahedral crystal structure.

This has been merely a bird's eye overview of how quantum mechanics can explain the basic principles behind chemistry. The calculations that must be performed to obtain precise numbers corresponding to the results found in experiments

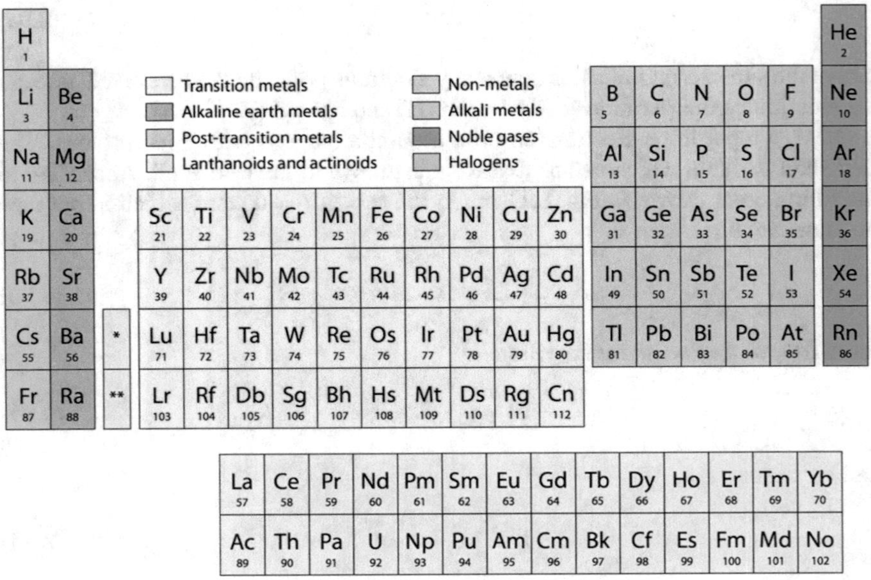

Fig. 8.12 The periodic table of elements (see supplementary material 12)

are increasingly complex. For elements heavier than hydrogen we cannot even find general formulas that describe the atom, and we are forced to use approximation methods such as perturbation theory. This still allows us to make calculations to any required precision, and the agreement with experiments is remarkable. To date, we have not found any case where quantum mechanics cannot explain the results we see in experiments, except for the situation where the calculations become too big to run on a computer.

And all this we obtain from the basic theory we set out in this book. We can say with confidence that quantum mechanics is the most successful theory of nature that we have come up with so far, both in its scope of applicability and in its precision with which it predicts physical processes and quantities.

Exercises

1. A particle with mass m in a one-dimensional tube of length L has a Hamiltonian H that is given by

$$H = -\frac{\hbar^2}{2m}\frac{d^2}{dx^2} + V(x),$$

where $V(x) = 0$ when $0 \leq x \leq L$, and $V(x) = \infty$ otherwise (the value of ∞ means that the walls are impenetrable).

(a) Show that the wave functions

$$\psi_n(x) = \sqrt{\frac{2}{L}} \sin\left(\frac{n\pi x}{L}\right)$$

are eigenfunctions of the Hamiltonian such that $H\psi_n = E_n\psi_n$, and $\psi_n = 0$ at the edges of the tube. Sketch the functions $\psi_n(x)$ for $n = 1, 2, 3, 4$.

(b) Show that the energy eigenvalues E_n are given by

$$E_n = \frac{n^2\pi^2\hbar^2}{2mL^2}.$$

(c) Are the wave functions $\psi_n(x)$ eigenfunctions of the *momentum* operator?

(d) Show that functions of the form $\phi_k(x) = \exp(\pm ikx)$ are solutions to the eigenvalue equation for momentum

$$-i\hbar\frac{d}{dx}\phi_k(x) = p\,\phi_k(x).$$

Determine the relationship between k and p.

(e) Construct the energy eigenfunctions $\psi_n(x)$ from the momentum eigenfunctions $\phi_k(x)$. What is the physical meaning of this relationship?

2. Rather than working out the energy eigenstates of a particle in a one-dimensional box, we may want to study more general time evolutions of the particle. A general wave function for the particle can be written as a superposition over all the energy eigenfunctions:

$$\psi(x) = \sum_n c_n\psi_n(x).$$

Determine the state $\psi(x, t)$ after some time t has elapsed.

3. Calculate the position uncertainty Δx for a momentum eigenstate $|p\rangle$, and the momentum uncertainty Δp for a position eigenstate $|x\rangle$. Together with the De Broglie relation $p = h/\lambda$, this implies that wave behaviour is manifested when momentum states are prepared or measured, while particle behaviour is manifested when position states are prepared or measured. What happens when the position is measured for particles prepared in a momentum eigenstate?

4. By using the chain rule for differentiation, calculate an expression for the commutator $[\hat{x}, \hat{p}]$ between the position and momentum operator. You will have to consider how the commutator acts on an arbitrary wave function $\psi(x)$.

5. Find the normalisation constant in (8.49).

6. Consider the wave function

$$\psi(x) = \frac{1}{\sqrt[4]{\pi\sigma^2}} \exp\left[-\frac{(x - x_0)^2}{2\sigma^2}\right],$$

with σ a positive real number and the position x running from $-\infty$ to $+\infty$.

(a) What is the probability of finding a particle in the position interval $[x, x + dx)$? Sketch the probability density function $|\psi(x)|^2$.

(b) Calculate the Fourier transform $\Psi(p)$ of $\psi(x)$. What is the probability of finding a particle with momentum in the interval $[p, p + dp)$? Sketch the probability density function $|\Psi(p)|^2$.

(c) By varying σ, show how a sharper position will lead to a more uncertain momentum, and vice versa.

7. Derive the reflection and transmission coefficients for quantum tunnelling in (8.83) and (8.84). Show that these quantities behave appropriately when $V_0 \to 0$ and $V_0 \to \infty$. What happens when $a \to 0$?

8. The quantum state of a wave with momentum p is given in position space by

$$|\psi\rangle = \frac{1}{\sqrt{2\pi\hbar}} \int dx \ e^{ipx/\hbar} |x\rangle \,.$$

express this state in momentum space.

Chapter 9
Quantum Uncertainty

In this chapter, we show how uncertainty relations arise naturally from quantum theory for position and momentum, time and energy, and so on. We explore the meaning of these relations and apply them to the quantum mechanical description of the pendulum. We then use our new-found knowledge of the pendulum to construct Schrödinger cat-like states in opto-mechanical systems. Finally, we show how we can use entanglement to enhance the precision of measurements.

9.1 Quantum Uncertainty Revisited

We have seen in Chaps. 1, 2, and 3 that a system in quantum mechanics can have a quantum uncertainty; the path of a photon in an interferometer may be uncertain, or an electron spin in the z-direction is fundamentally uncertain about the spin in the x- and y-directions. Moreover, in Chap. 7 we saw that this quantum uncertainty is different from classical uncertainty, in that nature itself does not have the information about the system. It is not a lack of knowledge on our part that makes the path of a photon inside a Mach–Zehnder interferometer uncertain, it is an inherent uncertainty in the path of the photon. When we take away this uncertainty via a QND detector we lose the interference in our photodetectors at the output of the Mach–Zehnder interferometer.

In this chapter, we will study in more detail how these uncertainties come about, and how the uncertainty in one observable of a system is related to the uncertainty

Supplementary Information The online version contains supplementary material available at https://doi.org/10.1007/978-3-031-16165-0_9.

P. Kok, *A First Introduction to Quantum Physics*, Undergraduate Lecture Notes in Physics, https://doi.org/10.1007/978-3-031-16165-0_9

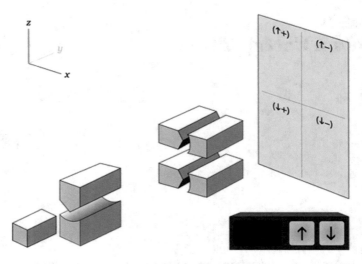

Fig. 9.1 Daisy-chaining two Stern-Gerlach experiments (see supplementary material 1)

of another observable. The famous Heisenberg Uncertainty Principle[1] says that it is impossible to measure both the position and the momentum of a particle with arbitrary precision. This principle was historically important in the development of quantum mechanics. Modern treatments of quantum mechanics tend to move away from the uncertainty principle and talk instead of "uncertainty relations", since they follow directly from the rules of quantum mechanics that we identified at the end of Chap. 5. There is no need for a fundamental "Principle" in addition to these rules. A second reason is that there are many ways in which we can define a (quantum) uncertainty. Heisenberg's Uncertainty Principle cannot tell the difference between these different definitions very well, but our modern formulation of the uncertainty relations is perfectly equipped to handle these complications.

To remind ourselves of the issue, consider the following experiment: We prepare an electron in the spin state $|\uparrow\rangle$ in the z-direction and send it into a Stern-Gerlach apparatus with magnets oriented in the z-direction. As a result, the electron will always be deflected upwards. If we create the electron spin in the state $|\downarrow\rangle$, the electron will be deflected downwards. However, instead of hitting a screen, we let the electron continue into a subsequent pair of Stern-Gerlach apparatuses, both oriented in the x-direction (see Fig. 9.1), before being detected on a fluorescent screen. We observe that the electrons indeed always appear either in the upper two quadrants or the lower two quadrants, as expected from the state preparation $|\uparrow\rangle$ or $|\downarrow\rangle$, but the outcome of the spin in the x-direction is completely uncertain. The electrons end up randomly with equal probability in the left or the right quadrant.

---

[1] W. Heisenberg, *Über den anschaulichen Inhalt der quantentheoretischen Kinematik und Mechanik*, Zeitschrift für Physik, **43** 172, 1927.

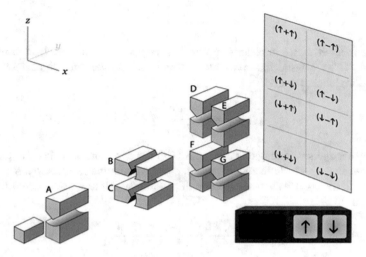

Fig. 9.2   Daisy-chaining three Stern-Gerlach experiments (see supplementary material 2)

Next, we replace the screen in the experiment with two subsequent Stern-Gerlach apparatuses with a third set of Stern-Gerlach apparatuses, this time again oriented in the z-direction. We collect the electrons again on a screen, as shown in Fig. 9.2. The eight positions on the screen provide enough possibilities that we can infer the original spin in the z-direction, the spin in the x-direction in the second apparatus, and the spin in the z-direction in the last apparatus. When we prepare electrons in the spin states $|\uparrow\rangle$ or $|\downarrow\rangle$ and let them pass through the three Stern-Gerlach apparatuses, we observe that the spin as recorded on the screen is no longer always in the same spin state as the preparation procedure: Even if we prepare the spin state in $|\uparrow\rangle$ initially, there is a 50:50 chance of finding the electron in the final spin state $|\uparrow\rangle$ or $|\downarrow\rangle$. *The measurement of the spin in the x-direction has disturbed the spin state in the z-direction.*

We can explain the measurement outcomes by calculating the state of the electron spin and the electron position at the screen, and show exactly how the interaction of the spin with the magnets will lead to the uncertainty in the measurement outcomes. First, let's assume that the spin state of the electron is selected as $|\uparrow\rangle$. The electron will always be deflected upwards by the magnets labeled "A". The state after magnet "A" is therefore

$$|\uparrow, \text{up}\rangle . \tag{9.1}$$

This is a state of two degrees of freedom for the electron, namely the spin value and the path information. We treat it in the same way as we did for composite systems in Chap. 6. Similarly, the electron spin $|\downarrow\rangle$ will become

$$|\downarrow, \text{down}\rangle \tag{9.2}$$

after the magnet. Next, in order to predict the behaviour of the electron in magnets "B" and "C", we write the spin state $|\uparrow\rangle$ in terms of the spin eigenstates in the x-direction, labeled $+$ and $-$:

$$|\uparrow, \text{up}\rangle = \frac{|+, \text{up}\rangle + |-, \text{up}\rangle}{\sqrt{2}} . \tag{9.3}$$

After the second set of magnets ("B" and "C") the component of the quantum state in the spin state $|+\rangle$ is deflected to the left, and the component in the spin state $|-\rangle$ is deflected to the right. When we include this in the description of the quantum state, we obtain

$$\frac{|+, \text{up}, \text{left}\rangle + |-, \text{up}, \text{right}\rangle}{\sqrt{2}} . \tag{9.4}$$

Notice that we now have entanglement between the spin state $|\pm\rangle$ and whether the electron is deflected to the left or to the right. The third set of magnets ("D" to "G") are oriented again in the z-direction, and we need to write the spin states $|\pm\rangle$ in terms of $|\uparrow\rangle$ and $|\downarrow\rangle$. This will determine whether the electron will be deflected upwards or downwards in the magnets. The state becomes

$$|\psi\rangle = \frac{1}{2}|\uparrow, \text{up}, \text{left}, \text{up}\rangle + \frac{1}{2}|\downarrow, \text{up}, \text{left}, \text{down}\rangle$$
$$+ \frac{1}{2}|\uparrow, \text{up}, \text{right}, \text{up}\rangle - \frac{1}{2}|\downarrow, \text{up}, \text{right}, \text{down}\rangle . \tag{9.5}$$

Notice that the spin components $|\downarrow, \text{up}, \text{left}, \text{down}\rangle$ and $-|\downarrow, \text{up}, \text{right}, \text{down}\rangle$ do not cancel, because the electron is in a different path in each case (right, instead of left). This lack of cancellation, or interference, means that we will see the electron appear both up and down on the screen in the upper two quadrants. Note also that there is entanglement between the spin and the path of the electron.

> We can retrieve the interference in $|\psi\rangle$ *when we use magnets to deflect the left and the right electron beams, so that they merge again. This is called "quantum erasure".*

9.2 Uncertainty Relations

It is the eigenstates of the observables S_z and S_x that are the cause of the uncertainty: the states $|\uparrow\rangle$ and $|\downarrow\rangle$ are different from $|+\rangle$ and $|-\rangle$, which is what causes the presence of both the upper and lower paths in (9.5). So in general, we expect that the uncertainty in a measurement of two observables A and B will be higher the more the eigenstates of A and B differ. Moreover, there should be no uncertainty if the eigenstates are the same, that is, when A and B commute.

Since physics is a quantitative science, we want to know precisely how big this uncertainty is. The so-called *Robertson relation*[2] relates the uncertainty in two observables A and B to the expectation value of their commutator:

$$\Delta A \ \Delta B \geq \frac{1}{2} |\langle [A, B] \rangle| \,. \tag{9.6}$$

This is what people usually mean by the uncertainty relation. It is a mathematical expression that is different from Heisenberg's Uncertainty Principle, which is the qualitative statement given in the previous section. We will now give the proof of the uncertainty relation, but it is not essential to the rest of the chapter, and if you wish you can skip to the next section.

✎ *Proof of the Uncertainty Relation*

One of the most useful theorems in algebra is the Cauchy–Schwarz inequality for scalar products. We will just state it here without proof:

$$\langle f|f \rangle \langle g|g \rangle \geq |\langle f|g \rangle|^2 \,. \tag{9.7}$$

Intuitively, you can see that this is true for real vectors $\mathbf{u}$ and $\mathbf{v}$ (with lengths u and v) that have a scalar product $\mathbf{u} \cdot \mathbf{v} = uv \cos \theta$, with θ the angle between the two vectors. The Cauchy–Schwarz inequality for these vectors becomes

$$u^2 \, v^2 \geq (uv \cos \theta)^2 \quad \text{or} \quad \cos^2 \theta \leq 1 \,. \tag{9.8}$$

The Cauchy–Schwarz inequality tells us that this is true for general complex vectors in arbitrary dimension as well.

The right-hand side of this equation is the modulus square of a complex number $\langle f|g \rangle$, and we can bound this by the imaginary part of $\langle f|g \rangle$:

$$|\langle f|g \rangle|^2 = |\text{Re}\langle f|g \rangle|^2 + |\text{Im}\langle f|g \rangle|^2 \geq |\text{Im}\langle f|g \rangle|^2 = \frac{1}{4} |\langle f|g \rangle - \langle g|f \rangle|^2 \,, \tag{9.9}$$

where we used that $\text{Im}\, z = (z - z^*)/2i$. The Cauchy–Schwarz inequality then becomes

$$\langle f|f \rangle \langle g|g \rangle \geq \frac{1}{4} |\langle f|g \rangle - \langle g|f \rangle|^2 \,. \tag{9.10}$$

---

[2] H. P. Robertson, *The Uncertainty Principle*, Phys. Rev. **34** 163, 1929.

This is the inequality that we will use to derive the Robertson relation.

Let $|f\rangle = (A - \langle A\rangle)\,|\psi\rangle$ and $|g\rangle = (B - \langle B\rangle)\,|\psi\rangle$. We calculate the scalar products $\langle f|f\rangle$ and $\langle g|g\rangle$ as

$$\langle f|f\rangle = \langle\psi|(A - \langle A\rangle)^2|\psi\rangle = (\Delta A)^2 \,,$$
$$\langle g|g\rangle = \langle\psi|(B - \langle B\rangle)^2|\psi\rangle = (\Delta B)^2 \,. \qquad (9.11)$$

The scalar product between $|f\rangle$ and $|g\rangle$ can be calculated as

$$\langle f|g\rangle = \langle\psi|AB|\psi\rangle - \langle A\rangle\,\langle B\rangle \quad \Rightarrow \quad \langle g|f\rangle = \langle\psi|BA|\psi\rangle - \langle A\rangle\,\langle B\rangle \,. \qquad (9.12)$$

This leads to

$$|\langle f|g\rangle - \langle g|f\rangle|^2 = |\langle\psi|AB - BA|\psi\rangle|^2 \,. \qquad (9.13)$$

Putting everything together, we obtain

$$(\Delta A)^2\,(\Delta B)^2 \geq \frac{1}{4}\,|\langle[A,\,B]\rangle|^2 \,, \qquad (9.14)$$

which gives the Robertson relation in (9.6) once we recognise that ΔA and ΔB are real positive numbers.

9.3 Position-Momentum Uncertainty

We can use the Robertson formula from the previous section to find the uncertainty relation between position $\hat{x}$ and momentum $\hat{p}$, with

$$\hat{x} = x \qquad \text{and} \qquad \hat{p} = -i\hbar\frac{d}{dx}\,. \qquad (9.15)$$

The commutator between $\hat{x}$ and $\hat{p}$ can be calculated by remembering that it is an operator acting on a state:

$$[\hat{x},\,\hat{p}]\,|\psi\rangle = -i\hbar\left(x\frac{d}{dx} - \frac{d}{dx}x\right)|\psi\rangle = -i\hbar x\frac{d}{dx}|\psi\rangle + i\hbar\frac{d}{dx}(x|\psi\rangle)$$
$$= -i\hbar x\frac{d}{dx}|\psi\rangle + i\hbar|\psi\rangle + i\hbar x\frac{d}{dx}|\psi\rangle = i\hbar|\psi\rangle\,, \qquad (9.16)$$

where we used the product rule for differentiation from the first to the second line. Since this is true for any state $|\psi\rangle$, we usually abbreviate this as

$$[\hat{x},\,\hat{p}] = i\hbar\,. \qquad (9.17)$$

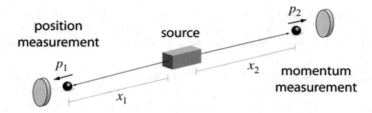

Fig. 9.3 Simultaneous position and momentum measurements on two particles

Substituting this into the Robertson relation in (9.6) gives the celebrated uncertainty relation between position and momentum:

$$\Delta x \ \Delta p \geq \frac{\hbar}{2}. \tag{9.18}$$

It tells us that the uncertainty Δx of the position of a particle means that the uncertainty in the momentum Δp in a subsequent measurement is at least $\hbar/(2\Delta x)$. These uncertainties are a combination of the inherent quantum uncertainty and the uncertainty associated with our classical lack of knowledge about the state of the particle (the uncertainties Δx and Δp are calculated with respect to the state of the system, which may be mixed, as we discovered in Chap. 7). Even when we know all there is to know about the state of a particle, there is an inherent quantum uncertainty in the position and the momentum of the particle due to the fact that the position and momentum operators do not have the same eigenstates, and therefore do not commute.

Albert Einstein was famously unhappy about the inherent quantum uncertainty in the position and the momentum of a particle, and tried to argue that the quantum mechanical description of physical systems is not complete. In 1935 he devised a thought experiment with Boris Podolski and Nathan Rosen that was supposed to prove the incompleteness of quantum mechanics. This has become known as the EPR paradox.[3]

The experiment goes as follows: Suppose we construct a source that sends two particles in opposite directions with equal but opposite momentum, so that the total momentum of the two particles is zero (see Fig. 9.3). We can then measure the position of particle 1, and the momentum of particle 2. There is no reason we cannot make each measurement arbitrarily precise: The measurements do not simultaneously measure the non-commuting observables $\hat{x}$ and $\hat{p}$ for each particle.

If the momentum of particle 1 is measured as p, we can infer that the momentum of particle 2 is $-p$ (since $p_1 + p_2 = 0$). Indeed, if we measured the momentum of particle 2 we would always find $-p$. However, instead of the momentum, we measure the position of particle 2 with respect to the position of the source, yielding x. If the

[3] A. Einstein, B. Podolsky, and N. Rosen, *Can Quantum-Mechanical Description of Physical Reality be Considered Complete?*, Phys. Rev. **47** 777, 1935.

source is located at the origin, we know that the distance from the origin of the other particle will also be x (in the opposite direction). Indeed, if we measured the position of the other particle we would find the distance x from the origin every time (that is, with probability 1).

Since the two particles are far away from one another, there is no way a signal from one could travel to the other particle in time to influence the measurement outcome. After all, nothing can travel faster than light. Einstein, Podolski and Rosen then make their profound argument: if you know with certainty what will happen in a measurement, then there should be an element of reality that determines the measurement outcome (having ruled out faster-than-light signalling). So the particles have inherently both a precise position and momentum, contrary to the claims of the uncertainty relation. This is the EPR paradox.

The spatial separation of the particles is crucial for the EPR thought experiment. Without it, we can imagine that the choice of measurement on particle 1 will influence the outcomes of the measurement on particle 2. We can introduce so-called "hidden variables" (the elements of reality of EPR) that are not part of the regular theory of quantum mechanics as it is developed in this book, but that dictate how the particles behave in measurements. The measurement outcome can then be determined by a signal from the other particle, instead of "elements of reality" already present in the particle. By placing the particles outside of each other's causal influence the EPR setup avoids this loophole.

The EPR paradox is one of the highlights of the famous debate between Einstein and Bohr about the meaning of quantum mechanics. To solve this paradox, we note that the two particles are entangled in position and momentum. This means that the position and momentum for each *individual* particle is completely uncertain, since the state of one component of a maximally entangled state is maximally mixed (i.e., maximally uncertain). The certainty of measurement outcomes that the EPR paradox identifies is in the *joint* properties $x_1 - x_2$ and $p_1 + p_2$ of the combined system of the two particles. This is fundamentally different from the properties of the individual particles, and the uncertainty relation for position and momentum as defined for individual particles still stands. The reason this works is that the two observables $x_1 - x_2$ and $p_1 + p_2$ commute:

$$\begin{aligned}
[\hat{x}_1 + \hat{x}_2, \hat{p}_1 - \hat{p}_2] &= [\hat{x}_1, \hat{p}_1 - \hat{p}_2] + [\hat{x}_2, \hat{p}_1 - \hat{p}_2] \\
&= [\hat{x}_1, \hat{p}_1] - [\hat{x}_1, \hat{p}_2] + [\hat{x}_2, \hat{p}_1] - [\hat{x}_2, \hat{p}_2] \\
&= [\hat{x}_1, \hat{p}_1] - 0 + 0 - [\hat{x}_2, \hat{p}_2] \\
&= i\hbar - i\hbar = 0,
\end{aligned} \tag{9.19}$$

where in the third line we used that operators on different particles commute.

In 1964, John Bell[4] showed that any theory of quantum mechanics that uses hidden variables to remove the inherent quantum uncertainty will have to involve superluminal signalling between particle 1 and particle 2. For many people this is too

---

[4] J. S. Bell, *On the Einstein Podolsky Rosen Paradox*, Physics **1** 195, 1964.

big a price to pay for regaining determinism. From a practical point of view, hidden variable theories are more complicated than orthodox quantum mechanics without any computational advantage, and that is why most people don't bother with them. Of course, this does not make the conceptual difficulties go away. We will explore those further in the next chapter.

9.4 Energy-Time Uncertainty

Another uncertainty relation that plays an important role in quantum mechanics is the energy-time uncertainty relation. Contrary to the position-momentum uncertainty relation, this is not a direct manifestation of the Robertson relation, because time is not an operator in quantum mechanics. Nevertheless, it takes a very similar form:

$$\Delta E \, \delta t \geq \frac{\hbar}{2} . \tag{9.20}$$

The question then becomes: what do we mean by ΔE and δt?

For operators such as $\hat{x}$ and $\hat{p}$ the uncertainties $(\Delta x)^2$ and $(\Delta p)^2$ are just the expectation value of the squared value away from the mean:

$$(\Delta x)^2 = \langle (x - \langle x \rangle)^2 \rangle . \tag{9.21}$$

It is a measure of the spread of the measurement outcomes, also known as the variance.

We have an operator for the energy, namely the Hamiltonian, and we can therefore unambiguously define ΔE in the same way as above. However, there is no (universal) time operator in quantum mechanics, so we cannot define the time uncertainty in the same way. That's why we write δt, to distinguish it from an operator uncertainty ΔA.

To find out what exactly we mean by δt, we first give a mathematical derivation. We start with the Robertson relation, where for B we substitute the Hamiltonian H:

$$\Delta A \, \Delta H \geq \frac{1}{2} |\langle [A, H] \rangle| . \tag{9.22}$$

Next, we want to relate the observable A to the time parameter t. For example, A may be the position of the second hand of a clock. Assume that the clock as a whole does not move in time, so the time variation of the expectation value $\langle A \rangle$ is entirely determined by the time evolution of the quantum state $|\psi\rangle$ of the pointer. We can then write, using the product rule, that

$$\frac{d\langle A \rangle}{dt} = \frac{d}{dt}\langle \psi | A | \psi \rangle$$

$$= \left(\langle \psi | \frac{\overleftarrow{d}}{dt} \right) A | \psi \rangle + \langle \psi | \left(\frac{dA}{dt} \right) | \psi \rangle + \langle \psi | A \left(\frac{d}{dt} | \psi \rangle \right)$$

$$= -\frac{1}{i\hbar}\langle \psi | HA | \psi \rangle + 0 + \frac{1}{i\hbar}\langle \psi | AH | \psi \rangle$$

$$= \frac{1}{i\hbar}\langle [A, H] \rangle , \tag{9.23}$$

where we used that the time derivative of A is zero, and we substituted the Schrödinger equation in the third line (twice). We can substitute this expression into the Robertson relation (9.22), which gives us

$$\Delta A \, \Delta H \geq \frac{\hbar}{2} \left| \frac{d\langle A \rangle}{dt} \right| . \tag{9.24}$$

Assuming that the expectation value $\langle A \rangle$ keeps changing over time, so that the right-hand side of (9.24) is never zero, we can define a time uncertainty as

$$\delta t \equiv \Delta A \left| \frac{d\langle A \rangle}{dt} \right|^{-1} . \tag{9.25}$$

This allows us to write an energy-time uncertainty relation

$$\delta t \, \Delta E \geq \frac{\hbar}{2} , \tag{9.26}$$

where we identified $\Delta E = \Delta H$.

We now have an expression for δt that will help us interpret the energy-time uncertainty relation. The right-hand side of (9.25) contains only well-defined terms: ΔA is the uncertainty in the measurement outcome of operator A, and $|d\langle A \rangle/dt|$ measures how fast the expectation value of A changes over time. When A is the position of a second hand on a clock, the denominator in (9.25) is the average (angular) speed of the second hand. The quantum uncertainty in the position of the second hand divided by the average speed is properly interpreted as the time it takes the second hand to move to a new position that is distinguishable from the position a time period δt earlier. This is a proper uncertainty, because it gives us the minimum time resolution of the clock. We can improve the precision of the clock by decreasing ΔA, which according to the uncertainty relation we can achieve only by choosing quantum states with a large uncertainty in energy.

In general, δt is the minimum time it takes for the system in a state $|\psi \rangle$ to evolve to a new state that is distinguishable from $|\psi \rangle$. It is a quantum uncertainty arising from the quantum uncertainty in the observable A. If the state is in an energy eigenstate, such that $\Delta E = 0$, the system merely accrues a global phase over time, and the system will never evolve to a distinguishable state. The time it takes for the system to evolve to a distinguishable state then becomes infinite: $\delta t \to \infty$.

Einstein raised an objection to the energy-time uncertainty relation by constructing the following paradox: Consider a device that emits a photon at a precise time t_0, a so-called photon gun. The photon carries energy away from the device, and via $E = mc^2$ we can relate this energy to a mass loss of the photon gun. So, if we measure the photon gun right before and right after t_0 we can get a precise determination of E from the mass difference, as well as a precise value for the time of emission t_0. This suggests that the product in the uncertainties ΔE and δt can be made arbitrarily small, violating the energy-time uncertainty relation.

Apparently, Bohr was quite agitated when Einstein presented this argument at the sixth Solvay conference in 1930. He did not present a counter-argument until the next day, but when he did it was a hailed as a triumph. First, Bohr created a precise model of Einstein's device. The photon gun is constructed using a slit in a box that briefly opens and lets a photon escape at time t_0, shown on a clock in the box next to the slit. The weighing is performed by suspending the box in a gravitational field g. After the photon has left the box, the box moves upwards and the experimenter starts adding mass m to the box to bring it back to its original position. The vertical position of the box is indicated by the pointer position x. We now have an actual model for Einstein's thought experiment that contains all the relevant bits.

Next, Bohr analysed what would happen when a photon leaves the box. After the release of the photon, the spring exerts a force on the box, moving it upwards and giving it some momentum p. By adding mass to the box we can return it to the equilibrium position with an uncertainty Δx, and consequently the uncertainty in the momentum of the box must increase according to the position-momentum uncertainty relation $\Delta x \Delta p \geq \hbar/2$. The time it takes to complete this procedure is given by T. Since the force on the mass added to the box is $F = mg$, we find from Newton's second law that

$$F = \frac{dp}{dt} = mg, \tag{9.27}$$

so the momentum at time T is $p(T) = mgT$. The uncertainty in momentum can be made small using

$$\Delta p < \Delta m \, gT. \tag{9.28}$$

This way we can get a very small uncertainty Δm in the mass by taking a long time T for the measuring procedure. Bohr then used Einstein's own theory of relativity to argue that after the photon emission, when the box is higher in the gravitational field, there is uncertainty δT in the reading of T due to the gravitational redshift. Since we do not know how high the box rises (this is related to the mass loss), we do not know how much gravitational redshift the clock experiences. This uncertainty is given by

$$\delta T = \frac{gT}{c^2} \Delta x. \tag{9.29}$$

We use our notation δT to indicate the uncertainty in a clock reading, as explained before. Using (9.28) and (9.29), we eliminate T and obtain

$$\delta T = \frac{gT}{c^2} \Delta x > \frac{g}{c^2} \frac{\Delta p}{g \Delta m} \Delta x = \frac{\Delta x \, \Delta p}{\Delta E} \geq \frac{\hbar}{2 \Delta E}, \tag{9.30}$$

where we used $\Delta E = c^2 \Delta m$. From this, it follows directly that

$$\delta T \Delta E > \frac{\hbar}{2}. \tag{9.31}$$

Thus Bohr demonstrated that in this case the energy-time uncertainty relation follows directly from the position-momentum uncertainty relation, and he constructed this argument using Einstein's own equivalence principle. Bohr's response to Einstein's paradox made a deep impression on the assembled physicists at the conference, since nobody had been able to find a flaw in Einstein's reasoning. Bohr's solution also underscores the importance of constructing a precise measurement device for any such situation, and how a quantity like time, which does not have an associated quantum mechanical observable, can be investigated by relating the time readout to a quantum measurement via a proper quantum observable.

9.5   The Quantum Mechanical Pendulum

How do the uncertainty relations affect real physical situations? To explore this we study one of the simplest and most interesting mechanical systems in physics: the pendulum. It has a regular periodic motion, which makes it an important component of mechanical clocks. It also is the prime example of harmonic motion, which is important in so many physical situations. The basic features of a classical pendulum are shown in Fig. 9.4.

Let us have a look at the way we describe the motion of a pendulum. First, we assume that all the mass m of the pendulum is concentrated in the bob, which is hanging from the ceiling by a thin wire of length l. Gravity pulls the pendulum towards the equilibrium position with a force $F = mg$. However, in quantum mechanics we prefer to deal with the energy of a system, so we first need to work out the potential energy of the (frictionless) pendulum as a function of the horizontal displacement x.

We choose the energy scale such that in the equilibrium position $x = 0$ the potential energy is zero. This means that for some displacement the potential energy becomes mgh. The period of a pendulum is given by

Fig. 9.4 The pendulum (see supplementary material 4)

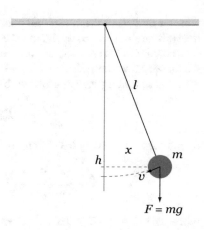

$$T = 2\pi\sqrt{\frac{l}{g}} = \frac{2\pi}{\omega},\qquad(9.32)$$

where ω is the angular frequency of the pendulum. We can therefore write $g = \omega^2 l$ and use this to eliminate g from the potential energy. Next, we express l in terms of x and h using Pythagoras' theorem:

$$(l - h)^2 + x^2 = l^2.\qquad(9.33)$$

Solving for l then gives us

$$l = \frac{h}{2}\left(1 + \frac{x^2}{h^2}\right).\qquad(9.34)$$

Since for small displacements $x^2 \gg h^2$, we can ignore the term 1 in this equation, and simplify to

$$h \simeq \frac{x^2}{2l}.\qquad(9.35)$$

Substituting g and h into mgh then gives

$$mgh \simeq \frac{1}{2}m\omega^2 x^2.\qquad(9.36)$$

The total kinetic and potential energy of the pendulum is then taken as

$$H = \frac{p^2}{2m} + \frac{1}{2}m\omega^2 x^2,\qquad(9.37)$$

where x and p are the *horizontal* position and momentum of the pendulum, respectively. Since x is small, we ignore the vertical movement.

In quantum mechanics, the total energy H becomes the Hamiltonian operator, where we now interpret x and p as operators. Finding the energy eigenstates then amounts to finding the solutions to the equation

$$H|E_n\rangle = E_n|E_n\rangle, \tag{9.38}$$

or equivalently solving the differential equation

$$\left(-\frac{\hbar^2}{2m}\frac{d^2}{dx^2} + \frac{1}{2}m\omega^2 x^2 - E_n\right)\psi_n(x) = 0, \tag{9.39}$$

where $\psi_n(x) \equiv \langle x|E_n\rangle$. This differential equation is a bit tricky to solve, so we take a different approach.

We define two new operators, $\hat{a}$, and its Hermitian adjoint $\hat{a}^\dagger$, such that

$$\hat{x} = \sqrt{\frac{\hbar}{2m\omega}}\left(\hat{a} + \hat{a}^\dagger\right) \tag{9.40}$$

and

$$\hat{p} = -i\sqrt{\frac{m\hbar\omega}{2}}\left(\hat{a} - \hat{a}^\dagger\right). \tag{9.41}$$

Now, this is a pretty strange step, and at this point completely unmotivated. However, it is entirely legal, and remember that we added operators before when we constructed the spin operator S_θ in an arbitrary direction in the xz-plane. Right now, I can only promise that this definition will have a grand pay-off in a few pages. You can check that the prefactor of $\hat{x}$ has dimensions of length and that of $\hat{p}$ has dimensions of momentum. This means that $\hat{a}$ and $\hat{a}^\dagger$ are dimensionless. Substituting these operators into (9.37), we obtain

$$H = \frac{1}{2}\hbar\omega\left(\hat{a}^\dagger\hat{a} + \hat{a}\hat{a}^\dagger\right). \tag{9.42}$$

Note that we do not assume that $\hat{a}^\dagger\hat{a} = \hat{a}\hat{a}^\dagger$. This is important. The operators $\hat{x}$ and $\hat{p}$ do not commute: $[\hat{x}, \hat{p}] = i\hbar$, and by imposing this commutation relation on the (9.40) and (9.41), we find that $\hat{a}$ and $\hat{a}^\dagger$ must obey the commutation relation

$$[\hat{a}, \hat{a}^\dagger] = 1. \tag{9.43}$$

We can use this to rewrite the Hamiltonian in (9.37) as

$$H = \hbar\omega \left(\hat{a}^\dagger \hat{a} + \frac{1}{2} \right) . \tag{9.44}$$

We will now use this form to extract the energy spectrum for the quantum mechanical pendulum.

The question that surely is going through your mind right now is: what do $\hat{a}$ and $\hat{a}^\dagger$ even mean? To answer this, we will use the fact that we can calculate commutation relations of functions of $\hat{a}$ and $\hat{a}^\dagger$, in particular H. Consider the expression $H\hat{a}|E_n\rangle$. In order to calculate what this is, we can commute $\hat{a}$ with H and apply (9.44):

$$H\hat{a} = \hat{a}H + \left[H, \hat{a} \right] , \tag{9.45}$$

and

$$\left[H, \hat{a} \right] = -\hbar\omega \hat{a} . \tag{9.46}$$

So we find that

$$H\hat{a}|E_n\rangle = \hat{a}H|E_n\rangle + \left[H, \hat{a} \right]|E_n\rangle = E_n\hat{a}|E_n\rangle - \hbar\omega \hat{a}|E_n\rangle$$
$$= (E_n - \hbar\omega)\,\hat{a}|E_n\rangle . \tag{9.47}$$

In other words, the (possibly unnormalised) state $\hat{a}|E_n\rangle$ is again an eigenstate of H with energy $E_n - \hbar\omega$. Similar reasoning leads to the relation

$$H\hat{a}^\dagger|E_n\rangle = (E_n + \hbar\omega)\hat{a}^\dagger|E_n\rangle . \tag{9.48}$$

The operators $\hat{a}$ and $\hat{a}^\dagger$ are therefore "ladder operators" that move up and down the ladder of energy eigenstates (it requires an additional argument to show that there are no energy eigenstates between E_n and $E_n \pm \hbar\omega$; the ladder operators do not skip energy eigenstates).

When the ground state of the quantum pendulum is denoted by $|E_0\rangle$, the lowering operator $\hat{a}$ on this state must return zero:

$$\hat{a}|E_0\rangle = 0 . \tag{9.49}$$

Otherwise it would create a state below the ground state, which is by definition impossible. Secondly, from (9.44) we see that $\hat{a}^\dagger \hat{a}$ must return the number n of the energy level E_n (the extra term $\hbar\omega/2$ is independent of n and can be subtracted by a redefinition of the energy scale). We therefore conclude that

$$\hat{a}^\dagger \hat{a}|E_n\rangle = n|E_n\rangle , \tag{9.50}$$

which we can write as

$$\hat{a}^{\dagger}\left(\hat{a}|E_n\rangle\right) = \hat{a}^{\dagger} f_n|E_{n-1}\rangle = f_n g_{n-1}|E_n\rangle = n|E_n\rangle . \tag{9.51}$$

Therefore, $f_n\, g_{n-1} = n$, which is satisfied for

$$f_n = \sqrt{n} \quad \text{and} \quad g_n = \sqrt{n+1}, \tag{9.52}$$

and we find the general relations

$$\hat{a}|E_n\rangle = \sqrt{n}|E_{n-1}\rangle \tag{9.53}$$

and

$$\hat{a}^{\dagger}|E_n\rangle = \sqrt{n+1}|E_{n+1}\rangle . \tag{9.54}$$

Note how this automatically satisfies the requirement that $\hat{a}|E_0\rangle = 0$. The energy spectrum of the quantum mechanical pendulum is therefore

$$E_n = \hbar\omega\left(n + \frac{1}{2}\right), \tag{9.55}$$

with energy eigenstates $|E_n\rangle$.

One interesting thing we can calculate is the average position and momentum in the ground state $|E_0\rangle$. For the position we find

$$
\begin{aligned}
\langle E_0|\hat{x}|E_0\rangle &= \sqrt{\frac{\hbar}{2m\omega}} \langle E_0|\left(\hat{a} + \hat{a}^{\dagger}\right)|E_0\rangle \\
&= \sqrt{\frac{\hbar}{2m\omega}} \left(\langle E_0|\hat{a}|E_0\rangle + \langle E_0|\hat{a}^{\dagger}|E_0\rangle\right) \\
&= \sqrt{\frac{\hbar}{2m\omega}} \left(0 + \langle E_0|E_1\rangle\right) = 0 ,
\end{aligned}
\tag{9.56}
$$

since the energy eigenstates $|E_0\rangle$ and $|E_1\rangle$ are orthogonal. Similarly, for the average momentum in the ground state we find

$$
\begin{aligned}
\langle E_0|\hat{p}|E_0\rangle &= -i\sqrt{\frac{m\hbar\omega}{2}} \langle E_0|\left(\hat{a} - \hat{a}^{\dagger}\right)|E_0\rangle \\
&= -i\sqrt{\frac{m\hbar\omega}{2}} \left(\langle E_0|\hat{a}|E_0\rangle - \langle E_0|\hat{a}^{\dagger}|E_0\rangle\right) \\
&= -i\sqrt{\frac{m\hbar\omega}{2}} \left(0 - \langle E_0|E_1\rangle\right) = 0 .
\end{aligned}
\tag{9.57}
$$

So the average position and momentum for a quantum mechanical pendulum in the ground state is zero, as expected.

However, what about the variances of x and p? We calculate $(\Delta x)^2$ and $(\Delta p)^2$ according to the standard formula:

$$
\begin{aligned}
(\Delta x)^2 &= \langle E_0|x^2|E_0\rangle - \langle E_0|x|E_0\rangle^2 \\
&= \frac{\hbar}{2m\omega}\langle E_0|\left(\hat{a}+\hat{a}^\dagger\right)^2|E_0\rangle \\
&= \frac{\hbar}{2m\omega}\langle E_0|\left(\hat{a}\hat{a}+\hat{a}\hat{a}^\dagger+\hat{a}^\dagger\hat{a}+\hat{a}^\dagger\hat{a}^\dagger\right)|E_0\rangle \\
&= \frac{\hbar}{2m\omega},
\end{aligned}
\tag{9.58}
$$

and for the momentum:

$$
\begin{aligned}
(\Delta p)^2 &= \langle E_0|p^2|E_0\rangle - \langle E_0|p|E_0\rangle^2 \\
&= -\frac{m\hbar\omega}{2}\langle E_0|\left(\hat{a}-\hat{a}^\dagger\right)^2|E_0\rangle \\
&= -\frac{m\hbar\omega}{2}\langle E_0|\left(\hat{a}\hat{a}-\hat{a}\hat{a}^\dagger-\hat{a}^\dagger\hat{a}+\hat{a}^\dagger\hat{a}^\dagger\right)|E_0\rangle \\
&= \frac{m\hbar\omega}{2}.
\end{aligned}
\tag{9.59}
$$

Therefore, the ground state uncertainties obey the relation

$$
\Delta x\,\Delta p = \frac{\hbar}{2},
\tag{9.60}
$$

and the ground state of the quantum pendulum is a minimum uncertainty state since it satisfies the equality in the uncertainty relation. Even though the position and momentum is zero on average, there are fluctuations of size Δx and Δp, and even in the ground state the pendulum is not completely still. This is fundamental quantum noise, and it cannot be removed from the system. This is a general aspect of quantum mechanics: no system can be completely still. There is always some *quantum noise* that cannot be reduced beyond a certain minimum that is determined by an uncertainty relation. In particle physics this manifests itself in the vacuum, which cannot be completely still and sees spontaneous generation and destruction of particles with energy E for a duration shorter than $\hbar/(2E)$.

There is one more thing we should say about the quantum mechanical pendulum. The most characteristic aspect of the pendulum is that it swings back and forth, while we have only determined the energy eigenstates $|E_n\rangle$. These states evolve over time as $e^{-\frac{i}{\hbar}E_n t}|E_n\rangle$, accumulating an unobservable global phase $e^{-\frac{i}{\hbar}E_n t}$. In other words, when a system is in an energy eigenstate, it will not show any "motion". Compare this to our discussion concerning the energy-time uncertainty relation, where we can have a clock timing resolution only if the clock state is not in an energy eigenstate.

We need some kind of motion to make a clock, like the pendulum in a grandfather clock.

The state of the pendulum in motion must therefore be a superposition of energy eigenstates. It is not so trivial to find out which state gives rise to the typical pendulum motion, but it turns out to be of the form

$$|\alpha(t)\rangle = e^{-\frac{1}{2}|\alpha|^2} \sum_{n=0}^{\infty} \frac{\alpha^n e^{-\frac{i}{\hbar} E_n t}}{\sqrt{n!}} |E_n\rangle , \tag{9.61}$$

where α is related to the maximum horizontal displacement (the amplitude) for the pendulum A via

$$A = |\alpha| \sqrt{\frac{2\hbar}{m\omega}} . \tag{9.62}$$

Choosing α real, we find that the expectation value of the position operator $\hat{x}$ defined in (9.40) with respect to $|\alpha(t)\rangle$ can be calculated as

$$\langle x \rangle = \langle \alpha(t)|\hat{x}|\alpha(t)\rangle = A \cos \omega t . \tag{9.63}$$

Similarly,

$$\langle p \rangle = \langle \alpha(t)|\hat{p}|\alpha(t)\rangle = -m\omega A \sin \omega t , \tag{9.64}$$

as you would expect. You should try to reproduce these expectation values using (9.53) and (9.54). The state $|\alpha(t)\rangle$ is called a *coherent state*. What happens when α is complex?

The expectation values of the quantum mechanical operators $\hat{x}$ and $\hat{p}$ follow the classical equations of motion for the pendulum. The probability density function $|\langle x|\alpha(t)\rangle|^2$ of finding the pendulum at displacement x at time t is shown as the blue line in Fig. 9.5, and clearly exhibits the harmonic motion.

Of course, the larger the amplitude A, the more energy there is in the pendulum. We can calculate this from the expectation value of the Hamiltonian in (9.44) with respect to the state $|\alpha(t)\rangle$:

$$\langle H \rangle = \hbar\omega \langle \alpha(t)| \left(\hat{a}^\dagger \hat{a} + \frac{1}{2} \right) |\alpha(t)\rangle = \hbar\omega|\alpha|^2 = \frac{1}{2}m\omega^2 A^2, \tag{9.65}$$

which is exactly what we would expect from (9.37), since classically $p = 0$ at the extreme displacement $x = A$. The uncertainty in the energy for the pendulum in the state of (9.61) is given by

$$\Delta E = \sqrt{\langle H^2 \rangle - \langle H \rangle^2} = \hbar\omega|\alpha| = A\sqrt{\frac{m\hbar\omega^3}{2}} . \tag{9.66}$$

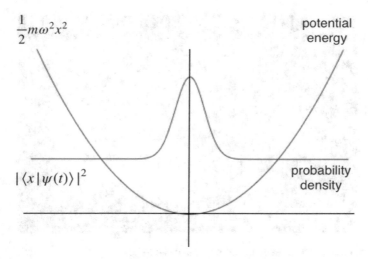

Fig. 9.5 The coherent state for the pendulum (see supplementary material 5)

From the uncertainty relation between energy and time we then have that

$$\delta t \geq \frac{\hbar}{2\Delta E} = \sqrt{\frac{\hbar}{2m\omega^3 A^2}} . \tag{9.67}$$

This means that if we use the pendulum as the oscillator in a clock, the highest precision we can achieve in principle is given by δt in (9.67).

You may find it strange that there are energy fluctuations in the swinging motion of the pendulum. What about energy conservation? It turns out that energy is strictly conserved in the quantum mechanical pendulum: potential energy is exactly converted into kinetic energy and vice versa (assuming no friction). However, the total value of the energy is uncertain and subject to quantum fluctuations. Energy conservation applies strictly to the *processes* governing physical systems, not the states themselves.

9.6 Schrödinger's Cat

In this section we will address a fundamental question about quantum states, and we will illustrate this with a system involving a quantum-mechanical pendulum. Quantum mechanical objects behave in a completely different way to the physical objects in our everyday experience. Typically, quantum mechanical objects such as photons, electrons or atoms are very small, and it is therefore not really surprising that their behaviour is different from what we naively expect. Our physical intuition evolved in order to exploit our environment for survival and to deal with physical

Fig. 9.6 Schrödinger's cat (see supplementary material 6)

threats, not to study atoms. Why should they behave in the same way as rocks and spears? On the other hand, since we think that quantum mechanics is a more fundamental theory of nature than classical (Newtonian) mechanics, it is certainly valid to ask how the classical world that we experience emerges from quantum mechanics. How do we lose the typical counterintuitive quantum behaviour in the every-day world around us?

In 1935, Erwin Schrödinger[5] came up with his famous thought experiment about a cat in a box (see Fig. 9.6). This argument pushed our question to the foreground with some urgency, so let us consider it in detail: A cat is put in a sealed box (with enough oxygen so it does not suffocate), together with a devious device consisting of a radioactive atom (for example the alpha particle emitter polonium-210), and a Geiger counter that is linked to a valve that opens a vial of poisonous gas. The device is rigged in such a way that when the polonium decays into a lead atom ($^{210}Po \rightarrow {}^{206}Pb$), the Geiger counter registers the emitted alpha particle and releases the valve. This briefly opens the vial with the poisonous gas, and the cat dies.

This all happens inside the sealed, soundproof box, so we do not know whether the atom has decayed or not. After some time, we have about a one in three chance of the atom being decayed (waiting longer increases this probability). Moreover, since the system inside the box is isolated, it will evolve quantum mechanically, turning the state of the cat (including the deadly contraption) into a superposition of alive and dead. At some specific time the superposition will be

$$|\text{cat}\rangle = \sqrt{\frac{2}{3}}|\text{alive}\rangle + \sqrt{\frac{1}{3}}|\text{dead}\rangle . \tag{9.68}$$

[5] E. Schrödinger, *Die gegenwärtige Situation in der Quantenmechanik*, Naturwissenschaften **23** 807, 1935.

According to the fifth postulate of quantum mechanics—the "projection postulate"—only when we open the box does the state of the cat collapse to "alive" or "dead".

This is clearly absurd. Classical objects like cats do not appear in quantum superpositions. Somewhere, we must pass from the quantum description to a classical description. Is the cat itself an observer of the valve on the vial, and already capable of triggering the projection postulate? Is the vial opening the classical observer? Or is it the Geiger counter? It is easy to see that we can in principle devise a chain of interacting systems of increasing size from the atom upwards (much like the internal workings of the Geiger counter), and it does not seem to be possible to say exactly at what point the quantum state collapses. This is known as the *measurement problem*. However, this paradox assumes that the quantum state corresponds directly to the reality of the physical system, something about which we were told to be cautious in Chap. 1. We will return to the precise meaning of the quantum state in Chap. 11.

If we wanted to perform the experiment with Schrödinger's cat in reality, we would quickly find that we made another crucial assumption: the cat and the contraption are completely sealed off from the environment. A moment's thought reveals that this is very difficult to achieve in practice. For example, while you are reading this, you may feel chilly, hot, or just right. This reflects the fact that the molecules in the air constantly interact with your skin. The same is true for the cat. And in the cat's case the air may have traces of the poisonous gas, depending on whether the atom decayed or not.

Before the box is opened, the joint quantum state $|\Psi\rangle$ of the air and the cat is given by

$$|\Psi\rangle \equiv \sqrt{\frac{2}{3}}|\text{pure}\rangle|\text{alive}\rangle + \sqrt{\frac{1}{3}}|\text{poison}\rangle|\text{dead}\rangle , \qquad (9.69)$$

where the different states of the air in the box are given by $|\text{pure}\rangle$ and $|\text{poison}\rangle$ for an undecayed and a decayed atom, respectively. In Chap. 7 we showed that we can remove the air from our description of the system via the partial trace. This leaves us with

$$\rho = \text{Tr}_{\text{air}}(|\Psi\rangle\langle\Psi|) = \frac{2}{3}|\text{alive}\rangle\langle\text{alive}| + \frac{1}{3}|\text{dead}\rangle\langle\text{dead}| . \qquad (9.70)$$

We see that the cat is no longer in the superposition of dead or alive. Instead, whether the cat is dead or alive is now subject to classical uncertainty. As we demonstrated in Chap. 7—where we considered a photon in a Mach–Zehnder interferometer interacting with the environment—we see the typical quantum behaviour only when the quantum system is isolated, and the larger the quantum system, the harder it is to keep it free from interactions with the environment. This transition from a quantum superposition to a state of classical uncertainty is called decoherence. The cat decoheres long before we open the box due to interactions with the environment that are beyond our control.

Why does the decoherence of the quantum state of the cat give us a classical uncertainty over the states "dead" and "alive", rather than some other superposition states such as

$$|+\rangle = \frac{1}{\sqrt{2}}|\text{alive}\rangle + \frac{1}{\sqrt{2}}|\text{dead}\rangle \,, \tag{9.71}$$

or

$$|-\rangle = \frac{1}{\sqrt{2}}|\text{alive}\rangle - \frac{1}{\sqrt{2}}|\text{dead}\rangle \,? \tag{9.72}$$

After all, quantum mechanics does not have any preference for the dead/alive basis over the $\pm$ basis. We saw this already in Chap. 3, where we could create the spin state of a spin-1/2 particle in any direction we want. This corresponds to different quantum states $|\uparrow\rangle$ and $|\downarrow\rangle$, or $|\pm\rangle$. So why does the cat decohere into the state in (9.70) and not, for example into

$$\frac{2}{3}|+\rangle\langle+| + \frac{1}{3}|-\rangle\langle-| \,? \tag{9.73}$$

Such a state would exhibit interference (caused by the off-diagonal terms in the density matrix) in the dead/alive basis, which is not what we expect in our classical description of the cat.

The (partial) answer to this question is in the *interaction* between the poisonous gas and the cat. The fact that, physically, it is the poisonous gas in the air that kills the cat means that the interaction between the cat and the air must take the form

$$|\text{poison}\rangle|\text{alive}\rangle \xrightarrow[H_I]{} |\text{poison}\rangle|\text{dead}\rangle \,, \tag{9.74}$$

where H_I denotes the interaction Hamiltonian (see Chap. 4). In other words, the interaction Hamiltonian H_I is diagonal in the poison/pure and dead/alive basis. If the interaction between the air and the cat had been different, the final state of the cat would be different, and possibly exhibit nonclassical interference between the states "alive" and "dead". But with this interaction we end up finding the cat either dead or alive, with a classical uncertainty. Therefore, the specific interaction between the quantum system and the environment causes rapid decoherence of large quantum systems, and this ensures that the macroscopic world looks classical, rather than quantum. However, this does not answer the question what the quantum state really means, only what we find in our measurements at the macroscopic, classical level. In Chap. 11 we will face the much more difficult question of the meaning of the quantum state head on.

There are currently many experimental efforts to overcome this decoherence and create ever larger physical systems in a quantum superposition, approaching Schrödinger's cat states. Not with real cats, obviously! For example, people are now per-

Fig. 9.7 The opto-mechanical cavity, where one mirror can vibrate, changing the length of the cavity

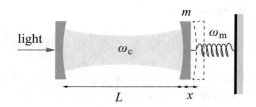

forming experiments similar to the double slit experiment in Chap. 8, in which we can see quantum behaviour in the interference pattern.[6] These experiments are not performed with single photons, but with increasingly larger molecules, and interference is indeed observed. These experiments give us vital evidence for the applicability of quantum theory to larger and larger systems, and so far no size limit has been found. It so far appears that quantum theory is generally applicable.

In the remainder of this section we will consider the creation of a cat-like state—a macroscopic quantum superposition—of a mirror in an optical cavity. This is called an opto-mechanical system, and the setup is shown in Fig. 9.7. An optical cavity is formed of two mirrors a distance L apart, and one mirror with mass m is fixed to a wall with a spring, allowing it to vibrate. Our aim is to put the macroscopic mirror into a superposition of two different vibrational quantum states. We treat the mirror as a quantum-mechanical harmonic oscillator, described in the same way as the quantum pendulum in the previous section, and the Hamiltonian H_m of the mirror is given by

$$H_m = \frac{\hat{p}^2}{2m} + \frac{1}{2}m\omega_m^2 x^2 , \qquad (9.75)$$

where $\hat{p}$ is the momentum operator for the mirror, x is the displacement of the mirror from the equilibrium position, and ω_m is the natural frequency of the mirror's oscillations. The Hamiltonian H_c for light in the cavity is given by

$$H_c = \hbar\omega_c \hat{n} , \qquad (9.76)$$

where $\hat{n}$ is the quantum observable for the number of photons in the cavity, and

$$\omega_c = \frac{\pi c}{L} \qquad (9.77)$$

is the frequency of the light that can exist in the cavity, and c is the speed of light. The total Hamiltonian for the system is then given by $H_m + H_c$.

What happens when the mirror is displaced by an amount x? This will change the length of the cavity to $L + x$, and therefore it will change the frequency of the

---

[6] S. Gerlich et al., *Quantum interference of large organic molecules*, Nature Communications **2** 263, 2011.

cavity:

$$\omega_c(x) = \frac{\pi c}{L + x} . \qquad (9.78)$$

Since $x \ll L$, we can write $\omega_c(x)$ more conveniently as a Taylor expansion around $x = 0$ and drop the terms that are quadratic in x and higher:

$$\omega_c(x) = \frac{\pi c}{L} - \frac{\pi c x}{L^2} = \omega_c \left(1 - \frac{x}{L}\right) . \qquad (9.79)$$

Substituting this form of $\omega_c(x)$ into H_c will give

$$H_c \rightarrow \hbar\omega_c\hat{n} - \hbar\frac{\omega_c}{L}\hat{n}x \equiv H_c - \hbar g\hat{n}x . \qquad (9.80)$$

We defined $g = \omega_c/L$ as the coupling strength between the light in the cavity and the mirror. This extra term $-\hbar g\hat{n}x$ is the interaction term of the Hamiltonian because it involves a *product* of the cavity observable $\hat{n}$ and the mirror observable x. We therefore finally arrive at the total Hamiltonian for the opto-mechanical system

$$H = H_m + H_c + H_{int}$$
$$= \frac{\hat{p}^2}{2m} + \frac{1}{2}m\omega_m^2 x^2 + \hbar\omega_c\hat{n} - \hbar g\hat{n}x , \qquad (9.81)$$

which completely determines the evolution of the joint cavity-mirror system in the absence of any photon losses from the cavity.

To get an idea of the evolution of the combined light-cavity system, we first note that we can create a coherent state of the mirror as given in (9.61) by displacing the mirror in the ground state $|0\rangle$ using the unitary operator

$$D(\alpha) = \exp\left[i\,\mathrm{Re}\,\alpha\sqrt{\frac{2m\omega}{\hbar}}\hat{x} + i\,\mathrm{Im}\,\alpha\sqrt{\frac{2}{\hbar m\omega}}\hat{p}\right] , \qquad (9.82)$$

such that $D(\alpha)|0\rangle = |\alpha\rangle$. Next, we consider the interaction Hamiltonian H_{int} and create a unitary transformation from it by exponentiation:

$$U_{int}(t) = \exp\left(-\frac{i}{\hbar}H_{int}t\right) = \exp\left(ig\hat{n}t\,x\right) . \qquad (9.83)$$

Note that $U_{int}(t)$ has a very similar form to $D(\alpha)$ when α is real. However, α would itself need to be an operator because of the photon number observable $\hat{n}$:

$$\alpha = gt\sqrt{\frac{\hbar}{2m\omega}}\hat{n} . \qquad (9.84)$$

This means that U_{int} acts not only on the mirror, but also on the cavity light field via $\hat{n}$. This makes sense, because U_{int} gives the change in the *total* state due to the interaction.

Assume that the mirror is in the ground state at the start of the experiment, and the light in the cavity is in some superposition state

$$|\psi\rangle = \sum_{n=0}^{\infty} c_n \, |n\rangle \, . \tag{9.85}$$

The state of the combined system due to the interaction for a duration t then becomes

$$|\Psi\rangle = U_{\text{int}}(t) \, |\psi, 0\rangle = \sum_{n=0}^{\infty} e^{ig\hat{n}tx} c_n \, |n, 0\rangle = \sum_{n=0}^{\infty} e^{igntx} c_n \, |n, 0\rangle = \sum_{n=0}^{\infty} c_n \, |n, \alpha_n\rangle \tag{9.86}$$

where α_n is the value of α in (9.84) with the observable $\hat{n}$ replaced by the photon number n. The third equality in the formula above uses the fact that we can replace an operator with its eigenvalue if it is applied to an eigenstate. Here $\hat{n}$ in the exponential acts on the eigenstate $|n\rangle$, so $\hat{n}$ can be replaced with n. The state $|\Psi\rangle$ in (9.86) is entangled, and if the mirror is a macroscopic object we can see some similarities with the cat state. However, there are potentially an infinite number of terms in this superposition, which is not quite a simple superposition between two macroscopically distinct states. In order to achieve this more obvious cat-like superposition, we have to modify the experiment a little.

First, we place a two-level atom in the cavity, as shown in Fig. 9.8. The atom interacts with the light in the cavity, and in turn the light interacts with the mirror. Without going into too much technical detail (which is really beyond the scope of this book), we can arrange the experiment such that the light mediates the interaction between the atom and the mirror with an interaction Hamiltonian

$$H_{\text{int}} = -\eta\hbar\sigma_z x \, , \tag{9.87}$$

where $\sigma_z = |g\rangle\langle g| - |e\rangle\langle e|$ is the Pauli z matrix for the atom with ground state $|g\rangle$ and excited state $|e\rangle$. The coupling η depends on the intensity of the light in the

Fig. 9.8 The opto-mechanical cavity with a two-level atom. The light inside the cavity mediates the interaction between the atom and the cavity mirror

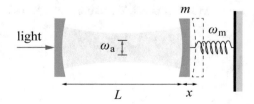

cavity and the coupling strength of the light with the atom and the mirror. The total Hamiltonian for this system, when we eliminated the light in the cavity, is

$$H = H_{\mathrm{m}} + H_{\mathrm{a}} + H_{\mathrm{int}} = \frac{\hat{p}^2}{2m} + \frac{1}{2}m\omega_{\mathrm{m}}^2 x^2 + \frac{1}{2}\hbar\omega_{\mathrm{a}}\sigma_z - \eta\hbar\sigma_z x , \qquad (9.88)$$

where $H_{\mathrm{a}} = \hbar\omega_{\mathrm{a}}\sigma_z$ is the Hamiltonian of the atom and $\hbar\omega_{\mathrm{a}} = E_e - E_g$ is the energy difference between the ground state and the excited state of the atom.

We can again consider the evolution due to the interaction Hamiltonian H_{int} and construct the unitary transformation

$$U_{\mathrm{int}}(t) = \exp\left(-\frac{i}{\hbar}H_{\mathrm{int}}t\right) = \exp\left(i\eta t \sigma_z x\right) . \qquad (9.89)$$

We apply this evolution to the joint atom-mirror state $|+, 0\rangle$, where $|+\rangle = (|g\rangle + |e\rangle)/\sqrt{2}$ is the initial state of the atom, and $|0\rangle$ is again the ground state of the mirror. The resulting state after a time t is then

$$\begin{aligned}
|\Psi\rangle = U_{\mathrm{int}}(t)\,|+, 0\rangle &= \exp\left(i\eta t \sigma_z x\right) \frac{|g, 0\rangle + |e, 0\rangle}{\sqrt{2}} \\
&= \frac{1}{\sqrt{2}} e^{i\eta t \sigma_z x} |g, 0\rangle + \frac{1}{\sqrt{2}} e^{i\eta t \sigma_z x} |e, 0\rangle \\
&= \frac{1}{\sqrt{2}} e^{+i\eta t x} |g, 0\rangle + \frac{1}{\sqrt{2}} e^{-i\eta t x} |e, 0\rangle \\
&= \frac{1}{\sqrt{2}} |g, +\alpha\rangle + \frac{1}{\sqrt{2}} |e, -\alpha\rangle ,
\end{aligned} \qquad (9.90)$$

where in the last line we used that the exponential operator in x again acts as the displacement operator $D(\alpha)$, with

$$\alpha = \eta t \sqrt{\frac{\hbar}{2m\omega}} . \qquad (9.91)$$

The state in (9.90) is again an entangled state, but now it has only two terms in the superposition. To take the final step and create a cat-like state for the mirror, we need to disentangle the atom from the mirror in such a way that the mirror remains in a quantum superposition of $|+\alpha\rangle$ and $|-\alpha\rangle$. We can achieve this by measuring the atom in the $|\pm\rangle$ basis. Write the joint state as

$$|\Psi\rangle = \frac{1}{\sqrt{2}} |g, +\alpha\rangle + \frac{1}{\sqrt{2}} |e, -\alpha\rangle$$

$$= \frac{1}{2} |+, +\alpha\rangle + \frac{1}{2} |-, +\alpha\rangle + \frac{1}{2} |+, -\alpha\rangle - \frac{1}{2} |-, -\alpha\rangle$$

$$= \frac{1}{\sqrt{2}} |+\rangle \left(\frac{|+\alpha\rangle + |-\alpha\rangle}{\sqrt{2}} \right) + \frac{1}{\sqrt{2}} |-\rangle \left(\frac{|+\alpha\rangle - |-\alpha\rangle}{\sqrt{2}} \right). \quad (9.92)$$

When we find the outcome $|+\rangle$, the mirror is projected into the state

$$|\text{cat}_+\rangle = \frac{|+\alpha\rangle + |-\alpha\rangle}{\sqrt{2}}, \quad (9.93)$$

which is clearly a superposition of two different states of motion of the mirror. When we find the outcome $|-\rangle$, the mirror is projected into the state

$$|\text{cat}_-\rangle = \frac{|+\alpha\rangle - |-\alpha\rangle}{\sqrt{2}}, \quad (9.94)$$

which is also a superposition of two different states of motion of the mirror, but with a relative phase shift. If the mirror is a macroscopic object, the state in (9.93) and (9.94) are macroscopic cat states. The sign of α determines which side of the equilibrium position the mirror starts, so the superposition is between two oscillations with the same amplitude α, but 180° out of phase.

The presentation here is somewhat simplified, since we did not include the complication of the mirror and atom or cavity Hamiltonians in U_{int}. That would require more advanced mathematical techniques. However, the basic entangling mechanism originates from the interaction Hamiltonian, and it shows that in principle cat states of macroscopic objects are possible in quantum mechanics. Indeed, people have created such cat states with mirrors that have masses up to milligrams. So far, there is no indication that quantum mechanics breaks down for larger systems.

9.7 ⚬ Precision Measurements

Quantum systems can evolve under the action of some interaction Hamiltonian, and we can make repeated measurements of the quantum system to figure out the strength of some of the physical quantities in the interaction Hamiltonian. For example, we can describe an electron spin in a magnetic field with the interaction Hamiltonian derived from (3.88)

$$H = -\frac{e}{m} \left(B_x S_x + B_y S_y + B_z S_z \right), \quad (9.95)$$

where S_x, S_y, and S_z are the spin observables of the electron; B_x, B_y, and B_z are the values of the magnetic field in the direction x, y, and z, respectively; and e and m are the electron charge and mass. We can then prepare the system repeatedly in the same state using the same preparation procedure, and for each preparation measure the electron spin to figure out the strength of the magnetic field.

Let's suppose that we send an electron through a region that has a magnetic field with unknown magnitude B in the z-direction ($B = B_z$), and $B_x = B_y = 0$. The state of the electron is given by

$$|\psi\rangle = \frac{|\uparrow\rangle + |\downarrow\rangle}{\sqrt{2}} = \frac{1}{\sqrt{2}} \begin{pmatrix} 1 \\ 1 \end{pmatrix}. \tag{9.96}$$

If we know the velocity of the electron and the size of the magnetic region, we can work out the time t the electron experiences the interaction H. The evolution U of the electron state in matrix form is calculated as

$$U = \exp\left(-\frac{i}{\hbar}Ht\right) = \exp\left(\frac{ieBt}{m\hbar}S_z\right) = \exp\left(\frac{ieBt}{2m}\sigma_z\right), \tag{9.97}$$

with

$$\sigma_z = \begin{pmatrix} 1 & 0 \\ 0 & -1 \end{pmatrix}. \tag{9.98}$$

The evolution U then becomes

$$U = \begin{pmatrix} e^{i\omega t/2} & 0 \\ 0 & e^{-i\omega t/2} \end{pmatrix} \quad \text{with} \quad \omega = \frac{eB}{m}, \tag{9.99}$$

and the state of the electron after interacting with the magnetic field is

$$|\psi(t)\rangle = \frac{e^{i\omega t/2}|\uparrow\rangle + e^{-i\omega t/2}|\downarrow\rangle}{\sqrt{2}} = \frac{1}{\sqrt{2}} \begin{pmatrix} e^{i\omega t/2} \\ e^{-i\omega t/2} \end{pmatrix}. \tag{9.100}$$

You should be able to verify this using the techniques developed in Chaps. 3 and 4. We can measure the spin of the electron in the x-direction using a Stern-Gerlach apparatus, and the probabilities of the measurement outcomes are

$$p_+ = |\langle +|\psi(t)\rangle|^2 = \frac{1}{2} + \frac{1}{2}\cos\left(\frac{eBt}{m}\right),$$

$$p_- = |\langle -|\psi(t)\rangle|^2 = \frac{1}{2} - \frac{1}{2}\cos\left(\frac{eBt}{m}\right). \tag{9.101}$$

By repeating this experiment N times we obtain N_+ outcomes "+" and N_- outcomes "−" with $N_+ + N_- = N$. This is shown in Fig. 9.9. When N is large, the relative frequencies of the measurement outcomes approach the probabilities:

Fig. 9.9 Measuring the spin direction on N separate spins (see supplementary material 9)

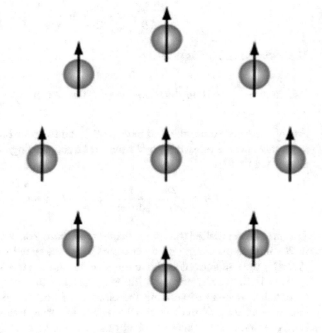

$$\frac{N_+}{N} \rightarrow p_+ \quad \text{and} \quad \frac{N_-}{N} \rightarrow p_- . \tag{9.102}$$

We can then estimate the value B for the magnetic field as

$$B = \frac{m}{et} \arccos (p_+ - p_-) \approx \frac{m}{et} \arccos \left(\frac{N_+ - N_-}{N} \right) , \tag{9.103}$$

which becomes exact as $N \rightarrow \infty$. Note the similarities with the gravitational wave detector in Chap. 2.

We would like to have some estimate of the error δB in this value of B, since for finite N the value for B will not be exact. We can relate δB to the variance in S_x via the standard formula for errors analogous to (9.24):

$$\delta B = \left| \frac{d \langle S_x \rangle}{dB} \right|^{-1} \Delta S_x . \tag{9.104}$$

We already calculated $\langle S_x \rangle$:

$$\langle S_x \rangle = \frac{\hbar}{2} p_+ - \frac{\hbar}{2} p_- = \frac{\hbar}{2} \cos \left(\frac{eBt}{m} \right) , \tag{9.105}$$

and the derivative becomes

$$\frac{d\langle S_x \rangle}{dB} = -\frac{e\hbar t}{2m} \sin\left(\frac{eBt}{m}\right), \qquad (9.106)$$

The variance $(\Delta S_x)^2$ is given by

$$(\Delta S_x)^2 = \frac{\hbar^2}{4} \langle \psi(t)|\sigma_x^2|\psi(t)\rangle - \frac{\hbar^2}{4}\langle \psi(t)|\sigma_x|\psi(t)\rangle^2 = \frac{\hbar^2}{4}\sin^2\left(\frac{eBt}{m}\right). \quad (9.107)$$

For N measurements, the variance $(\Delta S_x)^2$ adds up to become $N(\Delta S_x)^2$, and the expectation value of S_x is also N times as large, yielding $N\langle S_x \rangle$. When we combine all this we find

$$\delta B = \frac{2m}{Ne\hbar t}\frac{1}{|\sin \omega t|} \times \frac{\hbar}{2}\sqrt{N}|\sin \omega t| = \frac{m}{et}\frac{1}{\sqrt{N}}. \qquad (9.108)$$

This precision is called the *standard quantum limit*. You see that as $N \to \infty$ the error in B vanishes (assuming no other errors), and the speed with which this happens is given by the square root of the number of measurements N.

This is, however, not the ultimate precision that can be attained in quantum mechanics. Instead of sending the electrons through the magnetic region one by one, we can send them through all together in some suitably chosen quantum state. Suppose that we have again N electrons, with the first one prepared in the same state as before:

$$|\psi\rangle = \frac{|\uparrow\rangle + |\downarrow\rangle}{\sqrt{2}}. \qquad (9.109)$$

The rest of the electrons are prepared in the state $|\uparrow\rangle$. Next, we apply a so-called CNOT operation between the first electron and each one of the others, shown in Fig. 9.10. This will have the following effect on the state:

$$|\uparrow, \uparrow\rangle \xrightarrow[\text{CNOT}]{} |\uparrow, \uparrow\rangle \quad \text{and} \quad |\uparrow, \downarrow\rangle \xrightarrow[\text{CNOT}]{} |\uparrow, \downarrow\rangle,$$

$$|\downarrow, \uparrow\rangle \xrightarrow[\text{CNOT}]{} |\downarrow, \downarrow\rangle \quad \text{and} \quad |\downarrow, \downarrow\rangle \xrightarrow[\text{CNOT}]{} |\downarrow, \uparrow\rangle. \qquad (9.110)$$

In other words, the first electron flips the spin of the second if the first is in the state $|\downarrow\rangle$ (see also the XOR operation in section 6.4). After applying $N-1$ CNOT operations, the spin state of the N electrons becomes

$$|\psi\rangle = \frac{|\uparrow, \uparrow, \ldots, \uparrow\rangle + |\downarrow, \downarrow, \ldots, \downarrow\rangle}{\sqrt{2}}. \qquad (9.111)$$

This is a superposition of all electrons with spin up and all electrons with spin down, and this state has no classical analog. It is a special entangled state, called GHZ state,

Fig. 9.10 Measuring the spin direction on N entangled spins (see supplementary material 10)

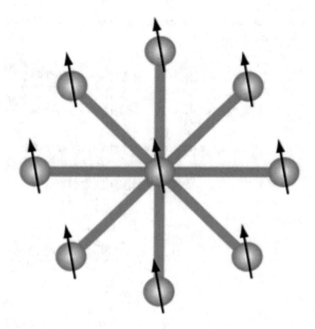

after Daniel Greenberger, Michael Horne and Anton Zeilinger, who first studied its non-classical properties.

As all electrons travel through the region with the magnetic field of unknown magnitude, each electron pics up the state dependent phase shift $e^{\pm i\omega t}$ according to (9.99), and the evolved state becomes

$$|\psi(t)\rangle = \frac{e^{iN\omega t/2}|\uparrow, \uparrow, \ldots, \uparrow\rangle + e^{-iN\omega t/2}|\downarrow, \downarrow, \ldots, \downarrow\rangle}{\sqrt{2}}. \tag{9.112}$$

Afterwards, we apply the same set of CNOT operations on the electrons, which turns the state into

$$|\Phi(t)\rangle = \frac{e^{iN\omega t/2}|\uparrow, \uparrow, \ldots, \uparrow\rangle + e^{-iN\omega t/2}|\downarrow, \uparrow, \ldots, \uparrow\rangle}{\sqrt{2}}. \tag{9.113}$$

We can interpret this state as $N - 1$ electrons in the state $|\uparrow\rangle$, and the first electron in the state

$$|\phi(t)\rangle = \frac{|\uparrow\rangle + e^{-iN\omega t}|\downarrow\rangle}{\sqrt{2}}, \tag{9.114}$$

up to a global phase. Notice how the phase shift is now N times larger than before: we have effectively teleported the phase shifts of all the $N - 1$ electrons onto the central electron. A measurement of this electron in the Stern-Gerlach apparatus then yields the probabilities

$$p_+ = |\langle +|\phi(t)\rangle|^2 = \frac{1}{2} + \frac{1}{2}\cos\left(\frac{NeBt}{m}\right),$$

$$p_- = |\langle -|\phi(t)\rangle|^2 = \frac{1}{2} - \frac{1}{2}\cos\left(\frac{NeBt}{m}\right). \tag{9.115}$$

Following the same procedure as before, we calculate

$$\delta B = \left|\frac{d\langle S_x\rangle}{dB}\right|^{-1} \Delta S_x, \tag{9.116}$$

with

$$\langle S_x\rangle = \frac{\hbar}{2}\cos\left(\frac{NeBt}{m}\right), \tag{9.117}$$

and

$$\frac{d\langle S_x\rangle}{dB} = -\frac{Ne\hbar t}{2m}\sin\left(\frac{NeBt}{m}\right). \tag{9.118}$$

The variance becomes

$$(\Delta S_x)^2 = \frac{\hbar^2}{4}\langle\psi(t)|\sigma_x^2|\psi(t)\rangle - \frac{\hbar^2}{4}\langle\psi(t)|\sigma_x|\psi(t)\rangle^2$$

$$= \frac{\hbar^2}{4}\sin^2\left(\frac{NeBt}{m}\right). \tag{9.119}$$

Note that we have only one measurement in this situation, so we multiply the variance and expectation value of S_x with 1. This yields for the error in B

$$\delta B = \frac{2m}{Ne\hbar t}\frac{1}{|\sin N\omega t|} \times \frac{\hbar}{2}|\sin N\omega t| = \frac{m}{et}\frac{1}{N}. \tag{9.120}$$

This is called the *Heisenberg limit*[7] of the precision in B, and you see it is much more precise than the standard quantum limit for large N. However, there is only a single measurement with a binary outcome $+$ or $-$, and this can never give us the full numerical value of B: there are many bits of information in a numerical value of the form $B = 5.2947 \times 10^{-7}$ T, but there is only one bit of information revealed in a binary measurement outcome. The Heisenberg limit therefore means that we

---

[7] M. Holland and K. Burnett, *Interferometric detection of optical phase shifts at the Heisenberg limit*, Phys. Rev. Lett. **71** 1355, 1993.

can detect the presence (a "yes/no" question) of a magnetic field of magnitude $B = m/eNt$. If the magnetic field is switched off (i.e., $B = 0$) the state does not evolve, and the measurement outcome will be $+$ with $p_+ = 1$. For a magnetic field magnitude at the Heisenberg limit the measurement outcome will be $-$ with probability $p_- = 1$. Any magnetic field that has a smaller magnitude will produce uncertainty in the measurement outcome, because $p_- < 1$.

It should also be noted that the CNOT operations are very hard to implement, and currently Heisenberg limited measurements have been demonstrated only for small N.

Exercises

1. Show that the variance of A vanishes when $|\psi\rangle$ is an eigenstate of A.
2. Calculate the maximum time precision for a 1.0 kg pendulum of length 1.0 m and amplitude 10 cm. Is the precision of a grandfather clock restricted by the laws of quantum mechanics?
3. Show that the precision of a clock cannot be improved by increasing the size of the face plate.
4. We can create entanglement by using a quantum erasure protocol. Consider two identical atoms with low lying energy states $|0\rangle$ and $|1\rangle$, and an excited state $|e\rangle$ that is coupled to $|1\rangle$. We can generate a photon if the atoms is in state $|1\rangle$ by exciting the atom to $|e\rangle$ with a laser, followed by the spontaneous emission of a photon.

   (a) both atoms are prepared in the state $(|0\rangle + |1\rangle)/\sqrt{2}$. Write down the state of the two atoms and any emitted photons after we excite both atoms and let them spontaneously emit a photon.
   (b) The modes that may contain photons are mixed on a beam splitter. What is the state of the atoms if we detect exactly one photon in the output modes of the beam splitter? You may assume our detectors are perfect.
   (c) What is the probability that we find exactly one photon after the beam splitter?

5. Prove (9.43) and (9.46).
6. Show that the momentum and the total energy can be measured simultaneously only when the potential is constant everywhere. What does a constant potential mean in terms of the dynamics of a particle?
7. A coherent state of a pendulum is defined by (9.61), and has a complex amplitude α. Calculate the expectation value for the position and momentum of the pendulum. When we write α in polar coordinates, such that $\alpha = r\,e^{i\phi}$, give a physical interpretation of ϕ.
8. Prove that the ladder operators obey the commutation relation $[\hat{a}, \hat{a}^\dagger] = 1$.

Chapter 10
The History of Quantum Mechanics

The history of quantum mechanics is full of interesting characters and stories. People often refer to the pioneers of quantum theory, so it is important that you have some familiarity with the history in order to participate in these shared stories. In this chapter I have collected some of the main results from the end of the nineteenth century to the end of the twentieth century, to give a broad overview of the development of the field.

The End of the 19th Century

Traditional classical physics consists of Newtonian mechanics, electromagnetism, and thermodynamics, which works remarkably well for most everyday phenomena. However, late 19th century physicists faced several questions that were deeply problematic for the classical world view held at the time.

Stability of atoms
Since the publication of Maxwell's theory of electromagnetism in 1864 and Heinrich Hertz' experiments with oscillating currents in the 1880s it was known that accelerating charges emit electromagnetic radiation (everything from radio waves to UV and X-rays). This caused problems for a simple "solar system" model of the atom, since orbiting electrons would undergo continual acceleration and therefore should lose energy and crash into the nucleus after only a short period of time. Other models of the atoms were proposed, such as J. J. Thomson's plum pudding model, that were also problematic. In short, physicists did not know how to explain the fact that atoms can exist. At the time this even led a fair number of physicists to believe that atoms did not exist at all, most notably Ernst Mach.

Supplementary Information The online version contains supplementary material available at https://doi.org/10.1007/978-3-031-16165-0_10.

P. Kok, *A First Introduction to Quantum Physics*, Undergraduate Lecture Notes in Physics, https://doi.org/10.1007/978-3-031-16165-0_10

Spectral lines

Another mystery was the origin of spectral lines. People understood early on that different materials emit light at different frequencies, and the technique of spectroscopy was and is used for the identification of different elements and molecules. In the 1860s, Gustav Kirchhoff formulated three laws of spectral lines (opaque objects emit a continuous spectrum, glowing gas emits spectral lines, and a cool gas in front of a hotter glowing object absorbs radiation at the frequencies of the emission lines of the gas). However, these are experimental facts that do not give a deeper understanding of the underlying mechanism. Why does a gas emit light at very specific frequencies, and why is it different for different gases?

The photo-electric effect

Observed by Heinrich Hertz in 1887, this was another head-scratcher. He noticed that currents would flow more easily across a spark gap when the material was illuminated with UV radiation. One would expect that more radiation would release more free charges, but the effect was highly frequency dependent. In particular, below a certain frequency it did not matter how intense the radiation was, it would not improve the conductivity.

Radioactivity

Towards the end of the nineteenth century, physicists and chemists started to notice that there are different types of radiation that behaved differently from electromagnetic radiation. In 1896, Henri Becquerel discovered that certain minerals left a print on photographic plates. In 1900 he was able to show that this radiation consists of fast negatively charged particles, which Ernest Rutherford later called "beta radiation". Rutherford also identified a different kind of radiation, "alpha radiation", which he found was easily stopped by only a small amount of material such as paper. In 1909, Thomas Royds proved that alpha particles are positively charged helium atoms, but they did not yet have a full understanding of the atomic structure. Marie Curie, the student of Becquerel, called the behaviour of materials that emit radiation *radioactivity*. She went on to use radioactivity to identify many new heavy elements, including radium. This element seemed to defy conservation of energy, and a proper understanding of the origin of radioactivity was badly needed.

Black body radiation

A black body is an object that emits electromagnetic radiation with a characteristic intensity for each wavelength in the spectrum. The curve depends only on the temperature of the body. When a black body is hot enough it glows very brightly, contrary to its name. The Sun is a good example of a black body. Lord Rayleigh (John William Strutt to his friends) and James Jeans worked out that for a black body the relationship between *spectral radiance* B as a function of wavelength λ and temperature T should go as

$$B(\lambda, T) = \frac{2ck_BT}{\lambda^4},$$

with c the speed of light and k_B Boltzmann's constant (check the units for a deeper understanding of spectral radiance). This Rayleigh-Jeans law worked out fine for longer wavelengths, but for shorter wavelengths B diverges. This is called the "UV catastrophe". Not surprisingly, this is not observed in nature (it would mean infinite energy). In reality, the spectral radiance gently returns to zero when $\lambda \rightarrow 0$. This dramatic discrepancy between theory and experiment demanded an explanation!

The Twentieth Century: The Old Quantum Theory

Quantum mechanics was developed in response to the puzzles encountered in the nineteenth century, but it took until the 1920s before a fully mature theory was created. In the early years of the twentieth century, several key discoveries were made that paved the way for Schrödinger, Heisenberg, and Dirac.

Planck and the black body spectrum
In 1900, Max Planck partially solved the puzzle of the black body spectrum, showing subsequent physicists the direction in which to go. The Rayleigh-Jeans law was derived by considering a volume and counting how many standing waves of electromagnetic radiation it can contain. In the classical case, each wave at a frequency f in the volume can have an arbitrary amplitude, leading to arbitrary energies ϵ in the waves. Integrating over all possible energies for waves with frequency f leads to the expression $B(\lambda, T)$ we encountered above. Planck, knowing from experimental data what the distribution should look like, chose *discrete* values of ϵ so that the integral changed into a discrete sum. Mathematically, this produced the desired result, but physically it was still a mystery why the oscillators producing the waves should be restricted like this. The allowed energies were now

$$\epsilon_n = nhf,$$

where $n = 1, 2, 3, \ldots$, and h was chosen to fit the experimental data. It became the famous Planck constant $h = 6.626 \times 10^{-34}$ J s.

Einstein and the photo-electric effect
While Planck assumed that the oscillators creating the waves were responsible for producing the discrete energies that give the right black body spectrum, Einstein realised that if the radiation *itself* occurred only with discrete energies, he could explain the photo-electric effect (much in the way you learned in highschool). Thus Einstein introduced the idea of *quantised* radiation. The name 'photon' was not introduced until 1916 by Leonard Troland, and popularised only in 1926 by Gilbert Lewis. Einstein received the Nobel Prize for his explanation of the photo-electric effect in 1921 (he did not receive the prize for his work on relativity!).

Rutherford's discovery of the atomic nucleus
As we have seen, Ernest Rutherford did important work in the experimental study of radioactivity. In 1909, with his assistants Hans Geiger and Ernest Marsden he fired alpha radiation at gold foil. Remember that the shape of atoms was unknown at this

time, and by studying how alpha particles scatter off the gold atoms he hoped to gain information about their shape. If the gold atoms were diffuse blobs of matter you would expect fairly gentle deviations in the path of the alpha particles, but instead they found that a small fraction of the alpha particles bounced right back towards their source. This could only be explained by assuming that atoms have a heavy positively charged nucleus, and hence the negatively charged electrons should orbit the nucleus. This was another key discovery that pointed ultimately to the quantum mechanical description of nature.

Bohr's model of the atom

In 1911 Niels Bohr became Rutherford's postdoc in Manchester, with his position funded by the Danish brewer Carlsberg. He learned about the shape of the atom from Rutherford's experiments, and set out to understand why the electrons in their highly accelerated orbits do not radiate their energy away. In 1913 he postulated, much in the style of Planck, that electrons can exist only in very specific orbits. They can jump between orbits by emitting or absorbing radiation. Since the orbits each have very specific energies, the radiation that is emitted or absorbed also has these energies. Thus Bohr explained in one go both the stability of atoms and the reason why spectral lines exist.

The work by Bohr, Einstein, and Planck became known as the 'old quantum theory'. It did not deal with physical systems at the atomic scale in a very systematic way, but some of the key ideas were already there, such as energy quantisation and the idea that systems are confined to allowable states. Bohr went on to become the 'father' of quantum mechanics by guiding a whole generation of physicists who would make tremendous progress in developing quantum theory, including Heisenberg, Dirac, Casimir, Oppenheimer, Wheeler, etc. He was the foremost proponent of the Copenhagen school of thought on quantum mechanics. Unfortunately he was notoriously difficult to understand, even by his peers, and as a result there is no single unique formulation of the so-called 'Copenhagen interpretation' of quantum mechanics.

Compton scattering and the existence of the photon

The idea that light consists of photons is nice and all, but it would be great if we had independent confirmation that photons behave as particles, scattering off other particles. In 1922, Arthur Compton scattered X-rays off electrons and found that their wavelength decreased as they were deflected by an angle θ:

$$\Delta\lambda = \frac{h}{m_e c}(1 - \cos\theta).$$

Here $\Delta\lambda$ is the change in wavelength of the scattered photon, m_e is the electron mass, and c is the speed of light. This formula can be derived directly from (relativistic) momentum conservation if we assume that the X-ray photons behave like particles. This was independent confirmation that photons were real. The discovery created new controversy, because the wave nature of light was well established, so how could light also be a particle? This question was the starting point of this book.

The New Quantum Theory

The work by Planck, Einstein, and Bohr inspired a new generation of mostly young physicists to come up with a new theory of quantum mechanics.

De Broglie's matter waves
One way of making sense of the strange behaviour of light and atoms according to the old quantum theory was to assign each particle a wavelength, and each wave a momentum. Louis de Broglie introduced this idea in his 1924 PhD thesis. He reasoned that Einstein's 'photons' were particles of light with a very specific momentum, which can be calculated from the classical radiation pressure of light. De Broglie expressed these thoughts in the form of his famous equation

$$p = \frac{h}{\lambda},$$

with p the momentum of the particle and λ its associated wavelength, as we have seen in Chap. 8. He generalised this idea to every particle and wave, introducing the so-called wave-particle duality. Bohr further developed the idea of this duality and stated that any physical system behaves either as a wave or as a particle, depending on the type of experiment you perform on it.

Schrödinger, Heisenberg, and Dirac
In 1925, the new quantum theory came of age. In this year first Erwin Schrödinger, and then Werner Heisenberg (with Born and Jordan) published their theories of how to describe systems quantum mechanically with a great degree of generality. Schrödinger's theory was referred to as wave mechanics, with the state of a quantum system described by a *wave function* (see Chap. 8). Heisenberg's theory was called matrix mechanics, which used the latest mathematical techniques to describe quantum systems.

Heisenberg's theory, which was developed together with Max Born and Pascual Jordan, was considerably more abstract than wave mechanics, and it was a conceptual hurdle for the more traditionally trained physicists at the time. As a result, wave mechanics initially gained more popularity. In 1926, Schrödinger proved that matrix mechanics and wave mechanics are in fact the same theory, formulated differently. In 1932, John von Neumann completed this programme with his unified description of wave mechanics and matrix mechanics in his book "Mathematical Foundations of Quantum Mechanics."

In 1925, Paul Dirac received Heisenberg's paper from his research supervisor Ralph Fowler. Dirac saw that Heisenberg's matrix mechanics was very similar to a not-so-well-known mathematical formulation of classical mechanics. From this insight he developed Heisenberg's theory in terms of so-called non-commuting observables. This way of thinking made it clearer what is going on and how quantisation works. He developed the notation of bras and kets that we still use today, including in this book.

In 1927, Dirac was staring into a fire, thinking about how to create a relativistic formulation of quantum mechanics. The problem was that the Schrödinger equation contains a first derivative in time, but a second derivative in space. Relativity treats space and time on an equal footing, so you would expect either first derivatives in both space and time, or second derivatives in both space and time. Dirac realised that he could make the mathematics work out, provided he introduced a new variable. This turned out to be spin, the mathematical description of which was independently given by Wolfgang Pauli in 1924. Neither knew what this meant physically, but it allowed for a relativistic description of the electron. The problem was that the theory also predicted particles with negative energy. Dirac postulated that all available negative energy states are filled, creating the so-called 'Dirac sea'. Today, the negative energy solutions to the Dirac equation are identified as anti-particles.

Spin, the Stern-Gerlach experiment, and magnetic moments

The problem of the physical meaning of spin was solved by Samuel Goudsmit and George Uhlenbeck in 1925. They worked out that the spin behaved like a magnetic moment based on the work by Otto Stern and Walter Gerlach. In 1921 Stern proposed a way to measure the magnetic moment of a particle and he related this to its angular momentum. Gerlach performed the experiment with silver atoms in 1922. Goudsmit and Uhlenbeck showed that spin is a kind of intrinsic angular momentum of the electron.

Probabilities of measurement outcomes

It took a while before people realised that quantum mechanics is at heart a theory about probabilities. This was figured out first by Max Born in July 1926. Born was a close friend of Einstein, defending his theory of special and general relativity in Germany against growing antisemitic sentiments, even among scientists. In one of his letters to Born, Einstein remarked that while quantum mechanics is an impressive theory, it is not yet the "real one". He added that "I, at any rate, am convinced that *He* does not play dice." Einstein nominated Heisenberg, Born and Jordan for the Nobel Prize in recognition of the development of matrix mechanics (Fig. 10.1).

The Einstein-Bohr debates

The fifth Solvay conference in Brussels in 1927 is probably the most famous conference in physics. It properly kicked off the debate between Einstein and Bohr about the meaning of quantum theory. The debate was to last decades, essentially until Einstein's death in 1955. Einstein was deeply disturbed that in particular in Heisenberg's formulation of matrix mechanics there was no mention anymore of an underlying space and time, opening the door to nonlocal effects. In addition, the fundamental probabilistic nature of quantum mechanics seems to undermine the notion of an objective underlying physical reality. Einstein constructed a series of very elegant thought experiments that were meant to show that quantum mechanics was either contradictory or incomplete (we have seen two of the most dramatic paradoxes in Chap. 9). At each turn, Bohr managed to parry the counterexamples, in the process showing everybody how quantum theory was to be interpreted (according to the Copenhagen interpretation). The debate culminated in the 1935 argument

Fig. 10.1 The Solvay conference in 1927, when it all kicked off between Einstein and Bohr. Back row left to right: Piccard, Henriot, Ehrenfest, Herzen, Den Donder, Schrödinger, Verschaffelt, Pauli, Heisenberg, Fowler, Brillouin. Centre row left to right: Debeye, Knudsen, Bragg, Kramers, Dirac, Compton, De Broglie, Born, Bohr. Front row left to right: Langmuir, Planck, Curie, Lorentz, Einstein, Langevin, Guye, Wilson, Richardson

by Einstein, Podolski and Rosen—known now as the EPR argument—that quantum mechanics cannot be a complete theory of nature. The debate about the interpretation of quantum mechanics continues to this day, and is briefly covered in Chap. 11.

Modern Quantum Physics

In the meantime, the world was tearing itself apart with the advent of the Second World War. This stopped most work on fundamental physics, and the focus was turned to practical applications to help the war effort instead.

The war years

Just before the war, nuclear physicists discovered a powerful new source of energy. Lise Meitner and her nephew Otto Frisch worked out the mechanism behind nuclear fission in 1938, while they were in Sweden as Jewish refugees from the nazis. Leo Szilárd realised later that the neutrons released in fission could induce a new fission event, creating a chain reaction. Given that a lot of energy is released in the process, this could have military applications. Szilárd and Einstein wrote a letter to president Franklin D. Roosevelt, urging the US to start a bomb project to counter a nazi bomb that may or may not be under development. This resulted in the Manhattan project, led by J. Robert Oppenheimer.

Many physicists fled Europe as the nazis conquered the continent, with a large number of European physicists joining the Manhattan project. Heisenberg stayed in Germany, heading the nazi bomb project (although the extent of his enthusiasm for this job is still unclear). This caused a rift between him and Bohr that never fully healed. Fundamental work on quantum theory came to a halt, but difficult questions remained: how to fully integrate quantum mechanics with relativity, or how to construct a quantum theory not just for particles, but also for the electromagnetic field? A new generation of physicists would take up these questions after the war, and they had a much more pragmatic outlook.

Relativistic quantum field theory

While Dirac had constructed a relativistic quantum theory, it was not perfect. In particular, it could not account for a small but significant shift in one of the spectral lines of hydrogen, discovered by Willis Lamb in 1947. This became known as the Lamb shift. In addition, naive attempts to describe the electromagnetic field quantum mechanically resulted in nonsensical results, giving the answer ∞ for many calculations. Clearly all was not well with relativistic quantum theory.

In 1948, Julian Schwinger gave a fully relativistic derivation of the Lamb Shift, using mathematical techniques that only a handful of people understood (including Freeman Dyson). Around the same time, Richard Feynman was approaching the problem in a completely different way, using clever diagrams to keep track of the calculations. Feynman and Dyson at one point took a road trip across the United States, which gave them the time to discuss Feynman's approach to calculating the Lamb shift. Dyson understood that Feynman's approach was equivalent to Schwinger's, but could be formulated in the more intuitive language of Feynman's diagrams. This showed the way to the development of quantum electrodynamics, the first quantum field theory.

Independent of Schwinger, Dyson and Feynman, Shin'ichiro Tomonaga (in a postwar Japan that was wrecked by the war) discovered methods to obtain finite results for relativistic quantum mechanics. We now call this *renormalisation*. Schwinger, Feynman and Tomonaga shared the 1965 Nobel prize. Their work led to the invention of the Standard Model for particle physics.

Gauge field theories and the Standard Model

In the sixties and seventies, new quantum field theories were developed that aimed to explain the plethora of fundamental particles found in particle accelerators. Particle accelerators had been built since the war years, notably by Enrico Fermi, and the embarrassment of riches of new types of matter beyond protons, neutrons and electrons that resulted meant that a deeper understanding of quantum field theory was necessary. Quantum electrodynamics became the model. Classical electrodynamics is a so-called gauge field theory. In simple terms, only potential differences have meaning, not the absolute value of the potential. Imposing this so-called *gauge invariance* on quantum physics resulted in the electromagnetic interaction between charged particles. Soon, new types of gauge invariance were constructed to create quantum field theories for the weak force (responsible for radioactive decay), and the strong force (the force that holds nuclei together against the repulsive electrical force between protons). Currently, quantum field theory underpins the Standard

Model, which describes to a very high degree of accuracy the physics of fundamental particles.

Yet, puzzles remain. The origin of the neutrino mass, and the imbalance between matter and anti-matter in the universe are some of the most pressing questions facing particle physicists today. And the biggest open question of all is how gravity fits into the picture.

Bell's theorem and nonlocality

The post-war development of relativistic quantum field theory, quantum condensed matter theory and particle physics was almost entirely pragmatic, and the culture discouraged speculation about the 'meaning' of quantum mechanics. The motto was 'Shut up and calculate!'. This attitude prevailed until well into the 1990s as this author knows from personal experience. However, in the 1960s there were a few notable exceptions. The particle physicist John Stuart Bell was sufficiently disturbed by the EPR argument that he formulated his famous *Bell's theorem*. It is a mathematical inequality that follows from the assumption that there exists a local deterministic reality. The inequality is violated by quantum mechanics, implying that there is no such thing as local deterministic reality. Experiments so far have all backed up the quantum mechanical prediction, and we say that quantum mechanics is *nonlocal*.

Quantum information

The properties of entanglement and Bell's theorem point to potential improvements when we use quantum systems for information processing. In 1984, Charles Bennett and Giles Brassard came up with a way to share a private key for cryptographic purposes that is unconditionally secure. In the same year, David Deutsch introduced the idea of a quantum bit, or *qubit* and showed how in principle it can perform calculations much faster than classical computers. In 1996, Peter Shor showed how qubits can factor large numbers into primes exponentially faster than conventional "classical" computers. This is an important application for code breaking, since most practical codes are based on the difficulty of factoring large numbers. Consequently, the race to build a quantum computer was on. In 2019, Google built the first quantum computer that outperforms any classical computer for a very specific task. It consisted of 53 qubits. We are currently in a research boom trying to work out how to build a fault tolerant general purpose quantum computer. This requires the control over millions of qubits.

Chapter 11
The Nature of Reality

In this chapter, we explore how quantum mechanics leads to some rather strange consequences for our notion of what is real and not real. What does quantum mechanics actually say about reality? Why does this theory make predictions that are so radically different from our everyday experience? We must answer these questions if quantum mechanics is to properly explain the world around us.

11.1   The Quantum State Revisited

What does the quantum state really mean? Some physicists hold that the state vector itself is a real property of the system, while others maintain that it is merely an expression of our knowledge of the system. Still others even argue that we should altogether drop the question of what $|\psi\rangle$ refers to in the real world. Niels Bohr[1] famously argued that we cannot meaningfully speak about what is going on at the quantum level at all:

> There is no quantum world. There is only an abstract physical description. It is wrong to think that the task of physics is to find out how nature is. Physics concerns what we can say about nature...

For many people this is unacceptable. Physical theories should have something to say about what nature really is, and not just be a prescription of what we can say about nature. This is what we identified as the second purpose of physical theories in Chap. 1: they should provide an explanatory framework to understand the natural world.

---

[1] A. Petersen, *The philosophy of Niels Bohr*, Bulletin of the Atomic Scientists **19** No. 7, 1963.

Supplementary Information   The online version contains supplementary material available at https://doi.org/10.1007/978-3-031-16165-0_11.

P. Kok, *A First Introduction to Quantum Physics*, Undergraduate Lecture Notes in Physics, https://doi.org/10.1007/978-3-031-16165-0_11

In this book, we developed quantum mechanics via careful reasoning about the probabilities of measurement outcomes. In particular, we did not ask the troublesome question what it all means at a "deeper" level. However, we cannot avoid this any longer. To attempt an answer to the question what $|\psi\rangle$ stands for, we need to introduce two concepts from philosophy:

- Epistemology: the study of what we can know.
- Ontology: the study of what exists.

We can broadly classify the different meanings of $|\psi\rangle$ into epistemic and ontic interpretations.[2] Epistemic interpretations of $|\psi\rangle$ say that the quantum state is an expression of our knowledge of the system, while ontic interpretations of $|\psi\rangle$ say that the quantum state corresponds to something real.

There are two types of epistemic interpretations:

- Quantum states reflect our knowledge of the system without referring to a deeper underlying reality. Call this ψ-epistemic of the first kind.
- Quantum states reflect our knowledge of the system with respect to a deeper underlying reality. We call this ψ-epistemic of the second kind.

Similarly, there are two types of ontic interpretations:

- Quantum states correspond to an underlying reality without any additional variables. We call this ψ-complete.
- Quantum states are supplemented by additional variables that correspond to an underlying reality. We call this ψ-supplemented.

The additional variables in the ψ-supplemented interpretation are typically hidden from us, and are commonly referred to as "hidden variables". They are not part of the theory of quantum mechanics as we have developed it here.

If you believe, like Niels Bohr, that there is no deeper underlying quantum reality (the epistemic interpretation of the first kind) then you are done. Quantum mechanics does not pose any conceptual problems for you. The phenomena are what they are, and quantum mechanics accurately predicts their probabilities.

However, if you believe that quantum mechanics says something about a microscopic physical reality, then three possibilities for the interpretation of the quantum state remain:

- complete,
- supplemented, and
- epistemic (of the second kind).

These three types of interpretations are collected in Fig. 11.1.

Which constraints do the predictions of quantum mechanics place on the nature of the underlying reality? This is a question that is currently actively researched, since it allows us to narrow down what $|\psi\rangle$ can mean. In the next two sections we

---

[2] N. Harrigan and R. W. Spekkens, *Einstein, Incompleteness, and the Epistemic View of Quantum States*, Found. Phys. **40** 125, 2010.

Fig. 11.1 A model of an underlying reality for quantum probabilities (see supplementary material 1)

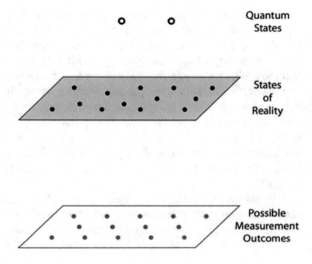

will set up two thought experiments that will limit the way reality can behave, given ψ-complete and ψ-supplemented interpretations. In the remainder of this section, we will deal a blow to the viability of ψ-epistemic interpretations of the second kind.

At first sight, ψ-epistemic interpretations of the second kind are rather attractive. The way we have developed quantum mechanics in this book shows that the entire theory can be constructed as a generalised probability theory. Probability theories are often interpreted as epistemic: the probabilities tell us something about our knowledge of a system. For example, when we say that a flipped coin can come up heads or tails—each with a probability of 1/2—we typically mean that we do not know the details of the movement of the coin and cannot calculate exactly how it is going to fall. If we could do this, we would know more about the coin flip, and we would not assign a 50:50 chance to the outcome. There are people who beat the odds at roulette by calculating how the ball moves, and where it will likely land. Seen in this light, quantum mechanics as a probability theory suggests a ψ-epistemic interpretation of the quantum state.

Another example is quantum teleportation (see Chap. 6): Alice sends an unknown quantum state to Bob by performing a two-qubit Bell measurement on her qubit in the unknown quantum state and one qubit of an entangled quantum system, which she shares with Bob. Before Alice sends the outcome of the measurement to Bob, he describes the state of his qubit as $\rho_B = \mathbb{I}/2$. This means that he is completely uncertain about the state of his qubit. After learning Alice's measurement outcome, he describes the state of his qubit as

$$|\psi'\rangle_B = U^\dagger |\psi\rangle_B \,, \tag{11.1}$$

where U is the corrective operation Bob has to perform in order to retrieve the quantum state $|\psi\rangle$.

Bob has not touched his system at all, but learning something about his qubit allows him to change the state of the qubit. In this situation it is natural to assume that there is no instantaneous physical process that changes the state of the system, Bob is merely updating his knowledge about the qubit. Again, this strongly suggests that the ψ-epistemic interpretation of the quantum state is the natural one.

Finally, the measurement process—where we learn something about the state of a system—is a discontinuous and non-deterministic process, while the unitary evolution determined by the Schrödinger equation is continuous and deterministic. Is it not strange that nature relies on two types of evolution, especially when the mechanism for the collapse of the wave function is not specified at all? On the other hand, wave function collapse as an update of our knowledge is not mysterious, but natural!

We now present a simple argument by Pusey, Barrett, and Rudolph[3] (PBR), which shows that a straightforward ψ-epistemic interpretation is a problematic view of the quantum state. First, we set up an experiment: we have two sources that can each create a qubit in the state $|0\rangle$ or $|+\rangle$. As usual, we define

$$|\pm\rangle = \frac{|0\rangle \pm |1\rangle}{\sqrt{2}}.$$

(11.2)

The qubits that are sent to the detector can therefore be in the states

$$|0, 0\rangle, \quad |0, +\rangle, \quad |+, 0\rangle, \quad \text{or} \quad |+, +\rangle.$$

(11.3)

The detector is set up to measure a special two-qubit observable with eigenstates

$$|\text{not } 00\rangle = \frac{1}{\sqrt{2}}|0, 1\rangle + \frac{1}{\sqrt{2}}|1, 0\rangle,$$

(11.4a)

$$|\text{not } 0+\rangle = \frac{1}{\sqrt{2}}|0, -\rangle + \frac{1}{\sqrt{2}}|1, +\rangle,$$

(11.4b)

$$|\text{not } +0\rangle = \frac{1}{\sqrt{2}}|+, 1\rangle + \frac{1}{\sqrt{2}}|-, 0\rangle,$$

(11.4c)

$$|\text{not } ++\rangle = \frac{1}{\sqrt{2}}|+, -\rangle + \frac{1}{\sqrt{2}}|-, +\rangle.$$

(11.4d)

These states from an orthonormal basis, and can therefore be used to create a valid observable. As the labels of the state indicate, a measurement outcome corresponding to the state in (11.4a) can happen only when the sources did *not* produce $|0, 0\rangle$, the state in (11.4b) indicates that the sources did *not* produce $|0, +\rangle$, and so on:

---

[3] M. Pusey, J. Barrett and T. Rudolph, *On the reality of the quantum state*, Nature Physics **8** 475, 2012.

$$\Pr(\text{not } 00|00) = |\langle \text{not } 00|00 \rangle|^2 = 0, \tag{11.5a}$$

$$\Pr(\text{not } 0+|0+) = |\langle \text{not } 0+|0+ \rangle|^2 = 0, \tag{11.5b}$$

$$\Pr(\text{not } +0|+0) = |\langle \text{not } +0|+0 \rangle|^2 = 0, \tag{11.5c}$$

$$\Pr(\text{not } ++|++) = |\langle \text{not } ++|++ \rangle|^2 = 0. \tag{11.5d}$$

This is what quantum mechanics predicts we would find in the experiment.

To relate this experiment to what "really" happens in nature, we assume that the quantum states $|0\rangle$ and $|+\rangle$ merely describe our knowledge about the underlying states of reality λ. In this description, the physical system is "really" in a state λ, independent of what we know about it or which observable we decide to measure. Each preparation procedure that creates a qubit in the state $|0\rangle$ will then result in some probability distribution $\mu_0(\lambda)$ over the states of reality λ. Similarly, the preparation procedure for $|+\rangle$ will result in a probability distribution $\mu_+(\lambda)$ over the same states of reality λ. These probabilities may overlap such that for some states of reality λ both $\mu_0(\lambda)$ and $\mu_+(\lambda)$ are nonzero. Alternatively the distributions may not overlap, such that any λ can give rise only to a nonzero $\mu_0(\lambda)$ or a non-zero $\mu_+(\lambda)$, but not both. Each λ then uniquely determines the state $|0\rangle$ or $|+\rangle$. However, the difference between the quantum state and the probability distribution over the states of reality is then merely a matter of words; the quantum state is just a shorthand for the possible real states λ, and can be considered a complete description of reality, leading to a ψ-complete interpretation. Therefore, ψ-epistemic interpretations imply distributions $\mu_0(\lambda)$ and $\mu_+(\lambda)$ with some overlap (see Fig. 11.2).

In this light, let us consider the above experiment again with a ψ-epistemic interpretation. When we create the state $|0\rangle$, there is a probability $p > 0$ that the actual state of reality λ lies in the overlap region where both $\mu_0(\lambda) > 0$ and $\mu_+(\lambda) > 0$. This implies directly that such a state could have been produced in either the $|0\rangle$ or the $|+\rangle$ preparation procedure (for keeping a record of the preparation procedure in the states of reality would effectively put λ outside the region of overlap).

Next, there is a probability $p^2 > 0$ that the state preparation $|0, 0\rangle$ by the two sources results in two states of reality (one for each qubit) that both lie in the overlap region where $\mu_0(\lambda) > 0$ and $\mu_+(\lambda) > 0$. However, if both states of reality lie in the overlap region, it is impossible to know whether the state preparation procedure was in fact $|0, 0\rangle$, $|0, +\rangle$, $|+, 0\rangle$, or $|+, +\rangle$. When we measure the observable with the eigenvectors from (11.4a)–(11.4d) the state of reality in the overlap region must occasionally produce measurement outcomes "not 00", even though the state preparation was in fact $|0, 0\rangle$. This contradicts the predictions of quantum mechanics unless $p = 0$, in which case the probability distributions $\mu_0(\lambda)$ and $\mu_+(\lambda)$ do not overlap. We argued above that non-overlapping distributions over the real states is not compatible with a ψ-epistemic interpretation of the quantum state, and we therefore have to seriously question its viability.

We should not be too hasty, though. This argument rests on two key assumptions. First, a physical system needs to have an objective real state ψ, independent of the observer. The argument depends crucially on it. Second, independently prepared sys-

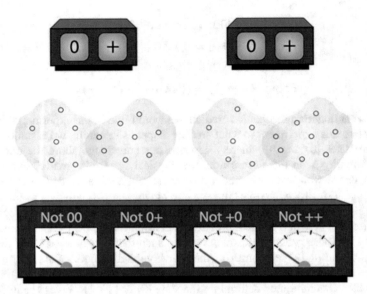

Fig. 11.2 The PBR theorem (see supplementary material 2)

tems have independent real states: we must assume that there is no grand conspiracy in the preparation of the two qubits, such that a preparation of $|0, 0\rangle$ will never trigger the outcome "not 00". These both seem natural assumption, and it is hard to see how we can save the ψ-epistemic interpretation without adding severe complications in the objective states of reality and how they interact with the rest of the universe.

11.2   Nonlocality

One of the problems in coming up with a realistic interpretation of the quantum state is how to deal with the probabilistic character of measurement outcomes. Any attempt at removing the probabilistic nature of quantum mechanics by introducing hidden variables (in a ψ-supplemented interpretation) inevitably leads to these hidden variables being nonlocal. To see how this comes about, we give a simple example, originally devised by Lucien Hardy.[4]

Alice and Bob, who are far apart from each other, each hold one qubit. They both have the choice of measuring their qubit either in the basis $\{|0\rangle, |1\rangle\}$ or in the basis

$$|+\rangle = \frac{|0\rangle + |1\rangle}{\sqrt{2}} \quad \text{and} \quad |-\rangle = \frac{|0\rangle - |1\rangle}{\sqrt{2}}, \quad (11.6)$$

---

[4] L. Hardy, *Nonlocality for two particles without inequalities for almost all entangled states*, Phys. Rev. Lett. **71** 1665, 1993.

corresponding to the eigenstates of the Pauli matrices σ_z and σ_x, respectively. For example, the qubits can be electrons with spin $1/2$, and Alice and Bob each have a Stern-Gerlach apparatus that they can orient in the z- or the x-direction.

We set up the state of the two qubits held by Alice and Bob in such a way that they never find the following joint measurement outcomes

$$\begin{aligned} &\text{Alice measures } 1 \quad \text{and} \quad \text{Bob measures } +\,, \\ &\text{Alice measures } + \quad \text{and} \quad \text{Bob measures } 1\,, \\ &\text{Alice measures } - \quad \text{and} \quad \text{Bob measures } -\,. \end{aligned}$$

We can express these outcomes with the probabilities

$$\Pr(1, +) = 0\,, \tag{11.7a}$$
$$\Pr(+, 1) = 0\,, \tag{11.7b}$$
$$\Pr(-, -) = 0\,. \tag{11.7c}$$

Later, we will see which specific quantum state of the two qubits will give these probabilities.

Before that, however, we expect that these probabilities allow us to infer that a measurement outcome "1" by Alice implies that if Bob measures σ_z, he must find measurement outcome "0". To see this, we note first that according to (11.7a), if Alice finds outcome "1", Bob must find outcome "$-$" in a measurement of σ_x:

$$1_A \xrightarrow[(10.7a)]{} -_B\,. \tag{11.8}$$

There are only two measurement outcomes for Bob's σ_x measurement, namely "+" and "$-$", and (11.7a) rules out the outcome "+".

Second, if Bob finds outcome "$-$", then according to (11.7c), Alice should always find outcome "+" in a σ_x measurement:

$$-_B \xrightarrow[(10.7c)]{} +_A\,. \tag{11.9}$$

Finally, if Alice finds outcome "+", then according to (11.7b), Bob should always find outcome "0" in a σ_z measurement:

$$+_A \xrightarrow[(10.7b)]{} 0_B\,. \tag{11.10}$$

Therefore, we have the chain of inference

$$1_A \xrightarrow[(10.7a)]{} -_B \xrightarrow[(10.7c)]{} +_A \xrightarrow[(10.7b)]{} 0_B\,, \tag{11.11}$$

In other words, the probability that Alice and Bob both find the measurement outcome "1" is zero:

$$\Pr(1, 1) = 0. \tag{11.12}$$

This is a perfectly reasonable conclusion when we assume that the properties of each qubit completely determine the outcomes of the measurements that are performed on it.

Next, we show how we can achieve the probabilities given in (11.7a)–(11.7c) in quantum mechanics. We prepare the two qubits held by Alice and Bob in the quantum state

$$|\psi\rangle = \frac{3|00\rangle + |01\rangle + |10\rangle - |11\rangle}{\sqrt{12}}. \tag{11.13}$$

You can verify that the probabilities are indeed

$$\Pr(1, +) = |\langle 1+|\psi\rangle|^2 = 0,$$
$$\Pr(+, 1) = |\langle +1|\psi\rangle|^2 = 0,$$
$$\Pr(-, -) = |\langle --|\psi\rangle|^2 = 0, \tag{11.14}$$

as required. However, when we calculate the probability that Alice and Bob both find the measurement outcome "1", we are in for a surprise:

$$\Pr(1, 1) = |\langle 11|\psi\rangle|^2 = \frac{1}{12}. \tag{11.15}$$

This directly contradicts our previous argument in (11.12) that $\Pr(1, 1) = 0$, which assumed that each qubit has properties that locally determine the outcomes of the measurements. Therefore, this assumption must be false!

The implications of this are profound: You may believe that the measurement outcomes for a quantum mechanical system are determined by extra hidden variables in order to "save" determinism in nature. To keep the predictions of quantum mechanics for our example—in particular (11.15)—the hidden variables for Alice's qubit must be influenced by the hidden variables of Bob's qubit, and vice versa. However, Alice and Bob can be arbitrary far apart, and can make their measurements at any time they like. This means that the influence of the hidden variables must travel faster than light! In this sense, quantum mechanics is often said to be nonlocal. Any hidden variable model must be nonlocal in order to reproduce the predictions of quantum mechanics. Moreover, these predictions are in fact verified in many experiments, and it therefore seems that nature really is nonlocal!

A note of caution: These results are often presented as contradicting Einstein's theory of relativity, which says that nothing can travel faster than light. However, quantum mechanics does not allow you to send signals faster than light. Any attempt to do so will ultimately fail due to the probabilistic nature of the measurement outcomes. The faster-than-light behaviour is restricted to the hidden variables, which are an addition to quantum mechanics, and not directly accessible to experimenters.

In 1964, John Bell proposed a very general proof that any hidden variable theory that reproduces the predictions of quantum mechanics *must* be nonlocal.[5] That is, even though it is still impossible to come up with schemes that allow faster than light communication, the hidden variables themselves must influence each other instantaneously. Bell derived a series of inequalities—now called Bell's inequalities—that relate the statistics of the possible measurement outcomes of an experiment in such a way that their violation demonstrates the nonlocal character of hidden variable models. To demonstrate a violation of Bell's inequalities, one has to make measurements on two entangled particles that lie outside of each other's light cone (to ensure no slower than light communication takes place). The first demonstration of such a violation for spatially separated measurements was carried out by Alain Aspect[6] and his co-workers in 1982. They measured the polarisation of two photons coming from a central caesium source. A possible loophole could still be found in the detection efficiencies of the photodetectors. This loophole was closed by David Wineland's group[7] in 2001, but without the spatial separation. In 2015, Ronald Hanson's group[8] demonstrated the violation of Bell's inequality with spatially separated measurements, while simultaneously closing the detection loophole. These experiments conclusively show that nature cannot be accurately described by local realistic theories. This is one of the most profound insights into the nature of reality that science has delivered in its long history.

11.3 Contextuality

We have seen that quantum mechanics is nonlocal, even though it does not allow us to exploit this for sending messages faster than light. Another way in which quantum mechanics is strange already shows up in a single system, and is called *contextuality*.

Before we give an example of contextuality in quantum mechanics, we consider a classical form of contextuality that was proposed by John Wheeler.[9] The group game of Twenty Questions is played as follows: one person—the player, let's call him Patrick—temporarily leaves the group, while the rest determine a secret word.

---

[5] J. S. Bell, in *Speakable and Unspeakable in Quantum Mechanics: Collected Papers on Quantum Philosophy*, Cambridge University Press, 1987.

[6] A. Aspect, J. Dalibard, and G. Roger, *Experimental Test of Bell's Inequalities Using Time-Varying Analyzers*, Phys. Rev. Lett. **49** 1804, 1982.

[7] M. A. Rowe et al., *Experimental violation of a Bell's inequality with efficient detection*, Nature **409** 791, 2001.

[8] B. Hensen et al., *Loophole-free Bell inequality violation using electron spins separated by 1.3 kilometres*, Nature **526** 682, 2015.

[9] J. A. Wheeler, *Information, physics, quantum: The search for links*, in W. H. Zurek, *Complexity, Entropy, and the Physics of Information*, Addison-Wesley 1990.

When Patrick returns, he may ask up to twenty questions that can be answered with either "yes" or "no". Patrick will address the questions to individual members of the group. For example, the group chose the word "steamboat", and Patrick asks the first group member, Alice: "Is it a person"? Alice will answer "no". Patrick then addresses Bob with the next question, and so on. If Patrick can guess the secret word using twenty questions or less, he wins. Otherwise the group wins.

At the next instance of the game, while a new player—Patricia—is waiting outside the room, the group feels mischievous and decides not to choose a word. Instead, each group member silently chooses a private word and will answer "yes" or "no" based on the word they have chosen. Every group member will have a different word in mind! Patricia first asks Alice: "Is it a person?", to which the answer is "yes". Bob had chosen the word "tree", which is clearly not a person. He therefore has to update his secret word such that it conforms to Alice's answer. Each successive group member will have to make sure that their private word conforms to all previous answers by the other members. Of course, this quickly becomes fiendishly difficult for the group members! Assume that Patricia wins with her final guessed word. It is unlikely that this word was thought of by any of the group members at the start, since everybody probably had to update their secret words at least once during the questions stage. So the final word was determined not only by the group members (as in the traditional way the game is played), but also by Patricia's choice of questions. The final outcome therefore depends not only on the choice of the group alone, but also on the context provided by the questions.

To see how this relates to quantum mechanics, we first set up a mathematical puzzle similar to Twenty Questions. We have five statements about a situation, $A_1, \ldots, A_5$, and use them to construct the following four joint questions, which all happen to have the answer "no":

$$\text{Are } A_1 \text{ and } A_2 \text{ both true? } \rightarrow \text{ No,} \qquad (11.16a)$$

$$\text{Are } A_2 \text{ and } A_3 \text{ both false? } \rightarrow \text{ No,} \qquad (11.16b)$$

$$\text{Are } A_3 \text{ and } A_4 \text{ both true? } \rightarrow \text{ No,} \qquad (11.16c)$$

$$\text{Are } A_4 \text{ and } A_5 \text{ both false? } \rightarrow \text{ No.} \qquad (11.16d)$$

We can naturally assume that the statements $A_1, \ldots, A_5$ each have pre-determined truth values (true or false, or "yes" or "no"), just like in the original game of Twenty Questions, and we construct the following chain of reasoning:

$$\text{If } A_1 \text{ is true } \xrightarrow{(10.6a)} A_2 \text{ is false,} \qquad (11.17a)$$

$$\text{If } A_2 \text{ is false } \xrightarrow{(10.6b)} A_3 \text{ is true,} \qquad (11.17b)$$

$$\text{If } A_3 \text{ is true } \xrightarrow{(10.6c)} A_4 \text{ is false,} \qquad (11.17c)$$

$$\text{If } A_4 \text{ is false } \xrightarrow{(10.6d)} A_5 \text{ is true,} \qquad (11.17d)$$

Therefore, we know the answer to the following question with certainty: If A_1 is true then A_5 is true, so

$$\text{Is } A_1 \text{ true and } A_5 \text{ false?} \rightarrow \text{No.} \tag{11.18}$$

This is what we expect if the statements $A_1, \ldots, A_5$ are individually true or false. If this game was played by the Twenty Questions group (the non-evil version), all members would agree on the truth values of the statements $A_1, \ldots, A_5$, and the answer to the player's question (11.18) must indeed be "no."

Next, we implement this experiment with a quantum mechanical system. Each experimental question is of the form "are A_1 and A_2 both true?" or "are A_4 and A_5 both false?", and corresponds to a different observable with a yes or no answer. To calculate the probability that A_1 is true given the quantum state $|\psi\rangle$, we first identify $|A_1\rangle$ as the state of the quantum system with the property that A_1 is true. The probability that A_1 is true is then given by the Born rule:

$$\Pr(A_1 \text{ is true}) = |\langle A_1|\psi\rangle|^2 = \langle\psi|P_1|\psi\rangle, \tag{11.19}$$

where $P_1 = |A_1\rangle\langle A_1|$ is the projector onto the state $|A_1\rangle$ (see Sect. 5.5). Similarly, we can calculate the probability that A_1 is false via

$$\begin{aligned}
\Pr(A_1 \text{ is false}) &= 1 - \Pr(A_1 \text{ is true}) \\
&= 1 - \langle\psi|P_1|\psi\rangle \\
&= \langle\psi|(\mathbb{I} - P_1)|\psi\rangle,
\end{aligned} \tag{11.20}$$

since A_1 is either true or false whenever we ask. Here, $\mathbb{I}$ is the identity operator. Measuring A_1 (that is, asking if A_1 is true or false) then amounts to the quantum state jumping into the eigenstate $|A_1\rangle$—indicating that A_1 is true—or in a state orthogonal to $|A_1\rangle$, indicating that A_1 is false. Finding that A_1 is true is therefore equivalent to saying that the quantum state is projected onto $|A_1\rangle$, by virtue of the projection postulate.

We set up a similar reasoning for A_2, and we associate the state "A_2 is true" with a quantum state $|A_2\rangle$ such that

$$\Pr(A_2 \text{ is true}) = |\langle A_2|\psi\rangle|^2 = \langle\psi|P_2|\psi\rangle \tag{11.21}$$

and

$$\Pr(A_2 \text{ is false}) = \langle\psi|(\mathbb{I} - P_2)|\psi\rangle. \tag{11.22}$$

To calculate the probability that both A_1 and A_2 are true, the observables must be compatible. This means that we must be able to ask the question "is A_1 true" and "is A_2 true" simultaneously in an experimental setting. Since a measurement of an observable leads to a state vector collapse into one of the eigenstates of the observable, the eigenstates for the observable A_1 and A_2 must be the same.

Therefore the projectors P_1 and P_2 must commute:

$$[P_1, P_2] = P_1 P_2 - P_2 P_1 = 0. \tag{11.23}$$

The probability that A_1 and A_2 are *both* true can then be calculated as

$$\Pr(A_1, A_2 \text{ both true}) = \langle\psi|P_1 P_2|\psi\rangle. \tag{11.24a}$$

When $|A_2\rangle$ is perpendicular to $|A_1\rangle$ we find that $P_1 P_2 = 0$, and the statement "A_1 and A_2 are both true" is always false, regardless of the quantum state $|\psi\rangle$.

We can follow this procedure also for the statements A_3, A_4 and A_5 with respective quantum states $|A_3\rangle$, $|A_4\rangle$, $|A_5\rangle$ and projectors P_3, P_4 and P_5. We then find that

$$\Pr(A_2, A_3 \text{ both false}) = \langle\psi|(\mathbb{I} - P_2)(\mathbb{I} - P_3)|\psi\rangle \tag{11.24b}$$
$$\Pr(A_3, A_4 \text{ both true}) = \langle\psi|P_3 P_4|\psi\rangle, \tag{11.24c}$$

and

$$\Pr(A_4, A_5 \text{ both false}) = \langle\psi|(\mathbb{I} - P_4)(\mathbb{I} - P_5)|\psi\rangle. \tag{11.24d}$$

This means that A_2 and A_3 must also be compatible observables, since we establish a joint truth value of A_2 and A_3 in (11.24b). Therefore P_2 and P_3 must commute: $[P_2, P_3] = 0$. Similarly, (11.24c) tells us that $[P_3, P_4] = 0$, and (11.24d) gives $[P_4, P_5] = 0$. However, just because $[P_1, P_2] = 0$ and $[P_2, P_3] = 0$ does not mean that $[P_1, P_3] = 0$, since we are not making joint statements about A_1 and A_3. The same is true for any other pair of observables that we do not wish to measure simultaneously. We will construct projectors with these commutation relations shortly.

First, let's summarise the probabilities that we have so far:

$$\Pr(A_1, A_2 \text{ both true}) = 0, \tag{11.25a}$$
$$\Pr(A_2, A_3 \text{ both false}) = 0, \tag{11.25b}$$
$$\Pr(A_3, A_4 \text{ both true}) = 0, \tag{11.25c}$$
$$\Pr(A_4, A_5 \text{ both false}) = 0. \tag{11.25d}$$

We saw in the classical case that this leads us to believe that

$$\Pr(A_1 \text{ true and } A_5 \text{ false}) = 0, \tag{11.25e}$$

as long as $A_1, \ldots, A_5$ have predetermined values. These predetermined values are the physical properties of the system that exist independent of any measurements of the system. Equation (11.25e) implies that $[P_1, P_5] = 0$.

Next, let's see what quantum mechanics makes of this. An explicit example of the game was proposed by Adàn Cabello[10] and co-workers, in which we consider a three-level quantum system. For example, the quantum system may be an atom with three energy levels, a particle with spin 1, or a photon that has the choice of three paths. It is convenient to solve this problem in vector form instead of kets, because the vectors are three-dimensional and can be made real, so they are easy to visualise.

For the five yes/no statements $A_1, \ldots, A_5$ we can choose anything we like that is allowed by quantum mechanics. In our case, we choose the eigenstates corresponding to the eigenvalue "true" for the five statements $A_1, \ldots, A_5$ as

$$|A_1\rangle = \frac{1}{\sqrt{3}} \begin{pmatrix} 1 \\ -1 \\ 1 \end{pmatrix}, \quad |A_2\rangle = \frac{1}{\sqrt{2}} \begin{pmatrix} 1 \\ 1 \\ 0 \end{pmatrix},$$

$$|A_3\rangle = \begin{pmatrix} 0 \\ 0 \\ 1 \end{pmatrix}, \quad |A_4\rangle = \begin{pmatrix} 1 \\ 0 \\ 0 \end{pmatrix}, \quad |A_5\rangle = \frac{1}{\sqrt{2}} \begin{pmatrix} 0 \\ 1 \\ 1 \end{pmatrix}, \tag{11.26}$$

with projectors $P_1 = |A_1\rangle\langle A_1|$, $P_2 = |A_2\rangle\langle A_2|$, and so on. Using matrix multiplication, you can verify that

$$P_1 = |A_1\rangle\langle A_1| = \frac{1}{3} \begin{pmatrix} 1 & -1 & 1 \\ -1 & 1 & -1 \\ 1 & -1 & 1 \end{pmatrix}, \tag{11.27a}$$

$$P_2 = |A_2\rangle\langle A_2| = \frac{1}{2} \begin{pmatrix} 1 & 1 & 0 \\ 1 & 1 & 0 \\ 0 & 0 & 0 \end{pmatrix}, \tag{11.27b}$$

$$P_3 = |A_3\rangle\langle A_3| = \begin{pmatrix} 0 & 0 & 0 \\ 0 & 0 & 0 \\ 0 & 0 & 1 \end{pmatrix}, \tag{11.27c}$$

$$P_4 = |A_4\rangle\langle A_4| = \begin{pmatrix} 1 & 0 & 0 \\ 0 & 0 & 0 \\ 0 & 0 & 0 \end{pmatrix}, \tag{11.27d}$$

and

$$P_5 = |A_5\rangle\langle A_5| = \frac{1}{2} \begin{pmatrix} 0 & 0 & 0 \\ 0 & 1 & 1 \\ 0 & 1 & 1 \end{pmatrix}. \tag{11.27e}$$

[10] A. Cabello et al., *Simple Hardy-Like Proof of Quantum Contextuality*, Phys. Rev. Lett. **111** 180404, 2013.

You should also check that the projectors pairwise commute according to our requirements above:

$$[P_1, P_2] = 0, \qquad [P_2, P_3] = 0, \qquad [P_3, P_4] = 0, \tag{11.28}$$

$$[P_4, P_5] = 0, \qquad \text{and} \qquad [P_5, P_1] = 0. \tag{11.29}$$

You can calculate the projectors $\mathbb{I} - P_2$ and $\mathbb{I} - P_3$ by using the matrix form of $\mathbb{I}$:

$$\mathbb{I} = \begin{pmatrix} 1 & 0 & 0 \\ 0 & 1 & 0 \\ 0 & 0 & 1 \end{pmatrix}. \tag{11.30}$$

Now we are finally ready to calculate the probabilities in (11.24d)–(11.24a). We have to prepare the system in a particular quantum state, and for our example we choose

$$|\psi\rangle = \frac{1}{\sqrt{3}} \begin{pmatrix} 1 \\ 1 \\ 1 \end{pmatrix}. \tag{11.31}$$

Using this state, we find indeed that the probabilities in (11.24d)–(11.24a) are zero. For example,

$$P_1 P_2 = \frac{1}{6} \begin{pmatrix} 1 & -1 & 1 \\ -1 & 1 & -1 \\ 1 & -1 & 1 \end{pmatrix} \begin{pmatrix} 1 & 1 & 0 \\ 1 & 1 & 0 \\ 0 & 0 & 0 \end{pmatrix} = \begin{pmatrix} 0 & 0 & 0 \\ 0 & 0 & 0 \\ 0 & 0 & 0 \end{pmatrix}, \tag{11.32}$$

and therefore $\langle\psi|P_1 P_2|\psi\rangle = 0$ since $P_1 P_2 = 0$. Similarly, $(\mathbb{I} - P_2)(\mathbb{I} - P_3)$ can be calculated as

$$(\mathbb{I} - P_2)(\mathbb{I} - P_3) = \begin{pmatrix} \frac{1}{2} & -\frac{1}{2} & 0 \\ -\frac{1}{2} & \frac{1}{2} & 0 \\ 0 & 0 & 2 \end{pmatrix} \begin{pmatrix} 1 & 0 & 0 \\ 0 & 1 & 0 \\ 0 & 0 & 0 \end{pmatrix} = \begin{pmatrix} \frac{1}{2} & -\frac{1}{2} & 0 \\ -\frac{1}{2} & \frac{1}{2} & 0 \\ 0 & 0 & 0 \end{pmatrix}, \tag{11.33}$$

and the probability $\langle\psi|(\mathbb{I} - P_2)(\mathbb{I} - P_3)|\psi\rangle = 0$. You can calculate the other probabilities in a similar way.

Finally, we calculate the probability of A_1 being true and A_5 being false: $\langle\psi|P_1(\mathbb{I} - P_5)|\psi\rangle$. This should be zero according to the classical thinking about physical properties that lead to (11.18). First, we determine

$$P_1(\mathbb{I} - P_5) = P_1 - P_1 P_5 = P_1, \tag{11.34}$$

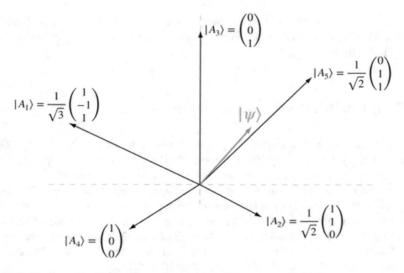

Fig. 11.3 Contextuality in three dimensions (see supplementary material 3)

since $P_1 P_5 = 0$. The probability $\Pr(A_1$ true and A_5 false) is then given by the expectation value

$$\Pr(A_1 \text{ true and } A_5 \text{ false}) = \langle\psi|P_1(\mathbb{I} - P_5)|\psi\rangle = \langle\psi|P_1|\psi\rangle = \frac{1}{9}. \qquad (11.35)$$

This is in contradiction to the earlier expected result that $\Pr(A_1$ true and A_5 false) $= 0$ if every proposition $A_1, \ldots, A_5$ is separately true or false. In other words, the propositions do not have separate truth values, and much like Wheeler's evil Twenty Questions, the outcome of the measurement is determined by the questions that you ask. In other words, whether $A_1, \ldots, A_5$ are real properties of the system depends on the context of the measurements that are being made! The geometric construction of this argument is shown in Fig. 11.3.

The contextuality of quantum mechanics was proved in 1967 by Simon Kochen and Ernst Specker, and it is now called the Kochen-Specker theorem.[11] It is an extremely important result in the foundations of quantum mechanics. It proves that it is impossible to assign definite values to all physical properties of a quantum system prior to a measurement without changing the experimental predictions. Since experiments have come out in favour of quantum mechanics, this means that you cannot say that quantum systems have properties in and of themselves. We can now appreciate Wheeler's original quote from Chap. 1 at an even deeper level:

No phenomenon is a physical phenomenon until it is an *observed* phenomenon

---

[11] S. Kochen and E. P. Specker, *The problem of hidden variables in quantum mechanics*, J. Math. Mech. **17** 59, 1967.

Our act of making a measurement—asking a question—brings physical properties into reality. We, as observers, are fundamentally instrumental in determining what reality looks like.

11.4 A Compendium of Interpretations

So how do we interpret the state of a physical system in quantum mechanics? Over the past century a great many attempts have been made to answer this question, and physicists and philosophers of physics have come up with all sorts of clever, crazy, outlandish and plausible solutions. At this point, no single interpretation of quantum mechanics clearly stands out as the generally accepted interpretation (although there are some front runners), and in the remaining pages of this book we will have a look at some of the most important ones at the time of writing. Detailed descriptions of these interpretations are given in the online Stanford Encyclopedia of Philosophy, and I have included footnotes to the various articles throughout.

11.4.1 The Copenhagen Interpretation

In the years immediately following the development of quantum mechanics, Niels Bohr and Werner Heisenberg (among many others) grappled with the meaning of the new theory. Their thinking resulted in a body of writing that we now call the Copenhagen interpretation.[12] At various points in time, the proponents of this interpretation have given contradictory statements as to what the interpretation entails, and as a result there is no clear unambiguous definition of the Copenhagen position on the quantum state.

Nevertheless, the key features of the Copenhagen interpretation seem to be that the wave function, or the state of a quantum system, is not a real property of the system, but a theoretical construction that we use to calculate measurement outcomes. The quantum state can therefore be classified as ψ-epistemic. There is a fundamental separation between the quantum and the classical world, where our direct experience and our measurement devices are described by the classical theories of physics. The microscopic quantum world is described by the quantum state, and somewhere between the quantum system and the measurement device the collapse of the wave function occurs. Historically, this prompted the thought experiment about Schrödinger's cat.

---

[12] See *Copenhagen Interpretation of Quantum Mechanics*, in: Stanford Encyclopedia of Philosophy, Fall 2008 edition.

In addition to the separation between the classical and the quantum world, Bohr introduced the Complementarity Principle[13] that must be invoked in the Copenhagen interpretation to explain how quantum objects can sometimes behave as particles, and other times as waves. In Bohr's own words:

> Evidence obtained under different experimental conditions cannot be comprehended within a single picture, but must be regarded as complementary in the sense that only the totality of the phenomena exhausts the possible information about the objects.

This principle provides an explanation of both de Broglie's wave-particle duality and Heisenberg's Uncertainty Principle, and also hints at the discovery of contextuality forty years later.

The Correspondence Principle says that for large systems the quantum description approaches the classical description. It is somewhat doubtful that this principle holds generally. For example, the aim of building a quantum computer is to create large systems that behave entirely differently from classical systems, thus violating the Correspondence Principle.

The Copenhagen interpretation is difficult to make philosophically precise while at the same time keeping to the key aspects that Bohr and Heisenberg put forward. One such attempt is the consistent histories interpretation, which specifies sets of questions about a quantum system that have consistent answers that do not lead to logical contradictions. In this interpretation, decoherence in the measuring device is enough to account for wave function collapse.

11.4.2 Quantum Bayesianism

Another refinement of the Copenhagen interpretation is Quantum Bayesianism.[14] To understand this philosophical position, we must first get to grips with regular Bayesianism. In this book we have talked about probabilities as if we know exactly what they mean. Indeed, we can measure probabilities by repeating an experiment a large number of times, and record the relative frequencies of the measurement outcomes. For example, if we measure the spin of an electron in the z-direction for N identically prepared spin states $a|\uparrow\rangle + b|\downarrow\rangle$, we expect to obtain the relative frequencies $f_\uparrow$ and $f_\downarrow$ defined by

$$f_\uparrow = \frac{N_\uparrow}{N} \simeq |a|^2 \quad \text{and} \quad f_\downarrow = \frac{N_\downarrow}{N} \simeq |b|^2, \tag{11.36}$$

when N becomes very large. Here $N_\uparrow$ and $N_\downarrow$ are the number of times we measure $\uparrow$ and $\downarrow$, respectively. However, $f_\uparrow$ and $f_\downarrow$ merely approach the probabilities $|a|^2$ and

---

[13] N. Bohr, *The Quantum Postulate and the Recent Development of Atomic Theory*, Nature **121** 580 1928.

[14] N. D. Mermin, *Physics: QBism puts the scientist back into science*, Nature **507** 421 2014.

$|b|^2$ as $N \to \infty$, they are not the same. As a result, we cannot define the probabilities as the relative frequencies. The question then becomes: what is a probability?

Bayesianism answers this question in the following way: A probability is a number assigned to a particular event by an agent (e.g., you or me), which expresses the degree of belief of the agent that such an event will happen. For example, I can say that the probability of rain today is 60%. This is an expression of my belief about the weather, and weather forecasters make statements like this all the time. This notion of probability does not require that an experiment is repeatable a large number of times. If we require that the rules for assigning and updating the values of the probabilities must be internally consistent, we recover the standard rules of probability theory!

Quantum Bayesianism adopts this view of probabilities, and interprets quantum mechanics as the theory that describes the degree of belief in the measurement outcomes according to an agent, such as an experimentalist. It does not say anything about the ontology of quantum systems, and is therefore a ψ-epistemic theory of the first kind. As we saw before, such a theory does not suffer from the interpretational problems in quantum mechanics, but at the same time lacks the explanatory power that we traditionally require from our theories about nature.

11.4.3   Quantum Logic

The problematic aspects of quantum mechanics such as nonlocality and contextuality (as described in the previous two sections) rely on specific chains of inference in (11.11) and (11.18). But what if logic itself is not valid in its ordinary form in quantum mechanics?

This notion is perhaps not as ridiculous as it may seem at first. Remember that geometry started as a branch of mathematics when Euclid formulated its fundamental rules in the third century BC. It turned out much later that there are many different geometries that are all mathematically consistent. The question is then: what is the geometry of the space around us? Consequently, geometry ceases to be a purely mathematical exercise, and becomes branch of physics with experimentally testable predictions. Not long after this realisation in the 19th century, general relativity showed that the geometry of space can change due to physical processes involving mass and energy.

We can apply the same argumentation to logic. Perhaps quantum mechanics can be understood if we use different logical rules, i.e., different rules of inference. Our everyday logic is then a bit like Euclid's geometry, while quantum logic is analogous to non-Euclidean geometry. We call this new type of logic quantum logic. It must replace some rules of classical logic with quantum versions and still form a consistent mathematical structure. On top of that, it must solve the conceptual puzzles of quantum mechanics. In most versions of quantum logic, the mathematical rule that is abandoned is the distributive law[15]:

$$A \text{ AND } (B \text{ OR } C) = (A \text{ AND } B) \text{ OR } (A \text{ AND } C). \qquad (11.37)$$

The standard example goes as follows: we make three statements A, B, and C about a particle,

A : the momentum is in the interval $[0, \hbar/4\ell]$,

B : the position is in the interval $[-\ell, 0]$,

C : the position is in the interval $[0, \ell]$,

where ℓ is some length. From statement A we see that the uncertainty in momentum is $\Delta p = \hbar/4\ell$, and from the statement B OR C we deduce that the uncertainty in position is $\Delta x = 2\ell$. Their product is therefore

$$\Delta p \Delta x = \frac{\hbar}{2}, \qquad (11.38)$$

which obeys the uncertainty relation for position and momentum, and is therefore allowed by quantum mechanics. Consequently, A AND $(B$ OR $C)$ can indeed be a true statement. However, the right-hand side of (11.37) can never be true. To see this, note that statements B and C both imply $\Delta x = \ell$, and the clauses A AND B and A AND C both lead to

$$\Delta p \Delta x = \frac{\hbar}{4}. \qquad (11.39)$$

This is not allowed by the uncertainty relation. Therefore, the left-hand side of (11.37) can be true, while the right-hand side must be false. Hence, the distributive law fails in quantum logic,[16] and has to be abandoned.

What exactly we should put into place depends on the particular version of quantum logic you wish to consider. It is safe to say that quantum logic is not an intuitive interpretation of quantum mechanics, and it is currently studied mostly for its mathematical structure.

---

[15] If you are not very familiar with formal logic, you can think of this as the logical equivalent of the distributive property of numbers: $x(y + z) = xy + xz$, where multiplication plays the role of AND, and addition plays the role of OR.

[16] G. Birkhoff and J. von Neumann, *The Logic of Quantum Mechanics*, Annals of Mathematics **37** 823, 1936; see also *Quantum Logic and Probability Theory*, in: Stanford Encyclopedia of Philosophy, Fall 2008 edition.

11.4.4   Objective Collapse Theories

Objective collapse theories[17] embrace the projection postulate, which says that the quantum state collapses after a measurement. These theories are ψ-complete, in that the quantum state is considered real, and add to quantum mechanics a physical mechanism that describes how the collapse happens.

One of the most well-known objective collapse theories is the version by Ghirardi, Rimini, and Weber[18] (GRW). It says that in addition to the standard evolution in quantum mechanics, particles very occasionally collapse their quantum state spontaneously. This happens on average once every hundred million years, so for individual particles this behaviour is never observed in practice. However, when we consider a large object that consists of 10^{23} particles, the spontaneous collapse of a few particles in the object triggers a cascade that causes the quantum state of the entire object to behave classically. The interaction of a small quantum system with a naturally collapsing, large measurement device is then enough to explain Schrödinger's cat: The cat is a macroscopic system, and will naturally decohere due to the spontaneous collapse mechanism of the GRW theory. In fact, the decoherence will already take place in the Geiger counter, since that is also a macroscopic object. The uneasy conclusion of the Schrödinger's cat paradox about counterintuitive macroscopic quantum superpositions is then circumvented.

Another well-known collapse theory is due to Roger Penrose,[19] who also views the quantum state as a real physical quantity. He notes that any massive particle in a superposition state of two or more locations creates a superposition of space-time curvature, since Einstein taught us that mass causes curvature in space. Penrose argues that there is an energy cost associated with creating a superposition of space-time, and if this energy becomes too big the system will spontaneously collapse into a lower energy system. This is a natural mechanism for the collapse of the wave function to a localised state. It explains how very light particles can be in a superposition, following the predictions of quantum mechanics, and why much heavier objects behave in a classical way.

These interpretations make predictions that are different from standard quantum mechanics. They put a limit on the size of the systems that are put in quantum superpositions, and this opens the way to experimental tests of the interpretation. Such experiments have been proposed, and are currently being implemented.

---

[17] See *Collapse Theories*, in: Stanford Encyclopedia of Philosophy, Fall 2008 edition.

[18] G. C. Ghirardi, A. Rimini and T. Weber, *Unified dynamics for microscopic and macroscopic systems*, Phys. Rev. D **34** 470, 1986.

[19] R. Penrose, *On Gravity's Role in Quantum State Reduction*, General Relativity and Gravitation **28** 581, 1996.

11.4.5   The de Broglie-Bohm Interpretation

The de Broglie-Bohm interpretation[20] is also known as the pilot-wave theory, or
Bohmian mechanics (after David Bohm), and it is a true hidden-variable theory (the
only one in our list). As such, it is a ψ-supplemented interpretation of quantum
mechanics.

In the de Broglie-Bohm interpretation, the wave function $\psi(x, t)$ is considered a
"pilot wave" that follows Schrödinger's equation, and which guides the trajectory of
the particle. The hidden variable is the actual trajectory $x(t)$ of the particle, which
obeys a separate equation from the Schrödinger equation. By claiming that a particle
has an actual trajectory, the problem of the indeterminism of position and momentum
in quantum mechanics goes away. However, the price we pay for this is two-fold:
First, we must solve an extra, often complicated equation for $x(t)$, such as

$$m\frac{d}{dt}x(t) = \hbar\frac{d}{dx}\text{Im}\left[\ln\psi(x, t)\right].\tag{11.40}$$

This means that we need to do extra work in Bohmian mechanics, compared to
standard quantum mechanics. Second, the trajectory $x(t)$ at time t in (11.40) depends
on the pilot wave $\psi(x, t)$ at the same time, and there is therefore an instantaneous
influence of ψ from all positions x at time t at the actual trajectory. This makes the
theory highly nonlocal. The de Broglie-Bohm interpretation is historically important,
because it prompted Bell to investigate the relationship between locality and hidden
variables.

11.4.6   Modal Interpretations

We know from the previous section on contextuality that we cannot assign actual
properties to a quantum mechanical system without running into serious contradictions, known as the Kochen-Specker theorem. Modal interpretations[21] circumvent
this theorem by assigning only part of all possible properties to the system. In other
words, modal interpretations say that certain physical quantities have real values
(they are properties of the system), while other physical quantities do not. It is easy
to see that the chain of reasoning in (11.35), which leads to the contradiction with
quantum mechanics, breaks down if we deny that one or more statements $A_1, \ldots, A_5$
do not have a definite truth value. The question then becomes: how do we choose
which statements have definite truth values and which do not?

---

[20] D. Bohm, *A Suggested Interpretation of the Quantum Theory in Terms of 'Hidden Variables' I*,
Phys. Rev. **85** 166, 1952; see also *Bohmian Mechanics*, in: Stanford Encyclopedia of Philosophy,
Fall 2008 edition.

[21] D. Dieks and P. Vermaas (eds.), *The Modal Interpretation of Quantum Mechanics*, Dordrecht:
Kluwer Academic Publishers, 1998; see also *Modal Interpretations of Quantum Mechanics*, in:
Stanford Encyclopedia of Philosophy, Fall 2008 edition.

Different modal interpretations give different answers to this question. They have in common that they all accept the standard formalism of quantum mechanics as we have developed it here, with the exception of the projection postulate, and with the addition of a property ascription rule. Modal interpretations therefore assign a certain reality to the quantum state without introducing hidden variables, and they can be classified as ψ-complete interpretations.

An example of a property ascription rule (in the Kochen-Dieks interpretation) is to look for the joint quantum state of the system S and the measurement device M. We can write this state in a unique way as a superposition of system states and measurement device states:

$$|\Psi\rangle = \sum_j d_j |\psi_j\rangle_S |M_j\rangle_M , \qquad (11.41)$$

where the index j runs from 1 to the dimension of the system. This is called the Schmidt decomposition. A theorem in linear algebra ensures that we can always find such a state. Next, we choose the states $|\psi_j\rangle$ as the eigenstates associated with the statements A_j that have real property values, along with statements that follow logically from these statements. For example, if A_1 and A_2 have definite property values, then (A_1 AND A_2) and (A_1 OR A_2) also have definite values. The interaction between the measurement device and the system therefore picks out the physical quantities that are the actual properties of the system. This construction ensures that the Kochen-Specker theorem can no longer be derived, while every measurement reveals a real property of the system.

In modal interpretations, quantum mechanics applies to all physical systems, large and small, and all measurements are ordinary physical interactions. In particular, there is no need for a "collapse of the wave function".

11.4.7   The Many Worlds Interpretation

In 1957, Hugh Everett[22] proposed a new interpretation of quantum mechanics that takes the quantum state at face value and assigns reality to each term in a quantum superposition. This means that an electron with spin in the quantum state $a|\uparrow\rangle + b|\downarrow\rangle$ has two real branches, or "worlds", one where the spin is $\uparrow$, and one where the spin is $\downarrow$. There are therefore as many real worlds as there are terms in a superposition (these are sometimes called parallel universes). No extra rules are required in this interpretation of quantum mechanics, and it is therefore ψ-complete.

The many worlds interpretation (also known as the Everett interpretation) traditionally elicits strong reactions, so it is important to give a good argument why this interpretation is to be taken seriously. Imagine that you want to send a signal to both

---

[22] H. Everett, *Relative State Formulation of Quantum Mechanics*, Rev. Mod. Phys. **29** 454, 1957; see also *Everett's Relative-State Formulation of Quantum Mechanics* and *Many-Worlds Interpretation of Quantum Mechanics*, in: Stanford Encyclopedia of Philosophy, Fall 2008 edition.

Alice and Bob. You can achieve this by encoding the signal in the amplitude of a laser beam, and divide the beam with a 50:50 beam splitter. The two outgoing beams are directed to Alice and Bob. At this point it is uncontroversial to say that both beams are real physical objects. Next, we slowly attenuate the intensity of the beam, making the signal weaker. Alice and Bob keep receiving the weaker signals, and we continue to regard them as real. Finally, we attenuate the laser so much that to a very good approximation there will be at most one photon in each pulse. The state of the photon can be written as

$$|\psi\rangle = \frac{|1, 0\rangle + |0, 1\rangle}{\sqrt{2}}, \tag{11.42}$$

where $|1, 0\rangle$ denotes a photon moving towards Alice, and $|0, 1\rangle$ denotes a photon moving towards Bob. If in the classical limit of a bright laser beam both signals to Alice and Bob are real, then in the quantum limit the two terms $|1, 0\rangle$ and $|0, 1\rangle$ should also both be real, since attenuation is a gradual process. Otherwise, at what point does the beam stop being real? However, Alice and Bob only ever see a single photon in either Alice's or Bob's detector. This means that the there are two "worlds" that are in some sense equally real, but in which different things happen.

The many worlds interpretation embraces the reality of each term in the quantum state, but it immediately raises an important question: we can always decompose a quantum state into a different basis, so which decomposition leads to the real branches? To answer this, consider a similar (but unproblematic) situation in classical electrodynamics. We can construct a wave packet of light by superposing waves of different frequencies (shown in Fig. 11.4). Depending on the choice of measurement (frequency or time-of-arrival measurements) we can decompose the wave packet

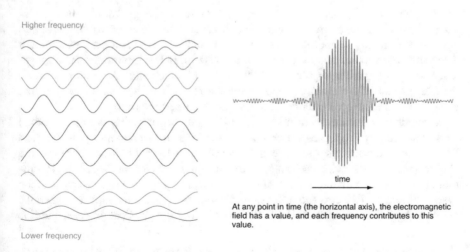

Higher frequency

Lower frequency

time

At any point in time (the horizontal axis), the electromagnetic field has a value, and each frequency contributes to this value.

Fig. 11.4 A localised wave packet as a superposition of plane waves (see supplementary material 4)

in different ways. The real underlying electromagnetic field does not exhibit any ambiguity. Moreover, every frequency component is part of the wave packet and is therefore "real" (you can make them visible using a prism), but so is the amplitude of the wave packet at every point in time. It is in this sense that the different branches of a superposition of frequencies are considered real.

The same happens with quantum mechanics in the many worlds interpretation: the quantum state $|\psi\rangle$ is unique and unambiguous and real just like the optical wave packet, but only when we wish to describe it in terms of measurements do we need to resolve the ambiguity of the measurement and the corresponding basis. This is resolved by the interaction Hamiltonian with the corresponding decoherence. Just like all the frequencies exist in the wave packet of the electromagnetic field, all terms in the superposition of a quantum state exist. Each branch is called a different "world", but you see that there is actually only one universe with a multiplicity of instantiations.

An often-heard objection to the many worlds interpretation is that introducing all those worlds violates energy conservation. The standard response is that energy conservation applies only within each world, and not across what may be an infinite number of worlds. Similarly, it is said that the many worlds interpretation is immune to Ockham's razor, since the number of assumptions in the interpretation is actually very small, even if the number of worlds aren't. For example, it does not require any hidden variables or collapse mechanisms that are not already present in standard quantum mechanics.

11.4.8 Relational Quantum Mechanics

In relational quantum mechanics,[23] the quantum state is an expression of the correlations between the quantum system and an observer. This observer may be alive and conscious, or it may be another physical system such as a macroscopic measurement device or even another microscopic quantum system. In this interpretation, all physical systems are quantum systems, and there is no fundamental distinction between the classical and the quantum world. A measurement is a physical interaction like any other, and there are no hidden variables in this interpretation.

In relational quantum mechanics, the basic elements of reality are the so-called "actualisations" of properties as they come about in the measurement by an observer. We can see that this is in some sense plausible when we look at Schrödinger's cat: the classical states of the cat, $|dead\rangle$ and $|alive\rangle$, are singled out by the interaction Hamiltonian in (9.74). An interaction Hamiltonian always requires at least two parties: the system of interest (the cat) and the system it interacts with (for example us, when we open the box). The interaction Hamiltonian describes the physical process, and it is

---

[23] C. Rovelli, *Relational quantum mechanics*, International Journal of Theoretical Physics **35** 1637, 1996; see also *Relational Quantum Mechanics*, in: Stanford Encyclopedia of Philosophy, Fall 2008 edition.

unique for any given interaction. Since the actualisations are defined in the interaction between the system and an observer, there are no absolute, observer-independent properties of quantum systems.

An immediate consequence of relational quantum mechanics is that two observers, Alice and Bob, can give different but equivalent descriptions. For example, Alice may measure the spin of an electron and find the measurement outcome $\uparrow$:

$$a \, |\uparrow\rangle + b \, |\downarrow\rangle \;\; \rightarrow \;\; |\uparrow\rangle \,, \tag{11.43}$$

while Bob, who describes the joint system of the electron spin and Alice, gives the following account:

$$(a \, |\uparrow\rangle + b \, |\downarrow\rangle)|A_0\rangle \;\; \rightarrow \;\; a \, |\uparrow\rangle \, |A_\uparrow\rangle + b \, |\downarrow\rangle \, |A_\downarrow\rangle \,. \tag{11.44}$$

Here, $|A_0\rangle$ is the initial state of Alice, and $|A_\uparrow\rangle$, $|A_\downarrow\rangle$ the state of Alice when she measured spin $\uparrow$ or $\downarrow$, respectively. These two descriptions are equally correct in relational quantum mechanics, and we just need to remember who the observer is when we choose which description we should use. Note that this is exactly what we did when we described the state of the teleported qubit according to Alice and Bob, earlier in Sect. 11.1.

In relational quantum mechanics it is meaningless to refer to the absolute, observer-independent state of any system. This echoes the situation in special relativity, where it is meaningless to call two events simultaneous without any reference to an observer. It also immediately solves the puzzle of Schrödinger's cat, since there is never any doubt how each "observer" (which includes us, the cat, the Geiger counter, etc.) describes the system they observe.

Alice cannot say anything about her own state, because that state is defined only relative to another observer, such as Bob or Charlie. They may interact differently with Alice, which leads to different quantum states of Alice. This impossibility of describing ones own state leads to the perceived non-unitary evolution in a measurement, since the measurement process involves both the states of the system and the observer. The quantum state of the universe is therefore a meaningless concept in this interpretation (unless you can find an observer that does not belong to this universe). Relational quantum mechanics is often compared with Everett's relative state description (see section 10.5.7 on the many worlds interpretation), but differs in a crucial way: where Everett assigns an element of reality to each term $|\uparrow\rangle$ and $|\downarrow\rangle$ in (11.43), thus positing parallel worlds in which an electron has incompatible spin values $\uparrow$ and $\downarrow$, relational quantum mechanics assigns either $\uparrow$ or $\downarrow$ as the real property of the electron *as seen be the observer who measures the spin observable* S_z.

Relational quantum mechanics is a ψ-epistemic theory of the second kind, and you may expect that it is subject to the no-go theorem in Sect. 11.1. However, the relational interpretation of quantum mechanics does not assume observer-independent states that led to the derivation of the theorem (quite the contrary). This means that we can have an epistemic interpretation of the second kind that is not ruled out by quantum mechanics itself.

11.4.9 Other Interpretations

The above-mentioned interpretations tend to have strong defenders and detractors. In addition to this short list, there are numerous other interpretations of quantum mechanics, such as cosmological interpretations, statistical interpretations, time-symmetric theories, transactional interpretations, and so on. Often, these interpretations share key aspects of the ones described here, and differ in subtle but important parts.

All good interpretations of quantum mechanics must stand up to detailed philosophical scrutiny regarding plausibility and internal consistency. It is currently not known if these considerations single out a unique interpretation, or leave open a family of possible interpretations. One thing to be wary of is quantum mysticism. There is no factual reason to believe that quantum mechanics has anything to do with religion, Eastern or otherwise. Likewise, interpretations that rely on some form of consciousness are inherently problematic, since we do not have a good understanding of what consciousness really is in the first place. Invoking one unknown to explain another is bad philosophy and bad physics.

If all these interpretations have left your head spinning, you are not alone. The philosophy of quantum mechanics is notorious for its many subtleties. The good news is that you do not need an interpretation of quantum mechanics in order to use it successfully. If you lack the patience to keep track of the various ways in which interpretations of quantum mechanics make statements about reality, you can always heed the advice of David Mermin,[24] and *just shut up and calculate!*

Exercises

1. Going back to your answer to the first exercise of Chap. 1 (What is the purpose of physical theories?), does quantum mechanics fulfil the requirements implied by your answer there?
2. The Pauli matrices σ_x, σ_y and σ_z can be considered observables, since they have real eigenvalues $+1$ and -1. Consider two qubits in some quantum state, and the following array of possible observables:

$$
\begin{array}{ccc}
\sigma_x^{(1)}\sigma_x^{(2)} & \sigma_x^{(1)}\mathbb{I}^{(2)} & \mathbb{I}^{(1)}\sigma_x^{(2)} \\
\sigma_z^{(1)}\sigma_z^{(2)} & \mathbb{I}^{(1)}\sigma_z^{(2)} & \sigma_z^{(1)}\mathbb{I}^{(2)} \\
\sigma_y^{(1)}\sigma_y^{(2)} & \sigma_x^{(1)}\sigma_z^{(2)} & \sigma_z^{(1)}\sigma_x^{(2)}
\end{array}
$$

---

[24] N. D. Mermin, *What's Wrong with this Pillow?*, Physics Today, April 1989.

where the superscript indicates the qubit the matrix operates on.

(a) Show that the three operators in every column and every row commute, and can therefore be measured simultaneously.
(b) Show that each operator has eigenvalues ± 1.
(c) Using the matrix form of the Pauli matrices, show that the operators in each row and column multiply to $+\mathbb{I}$, except the first column, which multiplies to $-\mathbb{I}$.
(d) Assuming that each entry in the array has a pre-determined value of ± 1, try to populate the array such that the conditions in part (c) are satisfied. What do you conclude?

3. Alice and Bob, positioned far away from each other, each receive an electron that is part of the entangled spin state

$$|\Psi^-\rangle = \frac{|\uparrow\downarrow\rangle - |\downarrow\uparrow\rangle}{\sqrt{2}},$$

where $\uparrow$ and $\downarrow$ define the positive and negative z-axis. Alice can freely choose to measure her electron spin in one of two directions: in the z-direction (denoted by $\mathbf{a}$) and in the direction an angle θ_A away from the z-axis towards the x-axis (denoted by $\mathbf{a}'$). Similarly Bob can freely choose to measure his electron spin in the z-direction (denoted by $\mathbf{b}$) and in the direction an angle θ_B away from the z-axis towards the x-axis (denoted by $\mathbf{b}'$).

(a) Calculate the probabilities $\Pr(a = \pm, b = \pm)$ of finding the $\pm\hbar/2$ eigenvalues in all possible measurement directions $\mathbf{a}$, $\mathbf{a}'$, $\mathbf{b}$ and $\mathbf{b}'$.
(b) Define $P(\mathbf{a}, \mathbf{b})$ as the expectation value over the ± 1 values of $\mathbf{a} \cdot \mathbf{b}$:

$$P(\mathbf{a}, \mathbf{b}) = \Pr(a = +, b = +) - \Pr(a = +, b = -)$$
$$- \Pr(a = -, b = +) + \Pr(a = -, b = -).$$

What are the minimum and maximum values that $P(\mathbf{a}, \mathbf{b})$ can take?
(c) Without superluminal signalling between Alice and Bob, the expectation values obey the so-called CHSH inequality

$$|P(\mathbf{a}, \mathbf{b})| + |P(\mathbf{a}, \mathbf{b}')| + |P(\mathbf{a}', \mathbf{b})| - |P(\mathbf{a}', \mathbf{b}')| \leq 2,$$

which is a special case of a *Bell inequality*. Show that the probabilities you calculated in part (a) violate this expression when $\theta_{\mathbf{a}} = 0$, $\theta_{\mathbf{a}'} = \pi/2$, $\theta_{\mathbf{b}} = \pi/4$, and $\theta_{\mathbf{b}'} = 3\pi/4$. What do you conclude?

Epilogue

In this book we have developed the basic theory of quantum mechanics, explored some of its consequences for physics, chemistry and technology, and studied what the theory says about the nature of reality. I hope that I have managed to convey some of the beauty and elegance of the theory and its applications, and at the same time made you feel a little uncomfortable about the things that quantum mechanics has to say about nature. Fundamentally, quantum mechanics tells us how we should reason about physical systems without falling into the trap of assuming that things behave the way we expect even when we are not looking at them. In this sense, quantum theory is a refinement of our understanding of probabilities. As such, the theory has a much broader application than it is traditionally given in atomic, solid state, and particle physics.

What I presented here is just the tip of the iceberg, and if you continue your studies in quantum mechanics, what can you expect to learn further? First of all, you will learn how to solve the Schrödinger equation for a range or problems in one, two, and three dimensions. We discussed here the particle in a one-dimensional box, a one-dimensional pendulum, and we sketched roughly the solutions for the hydrogen atom in three dimensions, but there is a whole family of interesting potentials $V(x, y, z)$ that we can substitute into the Hamiltonian. One important example is when we have a periodic potential, describing for example the (attractive) nuclei in a crystal. An electron with a certain energy in such a crystal can be bound to a particular nucleus, or it may be able to move freely through the crystal. The available energy levels for the electrons then determine the physical properties of the crystal, from insulators to metals and semiconductors, and they determine material properties like heat capacity and conductivity.

Solving partial differential equations is generally hard, and given a particular potential $V(x, y, z)$ it is unlikely that you can find solutions in closed form. However, the potential may be somewhat close to a different, simpler potential for which we do have solutions. In that case the solutions to the complicated problem will often be close to the solutions of a simple problem, and by studying how the potential changes

P. Kok, *A First Introduction to Quantum Physics*, Undergraduate Lecture Notes in Physics,
https://doi.org/10.1007/978-3-031-16165-0

between the two we can find approximations to the quantum state and the eigenvalues of various observables of interest, usually energy. This is an example of perturbation theory. For example, our system may have a potential in the Hamiltonian that has a quadratic term $\frac{1}{2}m\omega^2 x^2$ like the pendulum, but it may also contain a small term proportional to x^3 or x^4, etc. The motion of the system will be almost harmonic, but not quite. Perturbation theory can tell us how the motion will be different from the ideal pendulum.

There are many different perturbation theory techniques: first and second order, time-dependent, the variational method (which describes the above example), and the Wentzel-Kramers-Brillouin method that is suitable when the system behaves almost classically. In particle physics, perturbation theory takes the form of Feynman diagrams: all possible ways for the particles to interact are organised in an infinite series of increasingly complex—and unlikely—processes, represented graphically by a Feynman diagram. The more of these processes we calculate, the more precise the answer of our calculation will be. Problems in chemistry and solid state physics also almost always require us to use one or more methods from perturbation theory. While this is typically a very technical topic it is nevertheless essential, since it allows us to get meaningful answers from quantum mechanics in almost any application.

Sometimes perturbations are caused by external classical fields, such as electric and magnetic fields. Since electrons are both electrically charged and have a magnetic moment (their spin), we would expect that such fields will have an influence on the energy levels of the electron. Indeed, this is the case. You will learn more about so-called Stark shifts of energy levels that are created by electric fields, and Zeeman splittings of energy levels that are due to magnetic fields. Typically, these effects cannot be calculated exactly, and you will again have to use perturbation theory.

The next topic you may encounter is the complete description of angular momentum. We encountered angular momentum mostly in the form of the electron spin in this book, but there is also the orbital angular momentum associated with particles in a rotational motion. Angular momentum is important any time there is a central potential in a physics problem, such as the Coulomb potential in Sect. 8.7. Classically, the angular momentum is defined as the cross product $\mathbf{L} = \mathbf{r} \times \mathbf{p}$, where $\mathbf{r}$ is the position vector of the particle, and $\mathbf{p}$ is its momentum vector. Since $\mathbf{r}$ and $\mathbf{p}$ do not commute in quantum mechanics, the three components of $\mathbf{L}$ do not commute and cannot be measured simultaneously (just like we saw in chapter three that S_x, S_y and S_z cannot be measured simultaneously). The commutation relations between L_x, L_y, and L_z are ultimately responsible for the quantisation of angular momentum. We already encountered this in Sect. 8.7, (8.91) where we introduced two new integer quantum numbers, l and m, connected to angular momentum. In a more advanced course in quantum mechanics you will learn how to combine spin and orbital angular momentum into a combined total angular momentum, and in quantum optics you will encounter a very deep connection between angular momentum and the polarisation of light, but also, more surprisingly, with optical networks of beam splitters and phase shifters.

Another key topic in the study of quantum mechanics, particularly in the context of particle physics, is scattering. In a particle accelerator we prepare particles in states that are close to momentum eigenstates, typically with high energy. When the particles collide, this means that they will be deflected into a superposition of different momentum eigenstates. For each input momentum p there will be an amplitude for the scattering probability into an output momentum q. If we call this amplitude $S(p, q)$, then S is the so-called scattering matrix. The form of the scattering matrix tells us about the structure of the particles, and this is how, for example, we discovered that protons and neutrons are made up of three quarks. A simple one-dimensional example of scattering is the tunnelling example in section 8.6. A particle in a momentum eigenstate gets scattered back from the barrier or gets transmitted (note that we assumed the particle was in a momentum eigenstate), and by measuring the reflected and transmitted parts for different initial momentum states we can infer the height and width of the potential barrier. You can also think of the beam splitter matrix in chapter two as a scattering matrix, because it sends a photon with a specific momentum either into the reflected beam or into the transmitted beam. The scattering matrix is therefore basically the unitary transformation that changes the input state of a particle to the output state, written in the momentum basis. Another interesting example is Compton scattering, in which a photon scatters off an electron and changes frequency due to energy transfer between the electron and the photon.

We started this book by considering the definition of a photon. We could get quite some way towards our description of quantum theory, but a full description of the photon must take into account that two photons can be fundamentally indistinguishable from each other: Nothing can tell these photons apart, not their position, nor their polarisation, frequency, or shape of the wave packet. We cannot individually label the photons, and yet there are two of them. Similarly, two electrons can also be indistinguishable, but whereas photons can all happily sit together in the same quantum state, electrons do not want to share their quantum state with any other electron. This is the difference between bosons (like photons) and fermions (like electrons). The concept of identical particles plays a central role when we wish to extend quantum mechanics to the special relativistic case (excluding gravity; we still do not know how to reconcile that with quantum theory). In relativistic quantum mechanics you will learn how the spin of a particle determines whether it is a boson or a fermion. This is the so-called spin-statistics theorem: particles with integer spin $(0, 1, 2, \ldots)$ are bosons, while particles with half-integer spin $(\frac{1}{2}, \frac{3}{2}, \frac{5}{2}, \ldots)$ are fermions. This will lead to the relativistic Klein-Gordon equation for massive bosons and the Dirac equation for fermions, which are the relativistic analogs of the Schrödinger equation. These equations must allow negative energy solutions, which was a big puzzle until they were identified with anti-particles.

A comprehensive description of relativistic quantum particles requires quantum field theory, which is the basis for the Standard Model of particle physics. However, quantum field theory also has important applications in optical physics and solid state physics. A fully quantum mechanical treatment of an atom interacting with quantum light needs a quantum theory for electric and magnetic fields. The standard tool for this type of problem is the Jaynes-Cummings model, which describes the interaction of an atom in a cavity with the quantised electromagnetic field in the cavity. In condensed matter, we can treat crystal lattice vibrations as quantum fields at finite temperatures. This requires the concept of the density matrix that we encountered in chapter seven, but now for a very large number of identical particles. Generally, you will need non-relativistic quantum field theory to study the discipline of quantum statistical physics, and it will lead to such beautiful behaviour as Bose-Einstein condensation, superconductivity, and superfluidity, which have all been observed experimentally (in the case of superconductivity even before quantum theory was developed).

These are the more traditional topics in advanced textbooks on quantum mechanics. However, in recent years great progress has been made in controlling individual quantum systems. We can now control the interaction between a single atom and a single photon quite precisely (Serge Haroche and David Wineland won the Nobel prize for this in 2012), we can control single electrons in very small electrical circuits, and we are getting to the point where we can fully control electrons locked in small semiconductor boxes, called quantum dots. To make experimental progress on this front we need a theory that can cope with the imperfections that inevitably arise in the setup. Detectors will not always give the "right" answer, and we must develop a method that includes such errors. We can take a similar approach to the way we derived the density matrix: we construct a probability distribution that governs how likely it is that quantum state $|a_j\rangle$ triggers a measurement outcome m_k instead of m_j. This is enough to construct the most general measurement operators, which are no longer projectors but probability distributions over projectors called POVMs. This leads to a new definition of "generalised observables", where instead of a spectral decomposition in terms of projectors, the observable is given in terms of the POVM.

Similarly, it will be very difficult in most experiments to ensure that our system is completely isolated from the environment. As a result, the evolution of the system over time is no longer perfectly reversible since some information about the state of the system may leak into the environment. That means the evolution is no longer unitary. Generally, this is how a pure state turns into a mixed state, and the general method was described in chapter 7 via the tracing out of the environment. However, in practice, this will be a very difficult calculation for all but the simplest systems, and we need additional techniques to describe non-unitary evolution of quantum systems. We can again consider probability distributions over various possible unitary transformations, but this also has limited use. Instead we will have to introduce a replacement for the Schrödinger equation, called the Master equation. It is an equation for density matrices instead of vectors, and in the case of unitary evolution it reduces to the Schrödinger equation, as required. The Master equation is a very powerful tool, and requires sophisticated mathematical techniques to set up and solve.

Finally, you may study quantum mechanics in a completely different context, namely that of quantum information theory. This includes quantum computing and quantum communication, and a large effort is currently underway to come up with ways to implement error correction that allows us to actually build quantum computers and communication systems even in the presence of practical imperfections. The theory of quantum error correction is less a physical theory and more a quantum computer science research area, even though the errors will be physical in origin. Similarly, quantum mechanics can be used to study the capacity of communication channels for transmitting information. It asks questions like "how much information can Alice transmit to Bob if they have access to shared entanglement?" Quantum information theory is very much built upon the mathematics of density matrices and entropy that we encountered in chapter seven.

In all these topics, the basic principles described in this book still apply, even though the mathematics will get much more complicated. Hopefully, this book, along with its interactive figures will have given you a solid conceptual foundation of quantum mechanics so that you can focus on the technical aspects of the more advanced subjects.

Further Reading

The following is a list of recommended reading for anyone who wants to learn quantum mechanics in more detail. It is by no means an exhaustive list of all the excellent books on the topic, but these are the ones that are close to my heart.

Matrix Operations for Engineers and Scientists *Alan Jeffrey, Springer* (2010). Any serious study of quantum mechanics requires a good grounding in linear algebra, and especially the theory of matrices. Jeffrey takes a pedagogical approach similar to what I adopted here, and develops the material from example problems. I highly recommend *Matrix Operations* alongside this book.

Matrix Analysis *R. A. Horn and C. R. Johnson, Cambridge University Press* (2013). This book is a comprehensive textbook that covers everything you need to know and more. It is written in a mature mathematical style, and is a good follow-up from Jeffrey's book. It has a wealth of practice problems.

Quantum Mechanics *A. Rae, Fifth Edition, Taylor and Francis* (2008). This is an introductory textbook on quantum mechanics based on a more traditional approach. The book starts with the photoelectric effect, the Compton effect and wave-particle duality, and proceeds via the Schrödinger equation in one and three dimensions to angular momentum, perturbation theory, scattering theory and many-particle systems. It assumes a good working knowledge of calculus and complex numbers.

Quantum Systems, Processes and Information *B. Schumacher and M. Westmoreland, Cambridge University Press* (2010). This textbook is the natural next step from the present book, in that it approaches quantum mechanics as a theory of information and processes. After a brief introduction to the smallest physical systems—qubits—the book develops the mathematical formalism of quantum theory in terms of Hilbert spaces. It covers the topics in this book with more generality, and includes more dedicated chapters on quantum information processing, entropy, and error correction.

© The Editor(s) (if applicable) and The Author(s), under exclusive license to Springer Nature Switzerland AG 2023
P. Kok, *A First Introduction to Quantum Physics*, Undergraduate Lecture Notes in Physics, https://doi.org/10.1007/978-3-031-16165-0

Quantum Computer Science   *D. Mermin, Cambridge University Press* (2007). This is a textbook aimed at readers from different backgrounds, including mathematics and computer science. Written in his inimitable style, Mermin explores quantum mechanics with a focus on computation, rather than physics. It requires some basic knowledge of matrices and complex numbers, but is generally very accessible and heartily recommended to readers who enjoyed Chap. 6 and want to know more about the information theoretic applications of quantum mechanics.

Q is for Quantum   *Terry Rudolph, Amazon* (2017). In this highly accessible book Rudolph explores the fundamental puzzles that quantum mechanics poses to our classical world view. The style of the book is much like Chap. 1, but it penetrates deeply into the topics we covered in Chap. 10. Recommended reading for anyone who wants to explore the foundations of quantum mechanics further.

The Principles of Quantum Mechanics   *P. A. M. Dirac, Oxford University Press* (1988). Not often in science would one recommend reading the earlier textbooks on a topic, but this book by Paul Dirac is a notable exception. It is still one of the clearest expositions of quantum mechanics (including his famous treatment of the relativistic electron), and it is the original source for the bracket notation that almost everybody uses today.

Useful Formulas

Complex numbers:

$$(x + iy)^* = x - iy$$
$$(x + iy)(u + iv) = (xu - yv) + i(xv + yu)$$
$$(r_1 e^{i\phi_1})(r_2 e^{i\phi_2}) = r_1 r_2 e^{i(\phi_1 + \phi_2)}$$

Series expansion and trigonometry:

$$e^x = 1 + x + \frac{x^2}{2!} + \frac{x^3}{3!} + \cdots$$
$$e^{i\phi} = \cos \phi + i \sin \phi$$
$$\cos \phi = \frac{e^{i\phi} + e^{-i\phi}}{2}$$
$$\sin \phi = \frac{e^{i\phi} - e^{-i\phi}}{2i}$$

Dirac delta function:

$$\delta(x - x_0) = \frac{1}{2\pi} \int_{-\infty}^{+\infty} e^{i(x - x_0)y} \, dy$$

When $a \leq x_0 \leq b$ we have

$$\int_a^b f(x)\delta(x - x_0) \, dx = f(x_0)$$

Fourier transform of wave functions:

$$\Psi(p) = \frac{1}{\sqrt{2\pi\hbar}} \int_{-\infty}^{+\infty} dx \; \psi(x)e^{-ipx/\hbar}$$

Gaussian function with peak at x_0 and width σ:

$$f(x) = \frac{1}{\sqrt[4]{\pi\sigma^2}} \exp\left[-\frac{(x - x_0)^2}{2\sigma^2}\right].$$

Operator definitions:

Hermitian:	$A^\dagger = A$
Unitary:	$U^\dagger = U^{-1}$
Projector:	$P^2 = P$ and $P^\dagger = P$
Commutator:	$[A, B] = AB - BA$

Determinant and trace of 2×2 matrices:

$$\det \begin{pmatrix} a & b \\ c & d \end{pmatrix} = ad - bc$$

$$\mathrm{Tr}\left[\begin{pmatrix} a & b \\ c & d \end{pmatrix}\right] = a + d$$

© The Editor(s) (if applicable) and The Author(s), under exclusive license to Springer   285
Nature Switzerland AG 2023
P. Kok, *A First Introduction to Quantum Physics*, Undergraduate Lecture Notes in Physics,
https://doi.org/10.1007/978-3-031-16165-0

Cyclic property of the trace:

$$\mathrm{Tr}[ABC] = \mathrm{Tr}[CAB]$$

Eigenvalue equation:

$$\det(A - \lambda \mathbb{I}) = 0$$

The Pauli matrices:

$$\sigma_x = \begin{pmatrix} 0 & 1 \\ 1 & 0 \end{pmatrix}$$

$$\sigma_y = \begin{pmatrix} 0 & -i \\ i & 0 \end{pmatrix}$$

$$\sigma_z = \begin{pmatrix} 1 & 0 \\ 0 & -1 \end{pmatrix}$$

Anti-commutation relations of the Pauli matrices:

$$\sigma_x \sigma_y + \sigma_y \sigma_x = 0$$
$$\sigma_x \sigma_z + \sigma_z \sigma_x = 0$$
$$\sigma_y \sigma_z + \sigma_z \sigma_y = 0$$

Born rule:

$$\mathrm{Pr(outcome)} = |\langle \mathrm{outcome} | \psi \rangle|^2$$
$$\mathrm{Pr}([a, b]) = \int_a^b dx \ |\psi(x)|^2$$

Expectation value and variance:

$$\langle A \rangle = \langle \psi | A | \psi \rangle$$
$$(\Delta A)^2 = \langle \psi | A^2 | \psi \rangle - \langle \psi | A | \psi \rangle^2$$

Uncertainty relation for operators:

$$\Delta A \, \Delta B \geq \frac{1}{2} |\langle [A, B] \rangle|$$

Error propagation formula:

$$\delta \theta = \left| \frac{d \langle A \rangle}{d\theta} \right|^{-1} \Delta A$$

The Schrödinger equation:

$$i\hbar |\psi(t)\rangle = H |\psi t\rangle$$

Time evolution:

$$|\psi(t)\rangle = U(t) |\psi(0)\rangle$$
$$= \exp\left(-\frac{i}{\hbar} H t\right) |\psi(0)\rangle$$

Ladder operators:

$$\hat{a} = \sqrt{\frac{m\omega}{2\hbar}} \, \hat{x} + i \frac{1}{\sqrt{2m\hbar\omega}} \, \hat{p}$$

$$\hat{a}^\dagger = \sqrt{\frac{m\omega}{2\hbar}} \, \hat{x} - i \frac{1}{\sqrt{2m\hbar\omega}} \, \hat{p}$$

Bell states for qubits:

$$|\Phi^\pm\rangle = \frac{|00\rangle \pm |11\rangle}{\sqrt{2}}$$

$$|\Psi^\pm\rangle = \frac{|01\rangle \pm |10\rangle}{\sqrt{2}}$$

Von Neuman entropy:

$$S(\rho) = -\mathrm{Tr}[\rho \log_2 \rho]$$
$$= -\sum_j p_j \log_2 p_j$$

where the p_j are eigenvalues of ρ.

Answers to Selected Problems

1.2 $\Pr(D_1) = 0.3$ and $\Pr(D_2) = 0.7$.

1.5 A partial memory of the photon path will partially destroy the interference. So instead of detector D_1 being always silent, it now occasionally fires (but not as often as D_2).

2.1

$$AB = \begin{pmatrix} 27 & 60 \\ 3 & 30 \end{pmatrix},$$

$$BA = \begin{pmatrix} 54 & 39 \\ -12 & 3 \end{pmatrix}$$

2.2

$$z_1 = 5\,e^{i\arctan(4/3)}$$

$$z_2 = 4\sqrt{13}\,e^{-i\arctan(2/3)}$$

$$z_3 = 3\sqrt{41}\,e^{-i\arctan(4/5)}.$$

2.6 $\langle\psi|\phi\rangle = -(21 + 15i)$ and $\langle\phi|\psi\rangle = -(21 - 15i)$.

2.10 $\Pr(\text{right}) = 9/25$.

2.12 $\phi = \pi/3$.

2.17 For any state $e^{i\phi}\,|\psi\rangle$ and event a, the probability of a is given by $\Pr(a) = \left|e^{i\phi}\,\langle a|\psi\rangle\right|^2 = e^{i\phi}e^{-i\phi}\,|\langle a|\psi\rangle|^2$. But $e^{i\phi}e^{-i\phi} = 1$, so the probability does not depend on ϕ.

3.1 $\Pr(\downarrow) = 2/3,\qquad \Pr(+) = 1/2 + \sqrt{2}/3$.

3.3 The normalisation constant is $\sqrt{13}$, and $\Pr(\uparrow) = 2/13,\quad \Pr(+) = 1/2$. The expectation value in the z-direction is $\langle S_z\rangle = -5\hbar/26$.

3.11 $|\psi\rangle = 0.500\,|\!\uparrow\rangle + (0.433 + 0.750i)\,|\!\downarrow\rangle$.

3.13 $\langle S_x\rangle = 12\hbar/25$, and $\Delta S_x = 7\hbar/50$.

4.2 $|\psi(t)\rangle = (|g\rangle + \sqrt{2}e^{i\omega t}\,|e\rangle)/\sqrt{3}$, and $\Pr(+) = 1/2 + \frac{1}{3}\sqrt{2}\cos\omega t$.

4.4 a.

$$H = -\frac{e\hbar B}{2\sqrt{2}mc}\begin{pmatrix} 1 & 1 \\ 1 & -1 \end{pmatrix}$$

b. Use that $1/\sqrt{2} = \cos(\pi/4) = \sin(\pi/4)$ and use the addition formulas from trigonometry.

(d) No.

5.1 (a) -3, (b) $-5 + 14i$, (c) -69.

5.4 Both, neither, Hermitian.

5.6 a. $\cos\left(\frac{eBT}{2m}\right)|\!\uparrow\rangle + \sin\left(\frac{eBT}{2m}\right)|\!\downarrow\rangle$.

b. $\Pr(+) = \frac{1}{2} + \frac{1}{2}\sin\left(\frac{eBT}{m}\right)$.

© The Editor(s) (if applicable) and The Author(s), under exclusive license to Springer Nature Switzerland AG 2023
P. Kok, *A First Introduction to Quantum Physics*, Undergraduate Lecture Notes in Physics, https://doi.org/10.1007/978-3-031-16165-0

5.11 a. $P_a = |a\rangle \langle a|$, $P_b = |b\rangle \langle b|$,
and $P_c = |c\rangle \langle c|$. $\text{Pr}(a) = \langle \psi | P_a | \psi \rangle$, and so on.

b. $\text{Pr}(\text{not } a) = \langle \psi | P_{\text{not } a} | \psi \rangle$, with $P_{\text{not } a} = \mathbb{I} - P_a$.

c. $P_{a \text{ or } b} = P_a + P_b - P_a P_b$.

6.1 (a) and (b) anre entangled, (c) is separable.

6.2 $X = \sigma_x$, $\lambda_1 = +1$ with $(1, 1)/\sqrt{2}$, and $\lambda_2 = -1$ with $(1, -1)/\sqrt{2}$. $\langle X \rangle = 0$.

6.4 For all possible pairs of states calculate that their scalar product is 0.

$$\langle \phi_1 | \phi_2 \rangle = \frac{1}{2} (\langle 0, 1 | 0, - \rangle + \langle 1, 0 | 1, + \rangle)$$

$$= \frac{1}{2} \left(-\frac{1}{\sqrt{2}} + \frac{1}{\sqrt{2}} \right) = 0 .$$

6.6 $U |0\rangle = |0\rangle$ and $U |1\rangle = e^{i\omega t} |1\rangle$. Apply U to both qubits: $U_1 U_2 |00\rangle = |00\rangle$ and $U_1 U_2 |11\rangle = e^{2i\omega t} |11\rangle$. The phase $e^{2i\omega t}$ returns to 1 twice as fast as $e^{i\omega t}$.

7.1 Purity is $\frac{7}{9} < 1$, so a mixed state.

7.2 a. We can consider the Hamiltonian as a conditional operation that acts on qubit two only if the state of qubit 1 is $|1\rangle$. We then have $|\Psi(t)\rangle = \frac{1}{\sqrt{2}}(|00\rangle + \cos \omega T |10\rangle - i \sin \omega T |11\rangle)$.

b. Take the partial trace of the state in (a):

$$\rho_1 = \frac{1}{2} \begin{pmatrix} 1 & \cos \omega T \\ \cos \omega T & 1 \end{pmatrix},$$

$$\rho_2 = \frac{1}{1} \begin{pmatrix} 1 + \cos^2 \omega T & -i \sin 2\omega T \\ i \sin 2\omega T & \sin^2 \omega T \end{pmatrix}.$$

c. The entropy of ρ_1 and ρ_2 are the same. Eigenvalues of ρ_1 are $(1 \pm \cos \omega T)/2$, and the entropy becomes

$$S = \cos \omega T \log \sqrt{\frac{1 - \cos \omega T}{1 + \cos \omega T}} + \log \frac{2}{|\sin \omega T|} .$$

d. $T = \pi/2\omega$. This is when the entropy of the individual qubits is maximal.

7.8 Top view of equatorial plane of the Bloch sphere:

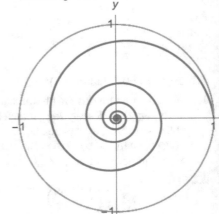

ω is the angular (rotation) frequency, while γ is the rate that determined how quickly the state moves to the origin.

7.10

$$S = -\sum_j p_j \log p_j$$

$$= -\sum_j p_j \log \frac{e^{-\beta E_j}}{Z}$$

$$= -\sum_j p_j \left(-\beta E_j - \log Z \right)$$

$$= \sum_j p_j \log Z + \beta \sum_j p_j E_j$$

$$= \log Z + \beta \langle E \rangle . \tag{.1}$$

7.14 $(\Delta A)^2 = \text{Tr}[A^2 \rho] - \text{Tr}[A\rho]^2$.

8.1 a. Differentiating ψ_n twice gives

$$\psi_n'' = -\frac{n^2\pi^2}{L^2}\psi_n \,.$$

In the region where $V = 0$, the energy eigenvalue equation becomes

$$\frac{n^2\pi^2\hbar^2}{2mL^2}\psi_n = E_n\psi_n \,.$$

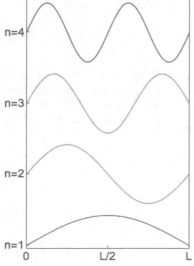

b. From (a) we see the result for E_n immediately.

c. Since the derivative of ψ_n is not proportional to ψ_n, it is not an eigenfunction of the momentum operator.

d. Differentiate and find $p = \pm\hbar k$.

e. $\psi_n(x) \propto \phi_k(x) + \phi_{-k}(x)$, which means two counter-propagating momentum functions. The position wave function ψ_n is a standing wave, and $k = n\pi/L$.

8.3 Both Δx for the state $|p\rangle$ and Δp for the state $|x\rangle$ diverge. When we measure the position of a particle in a momentum eigenstate, the wave function collapses to the position eigenstate corresponding to the measured position.

8.6 a. The probability is

$$\frac{\exp[-(x - x_0)^2/\sigma^2]}{\sqrt{\pi\sigma^2}}dx \,.$$

The probability density function looks like a bell curve with width σ.

b.

$$\Psi(p) = \frac{\exp\left[-\frac{p^2}{2\hbar^2/\sigma^2} - \frac{ipx_0}{\hbar}\right]}{\sqrt[4]{\pi\hbar^2/\sigma^2}} \,.$$

The probability density function looks like a bell curve with width $\hbar/\sigma$.

c. The widths of the bell curves are inverses of each other (up to a factor $\hbar$).

8.8

$$|\psi\rangle = \int \delta(p - p')|p'\rangle = |p\rangle \,.$$

9.1

$$\langle A \rangle = \sum_j a_j^2 |\langle a_j|\psi\rangle|^2$$

$$- \left(\sum_j a_j |\langle a_j|\psi\rangle|^2\right)^2$$

when $|\psi\rangle = |a_k\rangle$, $\langle A \rangle$ becomes

$$\langle A \rangle = \sum_j a_j^2 \delta_{jk} - \left(\sum_j a_j \delta_{jk}\right)^2$$

$$= a_k^2 - (a_k)^2 = 0$$

9.2 The precision of a grandfather clock is measured in milliseconds per cycle, whereas quantum mechanics puts the ultimate limit at

$$\delta t \geq \sqrt{\frac{\hbar}{2m\omega^3 A^2}} \approx 10^{-17} \text{ s.}$$

9.7

$$\langle x \rangle = A \cos(\omega t + \phi)$$
$$\langle x \rangle = -m\omega A \sin(\omega t + \phi)$$

where ϕ is the phase of the pendulum at $t = 0$.

10.3 a. For general angles θ for Alice and θ' for Bob:

$$Pr(++) = Pr(--)$$
$$= \frac{\sin^2[(\theta' - \theta)/2]}{2}$$

$$Pr(+-) = Pr(-+)$$
$$= \frac{\cos^2[(\theta' - \theta)/2]}{2}$$

b. Since $\cos 2x = \cos^2 x - \sin^2$, the range of values for P is between $+1$ and -1.

c. The CHSH inequality becomes

$$\left|\cos \frac{\pi}{4}\right| + \left|\cos \frac{\pi}{4}\right|$$
$$+ \left|\cos \frac{3\pi}{4}\right| - \left|\cos \frac{\pi}{4}\right| \leq 2.$$

However, the left-hand side is $2\sqrt{2}$, which is greater than 2. Any hidden variable model must admit superluminal signalling.

Index

© The Editor(s) (if applicable) and The Author(s), under exclusive license to Springer Nature Switzerland AG 2023
P. Kok, *A First Introduction to Quantum Physics*, Undergraduate Lecture Notes in Physics, https://doi.org/10.1007/978-3-031-16165-0

Printed in the United States
by Baker & Taylor Publisher Services